T0235158

Quantum Field Theory in Condensed Matter Physics

This book is a course in modern quantum field theory as seen through the eyes of a theorist working in condensed matter physics. It contains a gentle introduction to the subject and can therefore be used even by graduate students. The introductory parts include a derivation of the path integral representation, Feynman diagrams and elements of the theory of metals including a discussion of Landau Fermi liquid theory. In later chapters the discussion gradually turns to more advanced methods used in the theory of strongly correlated systems. The book contains a thorough exposition of such nonperturbative techniques as $1/N$-expansion, bosonization (Abelian and non-Abelian), conformal field theory and theory of integrable systems. The book is intended for graduate students, postdoctoral associates and independent researchers working in condensed matter physics.

ALEXEI TSVELIK was born in 1954 in Samara, Russia, graduated from an elite mathematical school and then from Moscow Physical Technical Institute (1977). He defended his PhD in theoretical physics in 1980 (the subject was heavy fermion metals). His most important collaborative work (with Wiegmann on the application of Bethe ansatz to models of magnetic impurities) started in 1980. The summary of this work was published as a review article in *Advances in Physics* in 1983. During the years 1983–89 Alexei Tsvelik worked at the Landau Institute for Theoretical Physics. After holding several temporary appointments in the USA during the years 1989–92, he settled in Oxford, were he spent nine years. Since 2001 Alexei Tsvelik has held a tenured research appointment at Brookhaven National Laboratory. The main area of his research is strongly correlated systems (with a view of application to condensed matter physics). He is an author or co-author of approximately 120 papers and two books. His most important papers include papers on the integrable models of magnetic impurities, papers on low-dimensional spin liquids and papers on applications of conformal field theory to systems with disorder. Alexei Tsvelik has had nine graduate students of whom seven have remained in physics.

Quantum Field Theory in Condensed Matter Physics

Alexei M. Tsvelik

Department of Physics

Brookhaven National Laboratory

CAMBRIDGE UNIVERSITY PRESS
Cambridge, New York, Melbourne, Madrid, Cape Town, Singapore, São Paulo

Cambridge University Press
The Edinburgh Building, Cambridge CB2 2RU, UK

Published in the United States of America by Cambridge University Press, New York

www.cambridge.org
Information on this title: www.cambridge.org/9780521822848

First published 1995

Reprinted 1996

First paperback edition 1996

Reprinted 1998

Second edition 2003

This digitally printed first paperback version 2006

A catalogue record for this publication is available from the British Library

Library of Congress Cataloguing in Publication data
Tsvelik, Alexei M.
Quantum field theory in condensed matter physics / Alexei M. Tsvelik. – [2nd ed.].
p.   cm.
Includes bibliographical references and index.
ISBN 0 521 82284 X (hardback)
1. Quantum field theory.   2. Condensed matter.   I. Title.
QC174.45.T79 2003
530.1′43 – dc21   2003043957

ISBN-13 978-0-521-82284-8 hardback
ISBN-10 0-521-82284-X hardback

ISBN-13 978-0-521-52980-8 paperback
ISBN-10 0-521-52980-8 paperback

To my father

Contents

Preface to the first edition

The objective of this book is to familiarize the reader with the recent achievements of quantum field theory (henceforth abbreviated as QFT). The book is oriented primarily towards condensed matter physicists but, I hope, can be of some interest to physicists in other fields. In the last fifteen years QFT has advanced greatly and changed its language and style. Alas, the fruits of this rapid progress are still unavailable to the vast democratic majority of graduate students, postdoctoral fellows, and even those senior researchers who have not participated directly in this change. This cultural gap is a great obstacle to the communication of ideas in the condensed matter community. The only way to reduce this is to have as many books covering these new achievements as possible. A few good books already exist; these are cited in the select bibliography at the end of the book. Having studied them I found, however, that there was still room for my humble contribution. In the process of writing I have tried to keep things as simple as possible; the amount of formalism is reduced to a minimum. Again, in order to make life easier for the newcomer, I begin the discussion with such traditional subjects as path integrals and Feynman diagrams. It is assumed, however, that the reader is already familiar with these subjects and the corresponding chapters are intended to refresh the memory. I would recommend those who are just starting their research in this area to read the first chapters in parallel with some introductory course in QFT. There are plenty of such courses, including the evergreen book by Abrikosov, Gorkov and Dzyaloshinsky. I was trained with this book and thoroughly recommend it.

Why study quantum field theory? For a condensed matter theorist as, I believe, for other physicists, there are several reasons for studying this discipline. The first is that QFT provides some wonderful and powerful tools for our research. The results achieved with these tools are innumerable; knowledge of their secrets is a key to success for any decent theorist. The second reason is that these tools are also very elegant and beautiful. This makes the process of scientific research very pleasant indeed. I do not think that this is an accidental coincidence; it is my strong belief that aesthetic criteria are as important in science as empirical ones. Beauty and truth cannot be separated, because 'beauty is truth realized' (Vladimir Solovyev). The history of science strongly supports this belief: all great physical theories are at the same time beautiful. Einstein, for example, openly admitted that ideas of beauty played a very important role in his formulation of the theory of general relativity, for which any experimental support had remained minimal for many years. Einstein is by no means alone; the reader is advised to read the philosophical essays of Werner Heisenberg, whose authority in the area of physics is hard to deny. Aesthetics deals with forms; it is not therefore suprising that a smack of geometry is felt strongly in modern QFT: for example, the idea that a vacuum, being an apparently empty space, has a certain symmetry, i.e. has a geometric figure associated with it. In what follows we shall have more than one chance to discuss this particular topic and to appreciate the fact that geometrical constructions play a major role in the behaviour of physical models.

The third reason for studying QFT is related to the first and the second. QFT has the power of *universality*. Its language plays the same unifying role in our times as Latin played in the times of Newton and Leibniz. Its knowledge is the equivalent of literacy. This is not an exaggeration: equations of QFT describe phase transitions in magnetic metals and in the early universe, the behaviour of quarks and fluctuations of cell membranes; in this language one can describe equally well both classical and quantum systems. The latter feature is especially important. From the very beginning I shall make it clear that from the point of view of calculations, there is no difference between quantum field theory and classical statistical mechanics. Both these disciplines can be discussed within the same formalism. Therefore everywhere below I shall unify quantum field theory and statistical mechanics under the same abbreviation of QFT. This language helps one

> To see a world in a grain of sand
> And a heaven in a wild flower,
> Hold infinity in the palm of your hand
> And eternity in an hour.[*]

I hope that by now the reader is sufficiently inspired for the hard work ahead. Therefore I switch to prose. Let me now discuss the content of the book. One of its goals is to help the reader to solve future problems in condensed matter physics. These are more difficult to deal with than past problems, all the easy ones have already been solved. What remains is difficult, but is interesting nevertheless. The most interesting, important and complicated problems in QFT are those concerning strongly interacting systems. Indeed, most of the progress over the past fifteen years has been in this area. One widely known and related problem is that of quark confinement in quantum chromodynamics (QCD). This still remains unresolved, as far as I am aware. A less known example is the problem of strongly correlated electrons in metals near the metal–insulator transition. The latter problem is closely related to the problem of high temperature superconductivity. Problems with the strong interaction cannot be solved by traditional methods, which are mostly related to perturbation theory. This does not mean, however, that it is not necessary to learn the traditional methods. On the contrary, complicated problems cannot be approached without a thorough knowledge of more simple ones. Therefore Part I of the book is devoted to such traditional methods as the path integral formulation of QFT and Feynman diagram expansion. It is not supposed, however, that the reader will learn these methods from this book. As I have

---

[*] William Blake, *Auguries of Innocence*.

said before, there are many good books which discuss the traditional methods, and it is not the purpose of Part I to be a substitute for them, but rather to recall what the reader has learnt elsewhere. Therefore discussion of the traditional methods is rather brief, and is targeted primarily at the aspects of these methods which are relevant to nonperturbative applications.

The general strategy of the book is to show how the strong interaction arises in various parts of QFT. I do not discuss in detail all the existing condensed matter theories where it occurs; the theories of localization and quantum Hall effect are omitted and the theory of heavy fermion materials is discussed only very briefly. Well, one cannot embrace the unembraceable! Though I do not discuss all the relevant physical models, I do discuss all the possible scenarios of renormalization: there are only three of them. First, it is possible that the interactions are large at the level of a bare many-body Hamiltonian, but effectively vanish for the low energy excitations. This takes place in quantum electrodynamics in $(3 + 1)$ dimensions and in Fermi liquids, where scattering of quasi-particles on the Fermi surface changes only their phase (forward scattering). Another possibility is that the interactions, being weak at the bare level, grow stronger for small energies, introducing profound changes in the low energy sector. This type of behaviour is described by so-called 'asymptotically free' theories; among these are QCD, the theories describing scattering of conducting electrons on magnetic impurities in metals (the Anderson and the s-d models, in particular), models of two-dimensional magnets, and many others. The third scenario leads us to critical behaviour. In this case the interactions between low energy excitations remain finite. Such situations occur at the point of a second-order phase transition. The past few years have been marked by great achievements in the theory of two-dimensional second-order phase transitions. A whole new discipline has appeared, known as conformal field theory, which provides us with a potentially complete description of all types of possible critical points in two dimensions. The classification covers two-dimensional theories at a transition point and those quantum $(1 + 1)$-dimensional theories which have a critical point at $T = 0$ (the spin $S = 1/2$ Heisenberg model is a good example of the latter).

In the first part of the book I concentrate on formal methods; at several points I discuss the path integral formulation of QFT and describe the perturbation expansion in the form of Feynman diagrams. There is not much 'physics' here; I choose a simple model (the $O(N)$-symmetric vector model) to illustrate the formal procedures and do not indulge in discussions of the physical meaning of the results. As I have already said, it is highly desirable that the reader who is unfamiliar with this material should read this part in parallel with some textbook on Feynman diagrams. The second part is less dry; here I discuss some miscellaneous and relatively simple applications. One of them is particularly important: it is the electrodynamics of normal metals where on a relatively simple level we can discuss violations of the Landau Fermi liquid theory. In order to appreciate this part, the reader should know what is violated, i.e. be familiar with the Landau theory itself. Again, I do not know a better book to read for this purpose than the book by Abrikosov, Gorkov and Dzyaloshinsky. The real fun starts in the third and the fourth parts, which are fully devoted to nonperturbative methods. I hope you enjoy them!

Finally, those who are familiar with my own research will perhaps be surprised by the absence in this book of exact solutions and the Bethe ansatz. This is not because I do not like these methods any more, but because I do not consider them to be a part of the *minimal* body of knowledge necessary for any theoretician working in the field.

Alexei Tsvelik
Oxford, 1994

Preface to the second edition

Though it was quite beyond my original intentions to write a textbook, the book is often used to teach graduate students. To alleviate their misery I decided to extend the introductory chapters and spend more time discussing such topics as the equivalence of quantum mechanics and classical statistical mechanics. A separate chapter about Landau Fermi liquid theory is introduced. I still do not think that the book is fully suitable as a graduate textbook, but if people want to use it this way, I do not object.

Almost 10 years have passed since I began my work on the first edition. The use of field theoretical methods has extended enormously since then, making the task of rewriting the book very difficult. I no longer feel myself capable of presenting a brief course containing the 'minimal body of knowledge necessary for any theoretician working in the field'. I strongly feel that such a body of knowledge should include not only general ideas, what is usually called 'physics', but also techniques, even technical tricks. Without this common background we shall not be able to maintain high standards of our profession and the fragmentation of our community will continue further. However, the best I can do is to include the material I can explain well and to mention briefly the material which I deem worthy of attention. In particular, I decided to include exact solutions and the Bethe ansatz. It was excluded from the first edition as being too esoteric, but now the astonishing new progress in calculations of correlation functions justifies its inclusion in the core text. I think that this progress opens new exciting opportunities for the field, but the community has not yct woken up to the change. The chapters about the two-dimensional Ising model are extended. Here again the community does not fully grasp the importance of this model and of the concepts related to it. For the same reason I extended the chapters devoted to the Wess–Zumino–Novikov–Witten model.

Alexei Tsvelik
Brookhaven, 2002

Fragmentation of the community.

Acknowledgements for the first edition

I gratefully acknowledge the support of the Landau Institute for Theoretical Physics, in whose stimulating environment I worked for several wonderful years. My thanks also go to the University of Oxford, and to its Department of Physics in particular, the support of which has been vital for my work. I also acknowledge the personal support of David Sherrington, Boris Altshuler, John Chalker, David Clarke, Piers Coleman, Lev Ioffe, Igor Lerner, Alexander Nersesyan, Jack Paton, Paul de Sa and Robin Stinchcombe. Brasenose College has been a great source of inspiration to me since I was elected a fellow there, and I am grateful to my college fellow John Peach who gave me the idea of writing this book. Special thanks are due to the college cellararius Dr Richard Cooper for irreproachable conduct of his duties.

Acknowledgements for the second edition

I am infinitely grateful to my friends and colleagues Alexander Nersesyan, Andrei Chubukov, Fabian Essler, Alexander Gogolin and Joe Bhaseen for support and advice. I am also grateful to my new colleagues at Brookhaven National Laboratory, especially to Doon Gibbs and Peter Johnson, who made my transition to the USA so smooth and pleasant. I also acknowledge support from US DOE under contract number DE-AC02-98 CH 10886.

I

Introduction to methods

1
QFT: language and goals

Under the calm mask of matter
The divine fire burns.

Vladimir Solovyev

The reason why the terms 'quantum field theory' and 'statistical mechanics' are used together so often is related to the essential equivalence between these two disciplines. Namely, a quantum field theory of a D-dimensional system can be formulated as a statistical mechanics theory of a $(D + 1)$-dimensional system. This equivalence is a real godsend for anyone studying these subjects. Indeed, it allows one to get rid of noncommuting operators and to forget about time ordering, which seem to be characteristic properties of quantum mechanics. Instead one has a way of formulating the quantum field theory in terms of ordinary commuting functions, more or less conventional integrals, etc.

Before going into formal developments I shall recall the subject of quantum field theory (QFT). Let us consider first what classical fields are. To begin with, they are entities expressed as continuous functions of space and time coordinates $(\mathbf{x}, t)$. A field $\Phi(\mathbf{x}, t)$ can be a scalar, a vector (like an electromagnetic field represented by a vector potential $(\phi, \mathbf{A})$), or a tensor (like a metric field g_{ab} in the theory of gravitation). Another important thing about fields is that they can exist on their own, i.e. independent of their 'sources' – charges, currents, masses, etc. Translated into the language of theory, this means that a system of fields has its own action $S[\Phi]$ and energy $E[\Phi]$. Using these quantities and the general rules of classical mechanics one can write down equations of motion for the fields.

Example

As an example consider the derivation of Maxwell's equations for an electromagnetic field in the absence of any sources. I use this example in order to introduce some valuable definitions. The action for an electromagnetic field is given by

$$S = \frac{1}{8\pi} \int dt d^3 x [\mathbf{E}^2 - \mathbf{H}^2] \tag{1.1}$$

where $\mathbf{E}$ and $\mathbf{H}$ are the electric and the magnetic fields, respectively. These fields are not independent, but are expressed in terms of the potentials:

$$\mathbf{E} = -\nabla\phi + \frac{1}{c}\frac{\partial \mathbf{A}}{\partial t}$$

$$\mathbf{H} = \nabla \times \mathbf{A} \tag{1.2}$$

The relationship between $(\mathbf{E}, \mathbf{H})$ and $(\phi, \mathbf{A})$ is not unique; $(\mathbf{E}, \mathbf{H})$ does not change when the following transformation is applied:

$$\phi \to \phi + \frac{1}{c}\frac{\partial \chi}{\partial t}$$

$$\mathbf{A} \to \mathbf{A} + \nabla\chi \tag{1.3}$$

This symmetry is called *gauge* symmetry. In order to write the action as a single-valued functional of the potentials, we need to specify the gauge. I choose the following:

$$\phi = 0$$

Substituting (1.2) into (1.1) we get the action as a functional of the vector potential:

$$S = \frac{1}{8\pi}\int dt d^3x \left[\frac{1}{c^2}(\partial_t \mathbf{A})^2 - (\nabla \times \mathbf{A})^2\right] \tag{1.4}$$

In classical mechanics, particles move along trajectories with minimal action. In field theory we deal not with particles, but with configurations of fields, i.e. with functions of coordinates and time $\mathbf{A}(t, \mathbf{x})$. The generalization of the principle of minimal action for fields is that fields evolve in time in such a way that their action is minimal. Suppose that $\mathbf{A}_0(t, \mathbf{x})$ is such a configuration for the action (1.4). Since we claim that the action achieves its minimum in this configuration, it must be invariant with respect to an infinitesimal variation of the field:

$$\mathbf{A} = \mathbf{A}_0 + \delta\mathbf{A}$$

Substituting this variation into the action (1.4), we get:

$$\delta S = \frac{1}{4\pi}\int dt d^3x [c^{-2}\partial_t \mathbf{A}_0 \partial_t \delta\mathbf{A} - (\nabla \times \mathbf{A}_0)(\nabla \times \delta\mathbf{A})] + \mathcal{O}(\delta\mathbf{A}^2) \tag{1.5}$$

The next essential step is to rewrite δS in the following canonical form:

$$\delta S = \int dt d^3x\, \delta\mathbf{A}(t, \mathbf{x})\mathbf{F}[\mathbf{A}_0(t, \mathbf{x})] + \mathcal{O}(\delta A^2) \tag{1.6}$$

where $\mathbf{F}[\mathbf{A}_0(t, \mathbf{x})]$ is some functional of $\mathbf{A}_0(t, \mathbf{x})$. *By definition*, this expression determines the function

$$\mathbf{F} \equiv \frac{\delta S}{\delta \mathbf{A}}$$

the *functional* derivative of the functional S with respect to the function $\mathbf{A}$. Let us assume that $\delta\mathbf{A}$ vanishes at infinity and integrate (1.5) by parts:

$$\delta S = -\frac{1}{4\pi}\int dt d^3x \left\{c^{-2}\partial_t^2 \mathbf{A}_0(t, \mathbf{x}) - [(\nabla \times \nabla) \times \mathbf{A}_0(t, \mathbf{x})]\right\}\delta\mathbf{A}(t, \mathbf{x}) \tag{1.7}$$

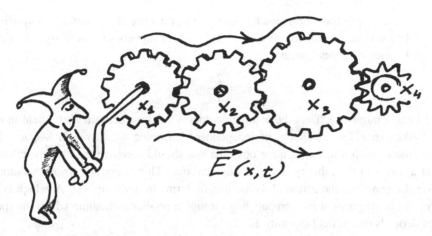

$$\vec{E}(x,t)$$

Figure 1.1. Maxwell's equations as a mechanical system.

Since $\delta S = 0$ for any $\delta \mathbf{A}$, the expression in the curly brackets (that is the functional derivative of S) vanishes. Thus we get the Maxwell equation:

$$c^{-2}\partial_t^2 \mathbf{A} - (\nabla \times \nabla) \times \mathbf{A} = 0 \tag{1.8}$$

Thus Maxwell's equations are the Lagrange equations for the action (1.4).

From Maxwell's equations we see that the field at a given point is determined by the fields at the neighbouring points. In other words the theory of electromagnetic waves is a mechanical theory with an infinite number of degrees of freedom (i.e. coordinates). These degrees of freedom are represented by the fields which are present at every point and coupled to each other. In fact it is quite correct to define classical field theory as the mechanics of systems with an infinite number of degrees of freedom. By analogy, one can say that QFT is just the quantum mechanics of systems with infinite numbers of coordinates.

There is a large class of field theories where the above infinity of coordinates is trivial. In such theories one can redefine the coordinates in such a way that the new coordinates obey independent equations of motion. Then an apparently complicated system of fields decouples into an infinite number of simple independent systems. It is certainly possible to do this for so-called linear theories, a good example of which is the theory of the electromagnetic field (1.4); the new coordinates in this case are just coefficients in the Fourier expansion of the field $\mathbf{A}$:

$$\mathbf{A}(\mathbf{x}, t) = \frac{1}{V} \sum_{\mathbf{k}} \mathbf{a}(\mathbf{k}, t) e^{i\mathbf{k}\mathbf{x}} \tag{1.9}$$

Substituting this expansion into (1.8) we obtain equations for the coefficients, which are just the Newton equations for harmonic oscillators with frequencies $\pm c|\mathbf{k}|$:

$$\partial_t^2 \mathbf{a}_i(\mathbf{k}, t) - (c\mathbf{k})^2 \left(\delta_{ij} - \frac{k_i k_j}{\mathbf{k}^2} \right) \mathbf{a}_j(\mathbf{k}, t) = 0 \tag{1.10}$$

where $\mathbf{a} = (a_1, a_2, a_3)$.

The meaning of this transformation becomes especially clear if we confine our system of fields in a box with linear dimensions L_i $(i = 1, \ldots, D)$ with periodic boundary conditions. Then our **k**-space becomes discrete:

$$k_i = \frac{2\pi}{L_i} n_i$$

(n_i are integer numbers). Thus the continuous theory of the electromagnetic field in real space looks like a discrete theory of independent harmonic oscillators in **k**-space. The quantization of such a theory is quite obvious: one should quantize the above oscillators and get a quantum field theory from the classical one. Things are not always so simple, however. Imagine that the action (1.4) has quartic terms in derivatives of **A**, which is the case for electromagnetic waves propagating through a nonlinear medium where the speed of light depends on the field intensity **E**:

$$c^2 = \left(c_0^2/n\right) + \alpha(\partial_t \mathbf{A})^2 \tag{1.11}$$

Then one cannot decouple the Maxwell equations into independent equations for harmonic oscillators.

We have mentioned above that QFT is just quantum mechanics for an infinite number of degrees of freedom. Infinities always cause problems, not only conceptual, but technical as well. In high energy physics these problems are really serious, but in condensed matter physics we are more lucky: here we rarely deal with systems where the number of degrees of freedom is really infinite. Numbers of electrons and ions are always finite though usually very large. If an infinity actually does appear, the first approach to it is to make it countable. We already know how to do this: we should put the system into a box and carry out a Fourier transformation of the fields. In condensed matter problems this box is not imaginary, but real. Another natural way to make the number of degrees of freedom finite is to put the system on a lattice. Again, in condensed matter physics a lattice is naturally present.

Usually QFT is concerned about *universal* features of phenomena, i.e. about those features which are independent of details of the lattice. Therefore QFT describes a continuum limit of many-body quantum mechanics, in other words the limit on a lattice with $a \rightarrow 0$, $L_i \rightarrow \infty$. We shall see that *this limit does not necessarily exist*, i.e. not all condensed matter phenomena have universal features.

Let us forget for a moment about possible difficulties and accept that QFT is just a quantum mechanics of systems of an infinite number of degrees of freedom. Does the word 'infinite' impose any additional requirements? It does, because this makes QFT a statistical theory. QFT operates with *statistically averaged quantum averages*. Therefore in QFT we average twice. Let us explain this in more detail. The quantum mechanical average of an operator $\hat{A}(t)$ is defined as

$$\langle \hat{A}(t) \rangle = \int \mathrm{d}^N x \, \psi^*(t, \mathbf{x}) \hat{A}(t) \psi(t, \mathbf{x}) = \sum_q C_q^* C_p \langle q | \hat{A}(t) | p \rangle \tag{1.12}$$

where $|q\rangle$ are eigenstates and the coefficients C_q are not specified. In QFT we usually consider systems in thermal equilibrium, i.e. we assume that the coefficients of the wave

Figure 1.2. Studying responses of a 'black box'.

functions follow the Gibbs distribution:

$$C_q^* C_p = \frac{1}{Z} e^{-\beta E_q} \delta_{qp} \tag{1.13}$$

where $\beta = 1/T$. In other words, the averaging process in QFT includes quantum mechanical averaging and Gibbs averaging:

$$\langle \hat{A}(t) \rangle_{QFT} = \frac{\sum_q e^{-\beta E_q} \langle q | \hat{A}(t) | q \rangle}{\sum_q e^{-\beta E_q}} = \frac{\text{Tr}(e^{-\beta \hat{H}} \hat{A})}{\text{Tr}(e^{-\beta \hat{H}})} \tag{1.14}$$

There is also another important language difference between quantum mechanics and QFT. Quantum mechanics expresses everything in terms of wave functions, but in QFT we usually express results in terms of correlation functions or generating functionals of these functions. It is useful to define these important notions from the very beginning. Let us consider a *classical* statistical system first. What is a correlation function? Imagine we have a complicated system where everything is interconnected appearing like a 'black box' to us. One can study this black box by its responses to external perturbations (see Fig. 1.2).

A usual measure of this response is a change in the free energy: $\delta F = F[H(\mathbf{x})] - F[0]$. In principle, the functional $\delta F[H(\mathbf{x})]$ carries all accessible information about the system. Experimentally we usually measure derivatives of the free energy with respect to fields taken at different points. The only formal difficulty is that the number of points is infinite. However, we can overcome this by discretizing our space as has been explained above. Therefore we represent our space as an arrangement of small boxes of volume Ω centred

Figure 1.3. Response functions are usually measurable experimentally.

around points $\mathbf{x}_n$ (recall the previous discussion!) assuming that the field $H(\mathbf{x})$ is constant inside each box: $H(\mathbf{x}) = H(\mathbf{x}_n)$. Thus our functional may be treated as a limiting case of a function of a large but finite number of arguments $F[H] = \lim_{\Omega \to 0} F(H_1, \dots, H_N)$.

Performing the above differentiations we define the following quantities which are called correlation functions:

$$\langle M(\mathbf{x}) \rangle \equiv \frac{\delta F[H]}{\delta H(\mathbf{x})}$$

$$\langle\langle M(\mathbf{x}_2) M(\mathbf{x}_1) \rangle\rangle \equiv \frac{\delta^2 F[H]}{\delta H(\mathbf{x}_2) \delta H(\mathbf{x}_1)} \cdots \qquad (1.15)$$

$$\langle\langle M(\mathbf{x}_N) M(\mathbf{x}_{N-1}) \cdots M(\mathbf{x}_1) \rangle\rangle \equiv \frac{\delta^N F[H]}{\delta H(\mathbf{x}_N) \cdots \delta H(\mathbf{x}_1)} \cdots$$

Recall that the operation $\delta F / \delta H$ thus defined is called a functional derivative. As we see, it is a straightforward generalization of a partial derivative for the case of an infinite number of variables. In general, whenever we encounter infinities in physics we can approximate them by very large numbers, so do not worry much about such things as functional derivatives and path integrals (see below); they are just trivial generalizations of partial derivatives and multiple integrals!

Response functions are usually measurable experimentally, at least in principle (see Fig. 1.3). By obtaining them one can recover the whole functional using the Taylor expansion:

$$\delta F[H(\mathbf{x})] = \sum_n \frac{1}{n!} \int d^D x_1 \cdots d^D x_N H(\mathbf{x}_1) \cdots H(\mathbf{x}_n) \langle\langle M(x_1) \cdots M(x_n) \rangle\rangle \quad (1.16)$$

In which way does the situation in QFT differ from the classical one? First of all, as we have seen, in QFT we average in both the quantum mechanical and thermodynamical sense, but what is more important is that the quantities $M(\mathbf{x})$ are now operators and the result of averaging depends on their ordering. As we know from an elementary course in quantum mechanics, operators satisfy the Heisenberg equation of motion:

$$i\hbar \frac{\partial \hat{A}}{\partial t} = [\hat{H}, \hat{A}] \tag{1.17}$$

where $\hat{H}$ is the Hamiltonian of the system. This equation has the following solution:

$$\hat{A}(t) = e^{-it\hbar^{-1}\hat{H}} \hat{A}(t = 0) e^{it\hbar^{-1}\hat{H}} \tag{1.18}$$

To describe systems in thermal equilibrium we usually use imaginary or the so-called Matsubara time

$$i\tau = t\hbar^{-1}$$

Its meaning will become clear later.

Suppose now that $\hat{A}$ is a perturbation to our Hamiltonian $\hat{H}$. Then this perturbation changes the energy levels:

$$E_n = E_n^{(0)} + \langle n|\hat{A}|n\rangle + \sum_{m \neq n} \frac{|\langle n|\hat{A}|m\rangle|^2}{E_n - E_m} + \cdots \tag{1.19}$$

and therefore changes the free energy:

$$F = -\beta^{-1} \ln \left(\sum_n e^{-\beta E_n} \right)$$

Now I am going to show that in the second order of the perturbation theory these changes in the free energy can be expressed in terms of some correlation function. Let me make some preparatory definitions. Consider an operator $\hat{A}(\mathbf{x})$ and its Hermitian conjugate $\hat{A}^+(\mathbf{x})$. Let us define their τ-dependent generalizations:

$$\begin{aligned} \hat{A}(\tau, \mathbf{x}) &= e^{\tau \hat{H}} \hat{A}(\mathbf{x}) e^{-\tau \hat{H}} \\ \hat{\bar{A}}(\tau, \mathbf{x}) &= e^{\tau \hat{H}} \hat{A}^+(\mathbf{x}) e^{-\tau \hat{H}} \end{aligned} \tag{1.20}$$

where the Matsubara 'time' belongs to the interval $0 < \tau < \beta$.

Then we have the following definition of the correlation function of two operators:

$$\begin{aligned} D(1, 2) &\equiv \langle\langle \hat{A}(\tau_1, \mathbf{x}_1) \hat{\bar{A}}(\tau_2, \mathbf{x}_2) \rangle\rangle \\ &= \begin{cases} \pm\{Z^{-1}\mathrm{Tr}[e^{-\beta\hat{H}} \hat{A}(\tau_1, \mathbf{x}_1)\hat{\bar{A}}(\tau_2, \mathbf{x}_2)] - \langle\hat{A}(\tau_1, \mathbf{x}_1)\rangle\langle\hat{\bar{A}}(\tau_2, \mathbf{x}_2)\rangle\} & \tau_1 > \tau_2 \\ \{Z^{-1}\mathrm{Tr}[e^{-\beta\hat{H}} \hat{\bar{A}}(\tau_2, \mathbf{x}_2)\hat{A}(\tau_1, \mathbf{x}_1)] - \langle\hat{\bar{A}}(\tau_2, \mathbf{x}_2)\rangle\langle\hat{A}(\tau_1, \mathbf{x}_1)\rangle\} & \tau_2 > \tau_1 \end{cases} \end{aligned} \tag{1.21}$$

The minus sign in the upper row appears if $\hat{A}$ is a Fermi operator. Here I have to make the following important remark. The terms Bose and Fermi are used in the following sense. Operators are termed Bose if they create a closed algebra under the operation of commutation, and they are termed Fermi if they create a closed algebra under anticommutation. The phrase 'closed algebra' means that commutation (or anticommutation) of operators of a certain set produces only operators of this set and nothing else. Thus spin operators on a lattice $\hat{S}^a(\mathbf{r})$ ($a = x, y, z$) create a closed algebra under commutation, because their commutator is either zero ($\mathbf{r} \neq \mathbf{r}'$) or a spin operator. One might think that $S = 1/2$ is a special case because the Pauli matrices on one site also satisfy the anticommutation relations:

$$\{\sigma^a, \sigma^b\} = 2\delta_{ab}$$

and it seems that one can choose alternative definitions of their statistics. It is not true, however, because the spin-1/2 operators from different lattice sites always *commute* and, on the contrary, their anticommutator is never equal to zero.

Imagine that we know all the eigenfunctions and eigenenergies of our system. Then we can rewrite the above traces explicitly using this basis. The result is given by

$$D(1, 2) = \sum_{n,m} \frac{e^{-\beta E_n}}{Z} |\langle n|\hat{A}(0)|m\rangle|^2 e^{i\mathbf{P}_{mn}\mathbf{x}_{12}} [\pm\theta(\tau_1 - \tau_2)e^{E_{nm}\tau_{12}} + \theta(\tau_2 - \tau_1)e^{E_{nm}\tau_{21}}]$$

(1.22)

where $\tau_{12} = \tau_1 - \tau_2$, $\mathbf{x}_{12} = \mathbf{x}_1 - \mathbf{x}_2$. Here we have used the following properties of eigenstates:

$$e^{\tau\hat{H}}|n\rangle = e^{\tau E_n}|n\rangle$$

$$\langle m|\hat{A}(\mathbf{x})|n\rangle = e^{i(\mathbf{P}_n - \mathbf{P}_m)\mathbf{x}}\langle m|\hat{A}(0)|n\rangle$$

The latter property holds only for translationally invariant systems where the eigenstates of $\hat{H}$ are simultaneously eigenstates of the momentum operator $\hat{\mathbf{P}}$. Now you can check that the change in the free energy can be written in terms of the correlation functions:

$$\beta\delta F = \int_0^\beta d\tau \langle A(\tau) \rangle + \frac{1}{2} \int_0^\beta d\tau_1 \int_0^\beta d\tau_2 D(\tau_1, \tau_2)$$

(1.23)

Therefore correlation functions are equally important in classical and quantum systems.

Let us continue our analysis of the pair correlation function defined by (1.21) and (1.22). This pair correlation function is often called the Green's function after the man who introduced similar objects in classical field theory. There are two important properties following from this definition. The first is that the Green's function depends on

$$\tau \equiv (\tau_1 - \tau_2)$$

which belongs to the interval

$$-\beta < \tau < \beta$$

The second is that for Bose operators the Green's function is a periodic function:

$$D(\tau) = D(\tau + \beta) \qquad \tau < 0$$

(1.24)

and for Fermi operators it is an antiperiodic function:

$$D(\tau) = -D(\tau + \beta) \qquad \tau < 0$$

(1.25)

These two properties allow one to write down the following Fourier decomposition of the Green's function:

$$D(\tau, \mathbf{x}_{12}) = \beta^{-1} \sum_{s=-\infty}^{\infty} \int \frac{d^D k}{(2\pi)^D} D(\omega_s, \mathbf{k}) e^{-i\omega_s \tau - i\mathbf{k}\mathbf{x}_{12}} \tag{1.26}$$

where

$$D(\omega_s, \mathbf{k}) = (2\pi)^D Z^{-1} \sum_{n,m} e^{-\beta E_n} (1 \mp e^{\beta - E_{mn}}) |\langle n|\hat{A}(0)|m\rangle|^2 \frac{\delta(\mathbf{k} - \mathbf{P}_{mn})}{i\omega_s - E_{mn}}$$

$$\equiv \sum_{n,m} \frac{\rho(n,m)(\mathbf{k})}{i\omega_s - E_{mn}} \tag{1.27}$$

and

$$\omega_s = 2\pi \beta^{-1} s$$

for Bose systems and

$$\omega_s = \pi \beta^{-1}(2s + 1)$$

for Fermi systems. Thus we get a function defined in the complex plane of ω at a sequence of points $\omega = i\omega_s$. We can continue it analytically to the upper half-plane (for example). Thus we get the function

$$D^{(R)}(\omega) = \sum_{n,m} \frac{\rho_{(n,m)}(\mathbf{k})}{\omega - E_{mn} + i0}$$

$$\rho_{(n,m)}(\mathbf{k}) = \frac{(2\pi)^D}{Z} e^{-\beta E_n} (1 \mp e^{\beta - E_{mn}}) |\langle n|\hat{A}(0)|m\rangle|^2 \delta(\mathbf{k} - \mathbf{P}_{mn}) \tag{1.28}$$

analytical in the upper half-plane of ω. This function has *two wonderful properties*. (a) Its poles in the lower half-plane give energies of transitions E_{mn} which tell us about the spectrum of our Hamiltonian. (b) We can write down our original Green's function in terms of the retarded one:

$$D(\omega_s, \mathbf{k}) = -\frac{1}{\pi} \int dy \frac{\Im m D^{(R)}(y)}{i\omega_s - y} \tag{1.29}$$

This relation is very convenient for practical calculations as will become clear in subsequent chapters.

We see that the quantum case is special due to the presence of the 'time' variable τ. What is specially curious is that the quantum correlation functions have different periodicity properties in the τ-plane depending on the statistics. We shall have a chance to appreciate the really deep meaning of all these innovations in the next chapters.

One should not take away from this chapter a false impression that in QFT we are doomed to deal with this strange imaginary time and are not able to make judgements about real time dynamics. The point is that the τ-formulation is just more convenient; for systems in thermal equilibrium the dynamic (i.e. real time) correlation functions are related to the thermodynamic ones through the following relationship:

$$D_{\text{dynamic}}(\omega) = \Re e D^{(R)}(\omega) + i \coth\left(\frac{\omega}{2T}\right) \Im m D^{(R)}(\omega) \qquad \text{(bosons)} \qquad (1.30)$$

$$D_{\text{dynamic}}(\omega) = \Re e D^{(R)}(\omega) + i \tanh\left(\frac{\omega}{2T}\right) \Im m D^{(R)}(\omega) \qquad \text{(fermions)} \qquad (1.31)$$

The proof of the above relations can be found in any book on QFT and I shall spend no time on it.

These relations are convenient if our calculational procedure naturally provides us with Green's functions in frequency momentum representation. This is not always the case, however. Sometimes we can work only in real space (see the chapters on one-dimensional systems). Then it is better not to calculate $D(i\omega_n)$ first and continue it analytically, but to skip this intermediate step and to express the retarded functions directly in terms of $D(\tau)$. In order to do this, we can use the relationship between the thermodynamic and the retarded Green's functions, which follows from (1.22) and (1.28):

$$D(\tau) = \theta(\tau)D_+(\tau) \pm \theta(-\tau)D_-(\tau)$$

$$D_+(\tau) = -\frac{1}{\pi} \int dx \, \Im m D^{(R)}(x) \frac{e^{-x\tau}}{1 \mp e^{-\beta x}} \qquad (1.32)$$

$$D_-(\tau) = -\frac{1}{\pi} \int dx \, \Im m D^{(R)}(x) \frac{e^{-x\tau}}{e^{\beta x} \mp 1}$$

(the upper sign is for bosons, the lower one for fermions). Then from (1.32) it follows that

$$D_+(\tau) - D_-(\tau) = -\frac{1}{\pi} \int_{-\infty}^{\infty} dx \, \Im m D^{(R)}(x) e^{-\tau x} \qquad (1.33)$$

from which we can recover $\Im m D^{(R)}(\omega)$:

$$\Im m D^{(R)}(\omega) = \frac{1}{2} \int_{-\infty}^{\infty} dt [D_-(it + \epsilon) - D_+(it + \epsilon)] e^{i\omega t} \qquad (1.34)$$

If you feel that the discussion of the correlation functions is too abstract, go ahead to the next chapter, where a simple example is provided. This is always the case with new concepts; at the beginning they look like unnecessary complications and it takes time to understand that, in fact, they make life much easier for those who have taken trouble to learn them. In order to make contact with reality easier, I outline below some experimental techniques which measure certain correlation functions more or less directly.

1. *Neutron scattering*. Being neutral particles with spin $1/2$, neutrons in condensed matter interact only with magnetic moments. The latter can belong either to nuclei (ions) or to electrons. Thus neutron scattering is a very convenient probe of lattice dynamics and electron magnetism. In experiments on neutron scattering one measures the differential cross-section of neutrons which is directly proportional to the sum of electronic and ionic dynamical structure factors $S_i(\omega, \mathbf{q})$ and $S_{\text{el}}(\omega, \mathbf{q})$. The ionic structure factor is the two-point *dynamical* correlation function of the exponents of ionic displacements $\mathbf{u}$ (see, for

example, Appendix N in the book by Ashcroft and Mermin in the select bibliography):

$$S_i(\omega, \mathbf{q}) = \frac{1}{N} \sum_{\mathbf{r}, \mathbf{r}'} e^{-i\mathbf{q}(\mathbf{r}-\mathbf{r}')} \int \frac{dt}{2\pi} \langle \exp[i\mathbf{q}\mathbf{u}(t, \mathbf{r})] \exp[-i\mathbf{q}\mathbf{u}(0, \mathbf{r}')] \rangle e^{i\omega t} \quad (1.35)$$

(N is the total number of ions in the crystal). In the case when the displacements are *harmonic* the expression for S_i can be simplified:

$$S_i(\omega, \mathbf{q}) = \sum_{\mathbf{r}} \int_{-\infty}^{\infty} \frac{dt}{2\pi} \exp \left\{ \frac{1}{2} q^a q^b [\langle u^a(t, \mathbf{r}) u^b(0, 0) \rangle - \langle u^a(0, 0) u^b(0, 0) \rangle] \right\} e^{i\omega t - i\mathbf{q}\mathbf{r}}$$

$$(1.36)$$

The electronic structure factor is the imaginary part of the dynamical magnetic susceptibility:

$$S_{el}^{ab}(\omega, \mathbf{q}) = \frac{1}{e^{\hbar\omega/kT} - 1} \Im m \chi^{(R), ab}(\omega, \mathbf{q})$$

$$(1.37)$$

$$\chi^{ab}(\omega_n, \mathbf{q}) = \int d^D r d\tau e^{-i\mathbf{q}\mathbf{r} - i\omega_n \tau} \langle \langle S^a(\tau, \mathbf{r}) S^b(0, 0) \rangle \rangle$$

where $S^a(\mathbf{r})$ is the spin density.

2. *X-ray scattering.* X-ray scattering measures the same ionic structure factor plus several other important correlation functions. In metals, absorption of X-rays with definite frequency ω is proportional to the single-electron density of states $\rho(\omega)$. The latter is equal to

$$\rho(\omega) = \frac{1}{\pi} \sum_{\sigma} \sum_{q} \Im m G_{\sigma\sigma}^{(R)}(\omega, \mathbf{q}) \quad (1.38)$$

where $G(\omega, \mathbf{q})$ is the single-electron Green's function. One can do even better than this, measuring X-ray absorption at certain angles. The corresponding method is called 'angle resolved X-ray photoemission' (ARPES); it measures $\Im m G_{\sigma\sigma}^{(R)}(\omega, \mathbf{q})$ directly.

3. *Nuclear magnetic resonance and the Knight shift.* A sample is placed in a combination of constant and alternating magnetic fields. Resonance is observed when the frequency of the alternating fields coincides with the Zeeman splitting of nuclei. The magnetic polarization of the electrons changes the effective magnetic field acting on the nuclei and thus changes the Zeeman splitting. The shift of the resonance line (the *Knight shift*) is proportional to the local magnetic susceptibility:

$$\Delta H / H \sim \sum_{q} F(q) \lim_{\omega \to 0} \Re e \chi^{(R)}(\omega, q) \quad (1.39)$$

where $F(q) = \sum_a \cos(\mathbf{q}\mathbf{a})$ is the structure factor of the given nuclei. A more detailed discussion can be found in Abrikosov's book *Fundamentals of the Theory of Metals.*

4. *Muon resonance.* This method measures internal local magnetic fields. Therefore it allows one to decide whether the material is in a magnetically ordered state or not. The problem of magnetic order may be very difficult if the order is complex, as in helimagnets or in spin glasses where every spin is frozen along its individual direction.

5. *Infrared reflectivity.* When a plane wave is normally incident from vacuum on a medium with dielectric constant ϵ, the fraction r of power reflected (the reflectivity) is given by

$$r = \left| \frac{1 - \sqrt{\epsilon}}{1 + \sqrt{\epsilon}} \right|^2 \tag{1.40}$$

In order to extract ϵ from the reflectivity one can use the Kramers–Kronig relations. This requires a knowledge of $r(\omega)$ for a considerable range of frequencies, which is a disadvantage of the method. The dielectric function $\epsilon(\omega, \mathbf{q})$ is directly related to the pair correlation function of charge density:

$$\Pi(\omega, \mathbf{q}) = \langle\langle \rho(-\omega, -\mathbf{q})\rho(\omega, \mathbf{q})\rangle\rangle \tag{1.41}$$

Its imaginary part is proportional to the electrical conductivity:

$$\Im m\epsilon = \frac{4\pi}{\omega}\Re e\sigma \tag{1.42}$$

Since photons have very small wave vectors $q = \omega/c$, the described method effectively measures values of physical quantities at zero q.

6. *Brillouin and Raman scattering.* In the corresponding experiments a sample is irradiated by a laser beam of a given frequency; due to the nonlinearity of the medium a part of the energy is re-emitted with different frequencies. Therefore a spectral dispersion of the reflected light contains 'satellites' whose intensity is proportional to the fourth-order correlation function of dipole moments or spins (light can interact with both). Scattering with a small frequency shift originates from gapless excitations (such as acoustic phonons and magnons) and is referred to as *Brillouin* scattering. For frequency shifts of the order of several hundred degrees the main contribution comes from higher energy excitations such as optical phonons; in this case the process is called *Raman* scattering. The practical validity of this kind of experimental technique is limited by the fact that measurements occur at zero wave vectors.

7. *Ultrasound absorption.* This measures the same density–density correlation function as light absorption, but with the advantage that q is not necessarily small, since phonons can have practically any wave vectors.

2
Connection between quantum and classical: path integrals

The efficiency of quantum field methods depends on convenient representation of the wave functions. Such representation exists; the wave function is written as the so-called *path integral*. Apart from being very convenient for practical calculations, this representation reveals a deep and rather unexpected relationship between quantum mechanics and classical thermodynamics.

To establish this connection I will use an example of a system of massless particles connected by springs subject to an external potential $aU(X)$. The particles are on a one-dimensional lattice with lattice constant a. The coordinate of the nth particle in the direction perpendicular to the chain is X_n. The total energy is the sum of the elastic energy and the potential energy (since the particles are massless there is no kinetic energy):

$$E = \sum_{n=1}^{N} \left[\frac{m}{2a}(X_{n+1} - X_n)^2 + aU(X_n) \right] \tag{2.1}$$

In what follows I intend to consider the limit $a \to 0$. The notation is adapted for this task.

Let us consider the thermodynamic probability distribution. From statistical mechanics we know that the probability of being in a state with energy E is given by the Gibbs distribution formula:[1]

$$dP(X_1, \ldots, X_N) = \frac{1}{Z} e^{-E(X)/T} d\Omega_X = \frac{e^{-E/T} d\Omega}{\int e^{-E(Y)/T} d\Omega_Y} \tag{2.2}$$

where $d\Omega_X$ is the volume which the state with coordinates $(X_1, \ldots, X_N) \equiv X$ occupies in the phase space. Here we obviously just need to integrate over the coordinates of all particles (in general there are also momenta, but here the energy is momentum independent) so that

$$d\Omega_X = dX_1 dX_2 \cdots dX_N$$

As I mentioned in Chapter 1, information available in statistical mechanics is provided by the partition function

$$Z = \int dX_1 \cdots dX_N \exp\{-E[X]/T\} \tag{2.3}$$

---

[1] The Boltzmann constant $k_B = 1$.

and the correlation functions such as

$$\langle X_n \rangle = \int dP(X_1, \ldots, X_N)X_n \qquad \langle X_n X_m \rangle = \int dP(X_1, \ldots, X_N)X_n X_m \qquad \text{etc.} \quad (2.4)$$

Since in expression (2.1) for the energy only neighbouring X_n are coupled, one can calculate the above integrals step by step integrating first over X_1, then over X_2 etc., up to X_N. For our purposes it will be more convenient to represent this integration as a recursive process. Therefore let us introduce the following functions:

$$\Psi(n, X_n) = (m/2T\pi a)^{n/2} \int dX_{n-1}dX_{n-2}\cdots dX_1$$

$$\times \exp\left\{-\frac{1}{T}\sum_{j=1}^{n}\left[\frac{m}{2a}(X_j - X_{j-1})^2 + aU(X_j)\right]\right\} \quad (2.5)$$

$$\bar{\Psi}(n, X_n) = (m/2T\pi a)^{(N-n)/2} \int dX_N dX_{N-1}\cdots dX_{n+1}$$

$$\times \exp\left\{-\frac{1}{T}\sum_{j=n+1}^{N}\left[\frac{m}{2a}(X_j - X_{j-1})^2 + aU(X_j)\right]\right\} \quad (2.6)$$

such that

$$Z = (m/2T\pi a)^{-N/2} \int dX_n \bar{\Psi}(n, X_n)\Psi(n, X_n) \quad (2.7)$$

for any n. The factors $(m/2T\pi a)^{n/2}$ and $(m/2T\pi a)^{(N-n)/2}$ in the definitions of Ψ and $\bar{\Psi}$ are introduced for later convenience. Rewriting (2.5) for $n+1$ we get

$$\Psi(n+1, X_{n+1}) = (m/2T\pi a)^{(n+1)/2} \int dX_n dX_{n-1}\cdots dX_1$$

$$\times \exp\left\{-\frac{1}{T}\sum_{j=1}^{n+1}\left[\frac{m}{2a}(X_j - X_{j-1})^2 + aU(X_j)\right]\right\}$$

$$= (m/2T\pi a)^{1/2} \exp\left[-aU(X_{n+1})/T\right] \int dX_n \exp\left[-\frac{m}{2a}(X_{n+1} - X_n)^2\right]$$

$$\times \left((m/2T\pi a)^{n/2} \int dX_{n-1}dX_{n-2}\cdots dX_1\right.$$

$$\left.\times \exp\left\{-\frac{1}{T}\sum_{j=1}^{n}\left[\frac{m}{2a}(X_j - X_{j-1})^2 + aU(X_j)\right]\right\}\right)$$

The expression in the large round brackets coincides with (2.5) and replacing it with $\Psi(n, X_n)$ we get the equation for $\Psi(n+1)$:

$$\Psi(n+1, X_{n+1}) = (m/2T\pi a)^{1/2} \exp[-aU(X_{n+1})/T]$$

$$\times \int dX_n \exp\left[-\frac{m}{2a}(X_{n+1} - X_n)^2\right]\Psi(n, X_n) \quad (2.8)$$

Let us now demonstrate that in the limit $a \to 0$ this equation becomes similar to the Schrödinger equation. At small a we can expand:

$$\Psi(n, X_n) \approx \Psi(n, X_{n+1}) + (X_n - X_{n+1})\frac{\partial \Psi(X_{n+1})}{\partial X_{n+1}} + \frac{1}{2}(X_n - X_{n+1})^2 \frac{\partial^2 \Psi(X_{n+1})}{\partial^2 X_{n+1}} \quad (2.9)$$

Substituting this into the integral in (2.8) and defining a new variable $y = X_n - X_{n+1}$ we get

$$(m/2T\pi a)^{1/2} \int dX_n \exp\left[-\frac{m}{2a}(X_{n+1} - X_n)^2\right]\Psi(n, X_n)$$

$$\approx (m/2T\pi a)^{1/2} \int_{-\infty}^{\infty} dy \left\{\Psi(n, X_{n+1}) + y\frac{\partial \Psi(X_{n+1})}{\partial X_{n+1}} + \frac{1}{2}y^2\frac{\partial^2 \Psi(X_{n+1})}{\partial^2 X_{n+1}}\right\}$$

$$\times \exp\left(-\frac{m}{2aT}y^2\right) = \Psi(n, X_{n+1})\left\{(m/2T\pi a)^{1/2} \int_{-\infty}^{\infty} dy\right.$$

$$\times \exp\left(-\frac{m}{2aT}y^2\right)\bigg\} + \frac{1}{2}\frac{\partial^2 \Psi(X_{n+1})}{\partial^2 X_{n+1}}\left\{(m/2T\pi a)^{1/2} \int_{-\infty}^{\infty} dy y^2 \exp\left(-\frac{m}{2aT}y^2\right)\right\}$$

The integral containing the first order of y vanishes.

Now let us use the formulas:

$$\int_{-\infty}^{\infty} dy \exp(-By^2) = \sqrt{\pi/B}$$

$$\int_{-\infty}^{\infty} dy y^2 \exp(-By^2) = \frac{1}{2B}\sqrt{\pi/B} \qquad (2.10)$$

Finally we get

$$(m/2T\pi a)^{1/2} \int dX_n \exp\left[-\frac{m}{2a}(X_{n+1} - X_n)^2\right]\Psi(n, X_n)$$

$$\approx \Psi(n, X_{n+1}) + \frac{aT}{2m}\frac{\partial^2 \Psi(n, X_{n+1})}{\partial^2 X_{n+1}} \qquad (2.11)$$

We can also expand

$$\exp[-aU(X_{n+1})/T] \approx 1 - aU(X_{n+1})/T \qquad (2.12)$$

Substituting this result and (2.11) back into (2.8) and keeping only terms linear in a, we get

$$\Psi(n + 1, X_{n+1}) - \Psi(n, X_{n+1}) = -[aU(X_{n+1})/T]\Psi(n, X_{n+1}) + \frac{aT}{2m}\frac{\partial^2 \Psi(n, X_{n+1})}{\partial^2 X_{n+1}}$$

$$(2.13)$$

Now we approximate

$$\frac{\Psi(n + 1, X_{n+1}) - \Psi(n, X_{n+1})}{a} \approx \frac{\partial \Psi(\tau, X)}{\partial \tau}$$

where $\tau = na$, and finally get

$$-T\frac{\partial \Psi(\tau, X)}{\partial \tau} = -\frac{T^2}{2m}\frac{\partial^2 \Psi}{\partial X^2} + U(X)\Psi(\tau, X) \qquad (2.14)$$

Similar considerations for $\bar{\Psi}$ yield

$$T \frac{\partial \bar{\Psi}(\tau, X)}{\partial \tau} = -\frac{T^2}{2m} \frac{\partial^2 \bar{\Psi}}{\partial X^2} + U(X)\bar{\Psi}(\tau, X) \tag{2.15}$$

Thus we see that functions Ψ and $\bar{\Psi}$ satisfy the Schrödinger equations, but in imaginary time! If we formally replace

$$T \to \hbar \qquad \tau \to it \tag{2.16}$$

in these equations one can identify Ψ with the wave function and $\bar{\Psi}$ with its complex conjugate.

Does this mean that quantum mechanics and classical thermodynamics are fully equivalent? Such a statement would contradict our basic intuition about quantum mechanics, namely the idea that it differs from classical thermodynamics fundamentally due to the phenomenon of interference of wave functions. I will return to this delicate matter later and discuss it in more detail.

Green's function for a harmonic oscillator

In order to illustrate how the established correspondence works in practice, let us consider a simple example of a harmonic oscillator (Fig. 2.1).

Let us start with the quantum mechanical calculation of its pair correlation function. The corresponding Hamiltonian has the following form:

$$\hat{H} = \frac{\hat{p}^2}{2M} + \frac{M\omega_0^2 \hat{x}^2}{2} \tag{2.17}$$
$$[\hat{x}, \hat{p}] = i\hbar$$

The quantum mechanical correlation function

$$D(1, 2) = \langle \hat{x}(\tau_1)\hat{x}(\tau_2) \rangle \tag{2.18}$$

can be easily calculated following the standard procedure of quantum mechanics. We introduce creation and annihilation operators defined as

$$\hat{x} = \sqrt{\frac{\hbar}{2M\omega_0}}(\hat{A} + \hat{A}^+)$$

$$\hat{p} = i\sqrt{\frac{\hbar M\omega_0}{2}}(\hat{A}^+ - \hat{A}) \tag{2.19}$$

satisfying the Bose commutation relations

$$[\hat{A}, \hat{A}^+] = 1 \tag{2.20}$$

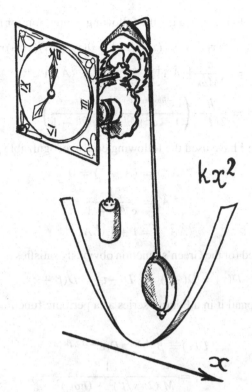

kx^2

Figure 2.1. The pendulum.

When expressed in terms of the above operators the Hamiltonian acquires the following form:

$$\hat{H} = \hbar\omega_0(A^+ A + 1/2)$$

and the normalized eigenstates are

$$|n\rangle = \frac{1}{\sqrt{n!}}(\hat{A}^+)^n|0\rangle \tag{2.21}$$

where the state $|0\rangle$ is defined as the state annihilated by the operator $\hat{A}$:

$$\hat{A}|0\rangle = 0$$

Now we can define the 'time'-dependent operators

$$\hat{A}(\tau) = e^{\tau\hat{H}}\hat{A}e^{-\tau\hat{H}} = e^{-\hbar\omega_0\tau}\hat{A}$$
$$\hat{A}(\tau) = e^{\tau\hat{H}}\hat{A}^+e^{-\tau\hat{H}} = e^{\hbar\omega_0\tau}\hat{A}^+ \tag{2.22}$$

(notice that $\hat{A}(\tau)$ is not a Hermitian conjugate of $\hat{A}(\tau)$!).

Substituting (2.22) into (1.21) we get the following expression for the Green's function:

$$D(1, 2) = \theta(\tau_1 - \tau_2)\langle \hat{x}(\tau_1)\hat{x}(\tau_2)\rangle + \theta(\tau_2 - \tau_1)\langle \hat{x}(\tau_2)\hat{x}(\tau_1)\rangle$$

$$= \frac{\hbar}{2M\omega_0}\left[\langle \hat{A}\hat{A}^+\rangle e^{-\hbar\omega_0|\tau_1-\tau_2|} + \langle \hat{A}^+\hat{A}\rangle e^{\hbar\omega_0|\tau_1-\tau_2|}\right]$$

$$= \frac{\hbar}{2M\omega_0}\left(\frac{e^{-\hbar\omega_0|\tau_1-\tau_2|}}{1 - e^{-\omega_0\hbar\beta}} + \frac{e^{\hbar\omega_0|\tau_1-\tau_2|}}{e^{\omega_0\hbar\beta} - 1}\right) \tag{2.23}$$

In the above derivation I have used the following easily recognizable properties:

$$\langle \hat{A}\hat{A}\rangle = \langle \hat{A}^+\hat{A}^+\rangle = 0$$

$$\langle \hat{A}^+\hat{A}\rangle = \frac{1}{e^{\beta\hbar\omega_0} - 1}$$

$$\langle \hat{A}\hat{A}^+\rangle = 1 + \langle \hat{A}^+\hat{A}\rangle$$

The expression obtained for the Green's function obviously satisfies the following relations:

$$D(\tau) = D(-\tau) \qquad D(-\tau) = D(\beta + \tau) \tag{2.24}$$

which allows us to expand it in a Fourier series as a periodic function of τ on the interval $(0, \beta)$:

$$D(\tau) = \beta^{-1}\sum_s D_s e^{2\pi i s\tau/\beta}$$

$$D_s = \frac{1}{M}\frac{1}{(2\pi s/\beta)^2 + (\hbar\omega_0)^2} \tag{2.25}$$

Now I am going to demonstrate that, in accordance with the general theorem established above, the same expression for D_s may be obtained from the solution of the problem of classical thermodynamics. The great advantage of the procedure which I am going to discuss is that it does not use such nasty things as time ordering, operators, etc. which seem to be unavoidable accessories of quantum theory.

The subsequent discussion of this and of the several following chapters is based, almost exclusively, on the properties of Gaussian integrals (2.10). Almost everything that follows is just a generalization of these identities for multi-dimensional integrals.

The classical counterpart of the harmonic oscillator problem is the problem of a classical string in a 'gutter'. Imagine that we have a *closed* string lying on a plane. Let us parametrize the position along the string by $0 < \tau < L$, and transverse fluctuations by $X(\tau)$, with the obvious boundary conditions:

$$X(\tau + L) = X(\tau)$$

The energy of the string in a parabolic potential is given by

$$E = \int_0^L d\tau \left[\frac{M}{2}\left(\frac{dX}{d\tau}\right)^2 + \frac{M\omega_0^2 X^2}{2}\right] \tag{2.26}$$

where M and ω_0 are just suitable notation for the coefficients.

As we already know, the two problems are related to each other. In the earlier derivation, however, I have discussed a discrete version of the classical problem. Now I will tackle the continuous version directly.

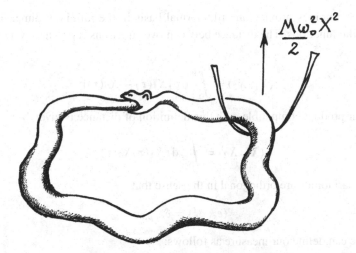

$$\frac{M\omega_o^2 X^2}{2}$$

Figure 2.2. A string in the gutter.

Let us consider the thermodynamic probability distribution for the system (2.26) given by the Gibbs distribution formula:

$$\mathrm{d}P(E) = \frac{1}{Z}\mathrm{e}^{-E/T}\mathrm{d}\Omega \tag{2.27}$$

where $\mathrm{d}\Omega$ is the volume which this state occupies in the phase space. The reader will understand that the main problem here is to define this volume or, in other words, the *measure of integration*. To do this we have to recall some facts about metric spaces.

A metric space is a space where one can define the distance between any two points. An N-dimensional Euclidean space with defined scalar product is a good example of a metric space. In such a space one can introduce an orthogonal basis of vectors

$$(\mathbf{e}_i, \mathbf{e}_j) = \delta_{i,j}$$

Then any vector function of coordinates can be represented as

$$\mathbf{X} = \sum_n x_n \mathbf{e}_n$$

and the element of volume is given by

$$\mathrm{d}\Omega = \mathrm{d}x_1 \cdots \mathrm{d}x_N$$

Does our function $X(\tau)$ belong to some metric space? It does: as a periodic function on the interval $(0, L)$ it can be expanded into Fourier harmonics:

$$X(\tau) = \sum_{s=-\infty}^{\infty} e_s(\tau)X_s$$
$$e_s(\tau) = \frac{1}{\sqrt{L}}\exp\left(2\mathrm{i}\pi s\tau/L\right) \tag{2.28}$$

The Fourier harmonics provide an orthonormal basis in the infinitely dimensional space of real periodic functions. The distance between two functions $X_1(\tau)$ and $X_2(\tau)$ is defined as

$$\rho^2(X_1, X_2) = \int_0^L d\tau \, [X_1(\tau) - X_2(\tau)]^2 \tag{2.29}$$

and the scalar product compatible with this definition of distance is given by

$$(X_1, X_2) = \int_0^L d\tau \, X_1(\tau) X_2(\tau) \tag{2.30}$$

The Fourier harmonics are orthogonal in the sense that

$$(e_s, e_p) = \delta_{s+p,0} \tag{2.31}$$

Therefore we can define our measure as follows:

$$d\Omega = \prod_s dX_s \tag{2.32}$$

where X_s are coefficients in the Fourier expansion of the function $X(\tau)$.
Substituting (2.28) into the expression for energy (2.26) we get:

$$E = \frac{M}{2} \sum_s X_{-s}[(2\pi s/L)^2 + (\omega_0)^2]X_s \tag{2.33}$$

and

$$dP[X] = \frac{1}{Z} e^{-E[X]/T} \prod_s dX_s \tag{2.34}$$

Now using the obtained probability distribution let us calculate the pair correlation function:

$$\langle X(\tau_1)X(\tau_2) \rangle = \int dP[X]X(\tau_1)X(\tau_2) = L^{-1} \sum_{s,s'} \exp\left[2i\pi(s\tau_1 + s'\tau_2)/L\right]$$

$$\times \left\{ \frac{\int_{-\infty}^{\infty} \prod dX_p X_s X_{s'} \exp\left(-\frac{M}{2T}\sum_p X_{-p}A_pX_p\right)}{\int_{-\infty}^{\infty} \prod dX_p \exp\left(-\frac{M}{2T}\sum_p X_{-p}A_pX_p\right)} \right\} \tag{2.35}$$

where

$$A_p = \left[(2\pi p/L)^2 + \omega_0^2\right]$$

The obtained Gaussian integral is just a generalization of (2.10). It is easy to see that it gives a nonzero answer only if $s + s' = 0$:

$$\left[\sum_{s,s'} \int X_s X_{s'} \exp\left(-\frac{M}{2T}\sum_p X_{-p}A_pX_p\right) \prod dX_p\right] \Big/$$

$$\left[\int \exp\left(-\frac{M}{2T}\sum_p X_{-p}A_pX_p\right) \prod dX_p\right] = \delta_{s,-s'}\frac{T}{MA_s}$$

Table 2.1. Equivalence between quantum and classical

Quantum oscillator	Classical string in a gutter
Green's function	Correlator
$\theta(\tau_{12})\langle \hat{x}(\tau_1)\hat{x}(\tau_2)\rangle$ $\theta(\tau_{21})\langle \hat{x}(\tau_2)\hat{x}(\tau_1)\rangle$	$\langle x(\tau_2)x(\tau_1)\rangle_{\text{Gibbs}}$
Hamiltonian	Energy
$\hat{H} = \dfrac{\hat{p}^2}{2M} + \dfrac{M\omega_0^2\hat{x}^2}{2}$	$E = \dfrac{M}{2}\left(\dfrac{dx}{d\tau}\right)^2 + \dfrac{M\omega_0^2 x^2}{2}$
Inverse temperature	Length of circle
$\hbar\beta$	L
Planck constant	Temperature
$\hbar$	T

Therefore we have

$$\langle X(\tau_1)X(\tau_2)\rangle = \frac{T}{ML}\sum_s \frac{\exp\left[2i\pi s(\tau_1 - \tau_2)/L\right]}{(2\pi s/L)^2 + \omega_0^2} \tag{2.36}$$

As we might expect, this expression is similar to the expression for the quantum Green function given by (2.25), which reflects the equivalence between classical and quantum mechanical descriptions. The reader has to remember, however, that these descriptions relate *different* objects (the oscillator and the string in the given case).

From what we have found out we can compose the 'dictionary' given in Table 2.1. In order to strengthen the case for the equivalence, let us calculate one more physical quantity, the partition function.

For the classical string we have

$$Z = \int \prod_s dX_s \exp\left[-\frac{M}{2T}X_{-s}A(s)X_s\right] \propto \prod_s A(s)^{-1/2} \tag{2.37}$$

$$\ln Z = -\frac{1}{2}\sum_s \ln\left[\omega_0^2 + (2\pi s/L)^2\right] \tag{2.38}$$

Since the above sum formally diverges, it is more convenient to calculate its derivative

$$\frac{\partial \ln Z}{\partial \omega_0^2} = -\frac{1}{2}\sum_{s=-\infty}^{\infty} \frac{1}{\omega_0^2 + (2\pi s/L)^2} \tag{2.39}$$

which converges.

To calculate the sum over s we can employ a standard trick: rewrite the sum as a contour integral in the complex plane surrounding the poles of $\coth(Lz/2)$:

$$\sum_s \frac{1}{\left[\omega_0^2 + (2\pi s/L)^2\right]} = \frac{L}{4i\pi}\oint_C \coth(Lz/2)\frac{dz}{\omega_0^2 - z^2}$$

Bending the contour C about the poles $z = \pm\omega_0$ we get

$$\frac{\partial \ln Z}{\partial \omega_0^2} = \frac{L}{4\omega_0} \coth(\omega_0 L/2) \qquad (2.40)$$

$$\ln Z = \ln Z_0 + \ln(1 - e^{-L\omega_0}) \qquad (2.41)$$

where Z_0 depends linearly on L. Now recall that L in the quantum language means inverse temperature and the thermodynamic potential is defined as

$$\Omega/\hbar = -\frac{T}{\hbar} \ln Z \rightarrow \frac{1}{L} \ln Z$$

Therefore from (2.25) we get

$$\Omega = \Omega_0 - \hbar T \ln(1 - e^{-\hbar\omega_0/T}) \qquad (2.42)$$

which is the well known expression for the harmonic oscillator.

So the equivalence holds.

Exercise

Consider a system of one-dimensional acoustic phonons. This system can be represented as a system of harmonic oscillators with frequencies $\omega(q) = c|q|$, where q is the wave vector of a phonon. The Green's function in this case depends not only on time but also on a spatial coordinate y. Calculate the correlation function of velocities:

$$\langle\langle\partial_{\tau_1}x(\tau_1, y)\partial_{\tau_2}x(\tau_2, 0)\rangle\rangle = -\int_{-\infty}^{\infty} \frac{dq}{2\pi} e^{iqy}\partial_\tau^2 D(\tau, q) \qquad (2.43)$$

where $\tau = \tau_1 - \tau_2$ and $D(\tau, q)$ is given by (2.23) with $\omega_0 = c|q|$. The answer is

$$\frac{\hbar T}{2Mc}\{\cot[\pi T(\tau + ix/c)] + \cot[\pi T(\tau - ix/c)]\} \qquad (2.44)$$

This result is intimately related to the material discussed further in Part IV.

3
Definitions of correlation functions: Wick's theorem

In the previous chapter we considered the equivalence between the quantum mechanics of a point oscillator and the classical statistical mechanics of a one-dimensional closed string. Now I shall generalize this equivalence as the equivalence between D-dimensional QFT and $(D + 1)$-dimensional statistical mechanics.

Suppose we have a quantum mechanical system at a temperature β^{-1}. (At this stage we shall not distinguish between the quantum mechanics of a finite number of particles and QFT, treating the latter as a limiting case.) On the classical level our system is described by the Lagrange function

$$L(\dot{x}_n, x_n)$$

where x_n are canonical coordinates. Then the quantum field theory for this system can be formulated in the language of *classical statistical mechanics* of *another* system at a temperature $\hbar$, whose energy functional is related to the classical Lagrangian of the original system:

$$E = -\int_0^{\hbar\beta} \mathrm{d}\tau\, L\left(-\mathrm{i}\frac{\mathrm{d}x_n}{\mathrm{d}\tau}, x_n\right) \tag{3.1}$$

This equivalence should be understood as the identity of the time ordered quantum correlation functions and the correlation functions of the classical system:

$$\langle \hat{T}\hat{x}_{n_1}(\tau_1)\cdots\hat{x}_{n_N}(\tau_N)\rangle_{\mathrm{QFT}} = \frac{\int \mathrm{D}x_n(\tau)x_{n_1}(\tau_1)\cdots x_{n_N}(\tau_N)\exp\left[\hbar^{-1}\int_0^{\hbar\beta}\mathrm{d}\tau L\right]}{\int \mathrm{D}x_n(\tau)\exp\left[\hbar^{-1}\int_0^{\hbar\beta}\mathrm{d}\tau L\right]} \tag{3.2}$$

Here we use the notation $\mathrm{D}x(\tau)$ to denote the infinite-dimensional integral (the path integral) introduced in the previous chapter. It is assumed that we integrate over all harmonics of the function $x(\tau)$ with the corresponding measure (for example, with the one given by (2.32)).

Now I can explain why the 'equivalence' outlined above does not breach a deep divide between the quantum and classical worlds. The reason is that in classical thermodynamics energy is a real quantity. However, not every Lagrangian remains real when time is made imaginary. Further in the text we shall encounter important examples of the theory of *free spin* (Chapter 16), and Chern–Simons (Chapter 15) and Wess–Zumino–Novikov–Witten models whose Lagrangians are *complex* in both Minkovsky and Euclidean space-time.

Therefore though the equivalence (3.1) holds on a formal level, in the sense that every quantum theory has a path integral representation, not every quantum theory has a meaningful classical counterpart with *real* energy.

For those who do not remember what the Lagrangian is I recall it briefly. In quantum mechanics we usually use the Hamiltonian which has the meaning of energy expressed in terms of coordinates and momenta:

$$H = E(p_n, x_n)$$

From the Hamiltonian one can find velocities:

$$\dot{x}_n = \frac{\partial H}{\partial p_n}$$

and re-express H in terms of *velocities* and coordinates:

$$H \to H(\dot{x}_n, x_n)$$

Then the Lagrangian is defined as follows:

$$L \equiv \sum_n p_n(\dot{x}_m, x_m)\dot{x}_n - H(\dot{x}, x)$$

As an example let us consider a system of nonrelativistic particles interacting via potential forces. Then we have

$$H = \sum_n \frac{p_n^2}{2M_n} + U(x_1, \ldots, x_N)$$

According to the definition of the Lagrangian we have

$$L = \sum_n \frac{M_n \dot{x}_n^2}{2} - U(x_1, \ldots, x_N)$$

Finally, substituting the imaginary time $\tau = it$ we have

$$E = -L(-i\dot{x}_n, x_n) = \sum_n \frac{M_n}{2}\left(\frac{dx_n}{d\tau}\right)^2 + U(x_1, \ldots, x_N)$$

This looks like the original Hamiltonian, but only for this particular simple case! Further in the text we shall encounter more difficult cases, especially in the chapters relating to spin models in Part III.

In what follows I shall assume

$$\hbar = k_B = 1$$

and call the quantity

$$S = -\int_0^\beta d\tau \, L(-i\dot{x}_n, x_n)$$

the 'thermodynamic action' or simply the 'action'. For those quantum theories which have meaningful classical counterparts, this action is real; however, as I have already mentioned, there are quantum theories where S is complex.

Let us study correlation functions. It was explained in Chapter 1 that the correlation functions can be introduced as functional derivatives of free energy with respect to an external field. Let us pursue this further. Suppose we are interested in correlation functions of some physical quantity M depending on the canonical coordinates $M = M(\mathbf{x})$. Then we can express its correlation functions as the functional derivatives of the so-called generating functional $Z[H]$:[1]

$$Z[H] = \int \mathrm{D}x_n(\tau) \exp\left\{-S + \int d\tau \sum_n H_n(\tau) M[x_n(\tau)]\right\} \tag{3.3}$$

Its derivatives are called reducible correlation functions:

$$\langle M_n(\tau) \rangle \equiv Z^{-1}[0] \frac{\delta Z[H]}{\delta H_n(\tau)}\bigg|_{H=0}$$

$$\langle M_{n_2}(\tau_2) M_{n_1}(\tau_1) \rangle \equiv Z^{-1}[0] \frac{\delta^2 Z[H]}{\delta H_{n_2}(\tau_2) \delta H_{n_1}(\tau_1)}\bigg|_{H=0} \cdots \tag{3.4}$$

$$\langle M_{n_N}(\tau_N) M_{n_{N-1}}(\tau_{N-1}) \cdots M_{n_1}(\tau_1) \rangle \equiv Z^{-1}[0] \frac{\delta^N Z[H]}{\delta H_{n_N}(\tau_N) \cdots \delta H_{n_1}(\tau_1)}\bigg|_{H=0} \quad \text{etc.}$$

Notice the difference between this definition and the definition of *irreducible* correlation functions given in Chapter 1. The latter are denoted as $\langle\langle \cdots \rangle\rangle$ and are defined by

$$\langle\langle M(1) \cdots M(N) \rangle\rangle = \frac{\delta^N \ln Z[H]}{\delta H(1) \cdots \delta H(N)}\bigg|_{H=0} \tag{3.5}$$

From the definitions (3.4) and (3.5) one can derive the relationship between these types of correlation functions. Let us do this step by step:

$$\langle\langle M(1) \rangle\rangle = \langle M(1) \rangle$$

$$\langle\langle M(1)M(2) \rangle\rangle = \langle M(1)M(2) \rangle - \langle M(1) \rangle \langle M(2) \rangle$$

The latter identity follows from the following one:

$$\frac{\delta^2 \ln Z[H]}{\delta H(1)\delta H(2)}\bigg|_{H=0} = \frac{\delta}{\delta H(1)}\left(\frac{1}{Z}\frac{\delta Z}{\delta H(2)}\right) = \frac{1}{Z}\frac{\delta^2 Z}{\delta H(1)\delta H(2)} - \frac{1}{Z^2}\frac{\delta Z}{\delta H(1)}\frac{\delta Z}{\delta H(2)} \tag{3.6}$$

Differentiating further we get:

$$\langle\langle M(1)M(2)M(3) \rangle\rangle = \langle M(1)M(2)M(3) \rangle - \langle M(1)M(2) \rangle \langle M(3) \rangle - \langle M(1)M(3) \rangle \langle M(2) \rangle$$
$$- \langle M(2)M(3) \rangle \langle M(1) \rangle + 2\langle M(1) \rangle \langle M(2) \rangle \langle M(3) \rangle$$

Let us now rewrite the third equation in terms of irreducible functions:

$$\langle M(1)M(2)M(3) \rangle = \langle\langle M(1)M(2)M(3) \rangle\rangle + \langle\langle M(1)M(2) \rangle\rangle \langle\langle M(3) \rangle\rangle$$
$$+ \langle\langle M(1)M(3) \rangle\rangle \langle\langle M(2) \rangle\rangle + \langle\langle M(2)M(3) \rangle\rangle \langle\langle M(1) \rangle\rangle + \langle\langle M(1) \rangle\rangle \langle\langle M(2) \rangle\rangle \langle\langle M(3) \rangle\rangle$$

---

[1] Putting the variable H into square brackets we emphasize that $Z[H]$ is a *functional* of a function $H(x)$.

Now we can conjecture the following general rule:

$$\langle M(1) \cdots M(N) \rangle = \langle\langle M(1) \cdots M(N) \rangle\rangle$$

$$+ \sum_{i=1}^{N} \langle\langle M(1) \cdots M(i-1)M(i+1) \cdots M(N) \rangle\rangle \langle\langle M(i) \rangle\rangle$$

$$+ \sum_{i>j} \langle\langle M(1) \cdots M(j-1)M(j+1) \cdots M(i-1)M(i+1) \cdots M(N) \rangle\rangle$$

$$\times \langle\langle M(i)M(j) \rangle\rangle + \cdots + \langle\langle M(1) \rangle\rangle \cdots \langle\langle M(N) \rangle\rangle \qquad (3.7)$$

This validity can be proved by induction.

Now let us apply the wisdom we have acquired to some practically important field theory. It is better to begin with something fairly simple. Let us consider, for example, the theory of a free massive scalar field described by the following action:

$$S = \frac{1}{2} \int d\tau \int d^D x \left[\left(\frac{d\Phi}{d\tau} \right)^2 + (\nabla \Phi)^2 + m^2 \Phi^2 \right] = \frac{1}{2\beta V} \sum_p \Phi(-p)(p^2 + m^2)\Phi(p)$$

$$(3.8)$$

where

$$p_0 = 2\pi n_0/\beta \qquad p_i = 2\pi n_i/L_i \qquad i = 1, \ldots, D$$

and

$$\Phi(p) = \int d\tau d^D x e^{i\omega_n \tau} e^{i\mathbf{p}\mathbf{x}} \Phi(\tau, \mathbf{x})$$

In high energy physics this model is known as the Klein–Gordon model. In condensed matter physics it describes long wavelength optical phonons. We shall study the action (3.8) in the Euclidean space-time ($t = i\tau, x_1, \ldots, x_D$).

Let us introduce a generating functional for Φ-fields:

$$Z[\eta] = \int \prod_p d\Phi(p) \exp \left[-\frac{1}{2} \sum_p \Phi(-p)(p^2 + m^2)\Phi(p) + \sum_p \eta(-p)\Phi(p) \right] \quad (3.9)$$

The above integral is just a product of simple Gaussian integrals and can be calculated by the shift of variables, which removes the term linear in Φ from the exponent:

$$\bar{\Phi}(p) = \Phi(p) + (p^2 + m^2)^{-1}\eta(p) \qquad (3.10)$$

$$\ln Z[\eta] = \ln Z[0] + \frac{1}{2\beta V} \sum_p \eta(-p)\frac{1}{(p^2 + m^2)}\eta(p) \qquad (3.11)$$

Applying here the definition of the irreducible correlation function (3.5) we find that the only nonvanishing irreducible correlation function is the pair correlator

$$\langle\langle \Phi(-p')\Phi(p) \rangle\rangle = \left. \frac{\delta^2 \ln Z}{\delta\eta(p)\delta\eta(p')} \right|_{\eta=0} = \delta_{p',p} \left(\frac{1}{p^2 + m^2} \right) \qquad (3.12)$$

Then according to the general theorem (3.7) any N-point *reducible* correlation function is represented as a sum of all possible products of two-point *irreducible* functions, i.e. *Wick's theorem*:

$$\langle \Phi(1) \cdots \Phi(2N) \rangle = \sum_P \langle\langle \Phi(p_{i_1}) \Phi(p_{j_1}) \rangle\rangle \cdots \langle\langle \Phi(p_{i_N}) \Phi(p_{j_N}) \rangle\rangle \qquad (3.13)$$

The validity of Wick's theorem is completely independent of how we define the correlation functions. We can do this using the path integral approach or use the old-fashioned definition with time ordered products – the theorem works all the same.

4

Free bosonic field in an external field

Let us now consider the situation which arises in many applications. Suppose the free scalar field introduced in the previous chapter interacts with some other field (or fields) f. The entire generating functional is given by

$$Z[\eta] = \int Df D\Phi \exp\left[-S[f] - \frac{1}{2}\Phi\bar{A}(f)\Phi + \int \eta\Phi\right] \tag{4.1}$$

where $\bar{A}(f)$ is some linear operator acting on Φ and $S[f]$ is an action for the field f. Since the path integral is a Gaussian one we can integrate over Φ (at least in principle) and get a result which depends only on f and η:

$$\int D\Phi \exp\left[-S[f] - \frac{1}{2}\Phi\bar{A}(f)\Phi + \int \eta\Phi\right] \equiv \exp\{-S_{\text{eff}}[f, \eta]\} \tag{4.2}$$

Now we can treat $S_{\text{eff}}[f, \eta]$ as a new action for the f field and forget about Φ.

Let us discuss the outlined procedure in more detail. First, let us recall how to calculate Gaussian integrals. In order to make things easier, let us discretize space-time. Then all operators become matrices and we have

$$\Phi\bar{A}(f)\Phi = \sum_{n,m} \Phi_n A(f)_{nm}\Phi_m \tag{4.3}$$

The path integral degenerates into the multi-dimensional integral:

$$\int \prod_n d\Phi_n \exp\left[-\frac{1}{2}\sum_{n,m}\Phi_n A(f)_{nm}\Phi_m + \sum_n \eta_n\Phi_n\right] \equiv I[f, \eta] \tag{4.4}$$

To calculate this integral we make a shift of variables which removes the term linear in Φ:

$$\Phi_n = \tilde{\Phi}_n + (A^{-1})_{nm}\eta_m \tag{4.5}$$

Since this shift does not affect the measure of integration we get

$$I[f, \eta] = \exp\left[\frac{1}{2}\sum_{n,m}\eta_n(A^{-1})_{nm}\eta_m\right] I[f, 0] \tag{4.6}$$

The remaining integral $I[f, 0]$ has the property

$$I[f, 0] \propto [\det \hat{A}(f)]^{-1/2} \tag{4.7}$$

In order to check this we represent Φ_n as a sum of eigenvectors of the matrix $\hat{A}$. In this representation the exponent in (4.4) becomes diagonal which trivializes the integral.

Using the remarkable identity

$$\ln \det \hat{A} = \mathrm{Tr} \ln \hat{A} \tag{4.8}$$

we get from (4.6) and (4.7) the following expression for the effective action:

$$S_{\mathrm{eff}}[f, \eta] = S[f] + \frac{1}{2} \mathrm{Tr} \ln \hat{A}(f) - \frac{1}{2} \eta \hat{A}^{-1} \eta \tag{4.9}$$

In fact, for the situation we have dealt with so far (I mean the free scalar field defined on a flat space-time), the above definition is quite rigorous and survives the transition to the continuum limit. Indeed, the continuous space-time becomes discrete in momentum representation and all differential operators become just matrices. One has to be careful, however, on a curved space-time. The corresponding problems are discussed in a later part of this chapter.

Let us consider a simple example of the scalar field on a flat space-time governed by the action (3.8) with an additional term

$$\int d\tau d^D x \, V(\xi) \Phi^2$$

where

$$\hat{A}(\xi, \xi') = \left[-\partial_a^2 + m^2 + V(\xi) \right] \delta^{(D+1)}(\xi - \xi') \tag{4.10}$$

and ξ denotes space-time coordinates. The matrix inverse to $\hat{A}$, usually called the Green's function of the operator $\hat{A}$, satisfies the following equation:

$$\left[-\partial_a^2 + m^2 + V(\xi) \right] G(\xi, \xi') = \delta^{(D+1)}(\xi - \xi') \tag{4.11}$$

At $V = 0$ this equation can be easily solved by the Fourier transformation:

$$G_0(\xi, \xi') = \int \frac{d^{D+1}p}{(2\pi)^{D+1}} \left[\frac{\exp(i\mathbf{p}\xi_{12})}{p^2 + m^2} \right] \tag{4.12}$$

where $\xi_{12} = \xi - \xi'$. At finite V we can rewrite (4.11) in the following form:

$$G(\xi, \xi') = G_0(\xi, \xi') - \int d^{D+1}\xi_1 G_0(\xi, \xi_1) V(\xi_1) G(\xi_1, \xi') \tag{4.13}$$

and then write the solution in the form of the expansion:

$$G(\xi, \xi') = G_0(\xi, \xi') - \int d^{D+1}\xi_1 G_0(\xi, \xi_1) V(\xi_1) G_0(\xi_1, \xi')$$

$$+ \int d^{D+1}\xi_1 \int d^{D+1}\xi_2 G_0(\xi, \xi_1) V(\xi_1) G_0(\xi_1, \xi_2) V(\xi_2) G_0(\xi_2, \xi') - \cdots \tag{4.14}$$

It is convenient to represent this expansion in a graphical form (see Fig. 4.1).

The wavy lines represent the V-field and the solid lines are the Green's functions $G_0(\xi_1 - \xi_2)$. We integrate over all coordinates of the vertices.

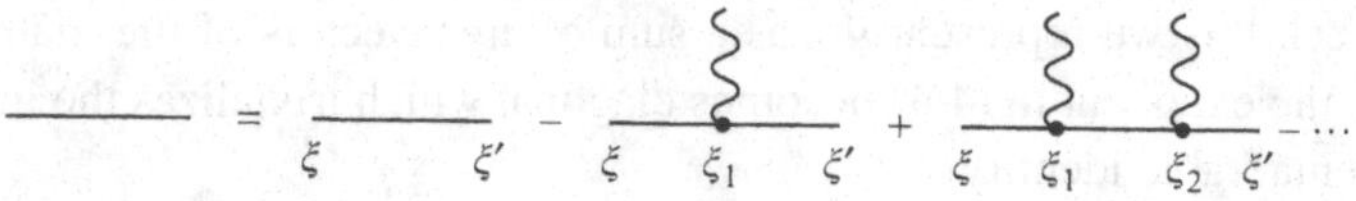

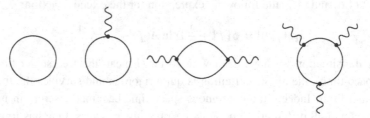

Figure 4.1.

Figure 4.2.

A similar expansion can be written for Tr ln G (see Fig. 4.2).

This sort of expansion is called a loop expansion. The function V is used as a small parameter.

Semiclassical approximation

In many applications we deal with situations where V changes strongly, but over a large integral, i.e. slowly (I shall make it clear below what the quantitative equivalent is to the word 'slowly'). In this case we have to expand not in powers of V, but in its gradients. This approach is called 'semiclassical approximation'. For constant $V = V_0$ the solution of (4.11) is

$$G_0(\xi_1 - \xi_2) = \int \frac{d^{D+1} p}{(2\pi)^{D+1}} \left[\frac{\exp(i\mathbf{p}\xi_{12})}{p^2 + m^2 + V_0} \right] \propto \exp(-M|\xi_{12}|)$$

which decays exponentially at large distances ($M^2 = m^2 + V_0$). Therefore there is a natural space-time scale M^{-1} in the theory and one can evaluate all changes with respect to this scale. A slow change of the potential means that its relative change on the distance M^{-1}, where the Green's function changes strongly, is small, i.e.

$$M^{-1} \left| \frac{\nabla V}{V} \right| \ll 1$$

Then one can suppose that $G(\xi_1, \xi_2)$ is a rapidly varying function of

$$x = \xi_1 - \xi_2$$

and a slowly varying function of

$$R = \frac{\xi_1 + \xi_2}{2}$$

In fact, since ξ_1 is present asymmetrically in the equation for the Green's function, it is more convenient to consider the Green's function as a function of $\xi_1 - \xi_2$ and $r \equiv \xi_1$ and represent it as follows:

$$G(\xi_1, \xi_2) = \int \frac{d^{D+1}p}{(2\pi)^{D+1}} \exp(i p \xi_{12}) G(\mathbf{p}, r) \qquad (4.15)$$

Substituting this expression into (4.11) we get the new equation for $G(\mathbf{p}, r)$:[1]

$$[p^2 + m^2 + V(r)]G(\mathbf{p}, r) + [2i\mathbf{p}\nabla_r - \Delta_r]G(\mathbf{p}, r) = 1 \qquad (4.16)$$

This equation can be solved by iteration in the same way as above, with the only difference that the 'tails' in the loop expansion now represent the operators

$$\Lambda = 2i\mathbf{p}\nabla_r - \Delta_r$$

and the bare Green's function is given by:

$$G_0(\mathbf{p}, r) = [p^2 + m^2 + V(r)]^{-1}$$

As an exercise let us calculate $\mathrm{Tr}G$ and $\mathrm{Tr}\ln G$ up to $(\nabla V)^2$. From (4.16) we find

$$G^{(1)} = -G_0[2i\mathbf{p}\nabla_r - \Delta_r]G_0(\mathbf{p}, r)$$
$$= 2iG_0^3\mathbf{p}\nabla_r V + G_0\Delta_r G_0$$
$$G^{(2)} = -G_0[2i\mathbf{p}\nabla_r - \Delta_r]G^{(1)} \qquad (4.17)$$
$$\approx 4G_0\mathbf{p}\nabla_r[G_0^3\mathbf{p}\nabla_r V]$$

Here we have used the fact that $\nabla G_0 = -G_0^2 \nabla V$. The first correction to $\mathrm{Tr}G$ is given by

$$\mathrm{Tr}G^{(1)} = \int dr \int \frac{d^{D+1}p}{(2\pi)^{D+1}}[2iG_0^3\mathbf{p}\nabla_r V + G_0\Delta_r G_0] = -\int dr \int \frac{d^{D+1}p}{(2\pi)^{D+1}}(\nabla G_0)^2$$

$$= -\int dr \int \frac{d^{D+1}p}{(2\pi)^{D+1}} \frac{(\nabla V)^2}{(p^2 + V + m^2)^4} \qquad (4.18)$$

The second correction is equal to

$$\mathrm{Tr}G^{(2)} = \int dr \int \frac{d^{D+1}p}{(2\pi)^{D+1}} 4G_0\mathbf{p}\nabla_r[G_0^3\mathbf{p}\nabla_r V] = 4\int dr \int \frac{d^{D+1}p}{(2\pi)^{D+1}} G_0^5(\mathbf{p}\nabla V)^2 \quad (4.19)$$

In a similar way we get

$$\mathrm{Tr}\ln G = \int dr \int \frac{d^{D+1}p}{(2\pi)^{D+1}}\left[-\ln(p^2 + m^2 + V) + \frac{4}{5}\frac{(\mathbf{p}\nabla V)^2}{(p^2 + m^2 + V)^4} - \frac{1}{4}\frac{(\nabla V)^2}{(p^2 + m^2 + V)^3}\right]$$

$$(4.20)$$

In Chapter 7 we shall need the expression for the effective action for $D + 1 = 4$, $m^2 = 0$, $V = \Phi_0^2$. The simple calculation gives for this case:

$$S_{\mathrm{eff}}(\Phi_0^2) = -\frac{1}{2}\mathrm{Tr}\ln G = \int dr \left[\frac{7}{960\pi^2}(\nabla\Phi_0)^2 + \frac{1}{8\pi^2}\Phi_0^4 \ln\left(\frac{1}{a\Phi_0^2}\right)\right] \qquad (4.21)$$

where a is of the order of the lattice spacing.

---

[1] Everywhere in this book we use the notation $\nabla^2 = \Delta$.

Elements of differential geometry

In what follows we shall need certain notions and expressions from differential geometry. For the sake of convenience I give all the necessary definitions here with brief explanations. The reader who feels that this section is too difficult can skip it and continue reading from Chapter 5.

As we have already seen above, a conscious use of path integrals requires a knowledge of the elements of the geometry of curved spaces. There are two possibilities: either our fields are defined on a curved space-time (see the example (4.37) below), or they belong to a curved manifold (see examples of nonlinear sigma models in later chapters). Naturally, both situations can coexist. One might think that curved spaces appear only in cosmology, and not in condensed matter physics; this is not the case. An elementary example of a curved space is a deformed crystal. There are less elementary examples; some of them are discussed in detail in later chapters, particularly in Part IV, where we shall use elements of the Riemann geometry in our analysis of $(1 + 1)$-dimensional conformal field theory. However, it is more often true that the background space is flat and the manifold where the fields live is curved, so that the second possibility occurs more frequently. Anyway, the Riemann geometry is necessary.

We have already discussed an example of a metric space in Chapter 2, when the concept of path integral was introduced. It has been stated that a metric space is a space where for any two points ξ_1, ξ_2 one can define a scalar function $\rho(\xi_1, \xi_2)$ called distance. Then the discussion continued with one particular metric space, the Euclidean space. Now we consider a general metric space. Its geometrical properties are characterized by the so-called metric tensor g_{ab}; it relates the distance between two infinitely close points ds to their coordinates:

$$ds^2 = g_{ab}dx^a dx^b \tag{4.22}$$

Examples

(i) A cylinder of radius R in polar coordinates:

$$ds^2 = dz^2 + R^2 d\phi^2$$

(ii) A two-dimensional sphere of radius R in spherical coordinates:

$$ds^2 = R^2(d\theta^2 + \sin^2\theta d\phi^2)$$

One can choose the coordinate system differently; each choice has its own metric tensor, but the distances remain invariant. Suppose y_a are new coordinates such that

$$x^a = X^a(y_1, \ldots, y_D) \tag{4.23}$$

Substituting this expression into (4.22), we get

$$ds^2 = g_{cd}\frac{\partial X^c}{\partial y^a}\frac{\partial X^d}{\partial y^b}dy^a dy^b \tag{4.24}$$

i.e. the new metric tensor is

$$\tilde{g}_{ab}(y) = g_{cd} \frac{\partial X^c}{\partial y^a} \frac{\partial X^d}{\partial y^b} \tag{4.25}$$

It is necessary to introduce the inverse metric tensor

$$g^{ab} \equiv [g^{-1}]_{ab} \qquad g^{ab} g_{bc} = \delta_c^a \tag{4.26}$$

The inverse metric tensor leaves invariant the following differential form:

$$g^{ab} \partial_a \Phi \partial_b \Phi \tag{4.27}$$

where Φ is an arbitrary scalar function.

As we already know, for metric spaces one can define not only distances, but also elements of volume. Again, an infinitesimal element of volume must be invariant with respect to general coordinate transformations ((4.23) and (4.25)). The following definition satisfies this requirement:

$$d\Omega = \sqrt{g} dx^1 \cdots dx^D \tag{4.28}$$

where $g = \det \hat{g}$.

Examples

(i) A cylinder of radius R in polar coordinates:

$$d\Omega = R dz d\phi,$$

(ii) A two-dimensional sphere of radius R in spherical coordinates:

$$d\Omega = R^2 \sin \theta \, d\theta d\phi$$

There is an important case when our curved space is a space of some group manifold, i.e. it is invariant with respect to transformations of some Lie group. For example, a one-dimensional unit circle $x_1^2 + x_2^2 = 1$ is invariant with respect to rotations

$$\begin{pmatrix} x_1' \\ x_2' \end{pmatrix} = \hat{R}(\phi) \begin{pmatrix} x_1 \\ x_2 \end{pmatrix}$$

where

$$\hat{R}(\phi) = \begin{pmatrix} \cos \phi & \sin \phi \\ -\sin \phi & \cos \phi \end{pmatrix} \tag{4.29}$$

Matrices $\hat{R}(\phi)$ create a representation of the U(1) group. The obvious choice of the coordinate system for this case is the polar system:

$$x_1 = \cos \phi \qquad x_2 = \sin \phi$$

Then the distance between two close points is given by

$$ds^2 = d\phi^2 \tag{4.30}$$

The important fact is that the same expression can be written in the form independent of the representation:

$$ds^2 = -\frac{1}{2}\text{Tr}[\hat{R}^{-1}d\hat{R}\hat{R}^{-1}d\hat{R}] \tag{4.31}$$

This expression is general and defines a metric of a group manifold in terms of the group variables. Let us consider another important example: the SU(2) group. By definition, the SU(2) group is composed of transformations which leave invariant a complex quadratic form

$$z_1^* z_1 + z_2^* z_2$$

The equation

$$R^2 = z_1^* z_1 + z_2^* z_2 \tag{4.32}$$

is an equation of a sphere in two-dimensional complex (four-dimensional real) space. One can parametrize the sphere as follows:

$$z_1 = Re^{i(\phi-\psi)/2}\cos(\theta/2)$$
$$z_2 = Re^{i(\phi+\psi)/2}\sin(\theta/2) \tag{4.33}$$

Substituting these expressions in the element of distance we get

$$ds^2 = dz_1^* dz_1 + dz_2^* dz_2 = \frac{1}{4}R^2(d\phi^2 - 2d\phi d\psi \cos\theta + d\psi^2 + d\theta^2) \tag{4.34}$$

The element of volume is given by

$$d\Omega = \frac{1}{8}R^3|\sin\theta|d\theta d\psi d\phi \tag{4.35}$$

as it must be for a four-dimensional sphere. Now let us derive the expression (4.34) from (4.31). For this purpose we choose the so-called Euler parametrization of the SU(2) group:

$$\hat{R} = \exp\left(\frac{i}{2}\phi\sigma^z\right)\exp\left(\frac{i}{2}\theta\sigma^x\right)\exp\left(\frac{i}{2}\psi\sigma^z\right)$$

$$= \begin{pmatrix} \cos(\theta/2)\exp[i(\phi+\psi)/2] & i\sin(\theta/2)\exp[i(\phi-\psi)/2] \\ i\sin(\theta/2)\exp[i(\psi-\phi)/2] & \cos(\theta/2)\exp[-i(\phi+\psi)/2] \end{pmatrix} \tag{4.36}$$

where σ^a are the Pauli matrices. The group coordinates ψ, θ, ϕ belong to the manifold which is a sphere of radius 2π whose boundary is equivalent to its central point. The matrix $\hat{R}$ is invariant with respect to the transformations of coordinates $\phi \to \phi + 2\pi$, $\psi \to \psi + 2\pi$, $\theta \to \theta$.

Exercise

Reproduce (4.34) with $R = 1$ substituting (4.36) into (4.31). In general, every group manifold is a metric space with the invariant metric (4.31).

One of the most important and general principles of modern physics is the principle of *general covariance*. According to this principle, laws of physics do not depend on the choice of the coordinate system. For classical physics this means that action is generally covariant. The principle of general covariance in QFT requires covariance of generating functionals. Therefore we have here a combined symmetry of action and a measure of functional integration. Let us discuss covariance of actions first. Consider the following action, which is a covariant generalization of the action (3.8):

$$\frac{1}{2}\Phi\bar{A}(g)\Phi = \frac{1}{2}\int d^D\xi\sqrt{g(\xi)}[g^{ab}\partial_a\Phi\partial_b\Phi + m^2\Phi^2] \tag{4.37}$$

It is covariant provided Φ does not change under coordinate transformations. Such functions are, by definition, scalars. Now let us rewrite this action in the form (4.1), i.e. in such a form where all derivatives act on the right-hand side. Integrating by parts we get

$$\Phi\bar{A}(g)\Phi = \int d^D\xi\sqrt{g(\xi)}\left[-\Phi\frac{1}{\sqrt{g(\xi)}}\partial_a\left(\sqrt{g(\xi)}g^{ab}\partial_b\Phi\right) + m^2\Phi^2\right] \tag{4.38}$$

from which we conclude that the generalization of (4.11) with $V = 0$ for a curved space-time is

$$-\frac{1}{\sqrt{g(\xi)}}\partial_a\left[\sqrt{g(\xi)}g^{ab}\partial_b G(\xi,\xi')\right] + m^2 G(\xi,\xi') = \frac{\delta(\xi,\xi')}{\sqrt{g(\xi)}} \tag{4.39}$$

Thus we have defined an invariant action for a scalar field. However, in physics we not only have scalars, but vectors and tensors of all sorts. Among the well known physical fields the vector potential A_μ is a vector and its strength $F_{\mu\nu}$ is a tensor. We have to learn how to write covariant actions for these fields. For this purpose let us consider a general definition of a covariant tensor of the Nth rank $\Phi_{a_1,\dots,a_N}$. By definition such a tensor transforms under coordinate transformations in such a way that the following form

$$\Phi_{a_1,\dots,a_N}dx^{a_1}\cdots dx^{a_N} \tag{4.40}$$

remains invariant. Each such tensor has its conjugate element, i.e. a *contravariant* tensor defined as

$$\bar{\Phi}^{a_1,\dots,a_N} = g^{a_1 b_1}\cdots g^{a_N b_N}\Phi_{b_1,\dots,b_N} \tag{4.41}$$

such that the product $\bar{\Phi}(\xi)\Phi(\xi)$ is a scalar, i.e. is invariant under coordinate transformations. One can move indices of tensors up and down, multiplying them by the inverse metric tensor g^{ab} (up) or by the metric tensor g_{ab} (down).

The measure of integration in the path integrals must be covariant; the proper definition is

$$D\mu = D\bar{\Phi}D\Phi \tag{4.42}$$

However, for a scalar field where $\bar{\Phi} = \Phi$, this simplifies to $D\Phi$.

Let us now discuss how to generalize the operation of differentiation. What we need is a linear operator D_a which satisfies the following two requirements. First, in the limit of flat space $D_a = \partial_a$. Second, if Φ_b is a covariant vector field, $D_a\Phi_b$ must be a covariant tensor of the second rank, i.e. it must transform under coordinate transformations as the metric tensor g_{ab} itself. The standard notation is

$$D_a \Phi_b = \partial_a \Phi_b - \Gamma_{ab}^c \Phi_c$$
$$D_a \Phi^b = \partial_a \Phi^b + \Gamma_{ca}^b \Phi^c$$

(4.43)

where Γ is a matrix independent of Φ (the operators D are linear!). In differential geometry this matrix is called the connection matrix or the Christoffel symbol. It can be shown that

$$\Gamma_{bc}^a = \frac{1}{2} g^{al} (\partial_b g_{cl} + \partial_c g_{bl} - \partial_l g_{bc})$$

(4.44)

Check that with this definition of Γ the expressions (4.43) have the necessary transformation properties. There are several properties of covariant derivatives which are important to remember: (i) the matrices Γ are not tensors and therefore one cannot raise and lower their indices by the metric tensor; (ii) for 'unscrewed' metric spaces (i.e. in the absence of dislocations) the connection matrices are symmetric with respect to permutation of their low indices, $\Gamma_{ab}^c = \Gamma_{ba}^c$; (iii) for a general tensor with P upper and Q lower indices we have

$$D_k T_{j_1 \cdots j_Q}^{i_1 \cdots i_P} = \partial_k T_{j_1 \cdots j_Q}^{i_1 \cdots i_P} + \sum_{a=1}^{P} \Gamma_{kl}^{i_a} T_{j_1 \cdots j_Q}^{i_1 \cdots i_{a-1} l i_{a+1} \cdots i_P} - \sum_{b=1}^{Q} \Gamma_{kj_b}^l T_{j_1 \cdots j_{b-1} l j_{b+1} \cdots j_Q}^{i_1 \cdots i_P}$$

(4.45)

(iv) covariant derivatives of metric (inverse metric) tensors are equal to zero:

$$D_a g^{bc} = 0$$

One can define a contravariant derivative acting by the inverse metric tensor:

$$D^a = g^{ab} D_b$$

Naturally, covariant (as well as contravariant) derivatives do not commute. Let us check this with the operators $D_a D_b$ and $D_b D_a$ acting on a scalar function f. Since f is a scalar, we have:

$$D_b f = \partial_b f$$

(4.46)

which is already a covariant vector. The corresponding contravariant vector is $g^{br} \partial_r f$ to which we can apply (4.43). Thus we have

$$D_a \partial_b f = \partial_a \partial_b f - \Gamma_{ab}^c \partial_c f$$

(4.47)

and

$$[D_a, D_b] f = R_{kab}^k f$$

(4.48)

where

$$R_{lab}^k = \left(\partial_a \Gamma_{lb}^k + \Gamma_{am}^k \Gamma_{bl}^m \right) - (a \leftrightarrow b)$$

(4.49)

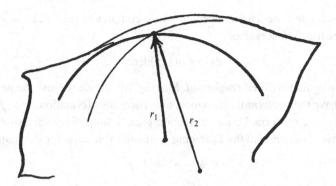

Figure 4.3. Two main radii of a two-dimensisonal surface at a given point A.

is called the Riemann tensor. This tensor generates two other important tensors: (i) $R_{ab} = R^k_{kab}$, the Ricchi tensor; and, (ii) $R = R^a_a$, the scalar curvature.

Now we can formulate a generalization of the action (4.37) for tensor fields $\Phi^{(n)} \equiv \Phi_{i_1,...,i_n}$:

$$\frac{1}{2}\bar{\Phi}^{(n)}\bar{A}(g)\Phi^{(n)} = \frac{1}{2}\int d^D\xi\sqrt{g(\xi)}\bar{\Phi}^{(n)}[-g^{ab}D_aD_b\Phi^{(n)} + m^2\Phi^{(n)}] \qquad (4.50)$$

Notice that on flat space the actions for scalars and tensors ((4.50) and (4.37)) coincide.

For two-dimensional spaces where $D = 2$ the situation is greatly simplified. The Riemann tensor can be written as follows:

$$R^k_{lab} = \frac{1}{2}\gamma^k_l\gamma_{ab}R \qquad (4.51)$$

where

$$\gamma_{ab} = \epsilon_{ab}/\sqrt{g} \qquad (4.52)$$

and ϵ_{ab} is the absolutely antisymmetric tensor whose nonzero components are $\epsilon_{12} = -\epsilon_{21} = 1$.

Imagine now that our two-dimensional space is a surface embedded in a three-dimensional flat space. Then the scalar curvature R has a very simple interpretation: $R = 2/r_1r_2$ where r_i are two main radii at this point (see Fig. 4.3).

Another great simplification comes from the fact that in two dimensions one always chooses (at least locally) such a coordinate system where the metric tensor is diagonal:

$$ds^2 = \rho(\xi)(d^2\xi^1 + d^2\xi^2) \qquad (4.53)$$

For example, for a sphere we have

$$\xi_1 = \phi, \ \xi_2 = \int \frac{d\theta}{\sin\theta}$$

and

$$\rho = \sin^2\theta$$

In such a coordinate system it is convenient to use complex coordinates $z = \xi_1 + i\xi_2$, $\bar{z} = \xi_1 - i\xi_2$ and rewrite the interval as

$$ds^2 = \rho(\bar{z}, z)d\bar{z}dz \tag{4.54}$$

This coordinate system is called *conformal*. Making this choice we lose the general covariance, but still have the *conformal* invariance: analytical transformations $z = f(z)$ preserve the form (4.54). The conformal parametrization greatly simplifies expressions for the covariant derivative. There is also the following valuable expression for the scalar curvature:

$$R = -\rho^{-1}\partial_a^2 \ln \rho \tag{4.55}$$

(it is valid in a conformal coordinate system only!)

In conformal notation the action (4.50) becomes particularly simple:

$$\int d\bar{z}dz\rho \bar{\Phi}^{(n)} \left[-\rho^{-(n+1)}\partial(\rho^n \bar{\partial}\Phi^{(n)}) + \frac{m^2}{4}\Phi^{(n)} \right] \tag{4.56}$$

(make sure that, for a scalar field $n = 0$, this expression is in agreement with (4.37)).

Obviously, there are not many examples of curved spaces where (4.39) can be solved in an explicit form. For $m^2 = 0$ it is possible to solve for a general metric tensor in two-dimensional space-time; we shall discuss this example later in Part IV, Chapter 22.

5

Perturbation theory: Feynman diagrams

This chapter is the first one in a sequence of chapters where the perturbation theory expansion is discussed. This expansion will be formulated in a graphical form, i.e. every term of the perturbation series will be represented by a picture – the Feynman diagram. In our derivation we shall follow the path integral approach. The final results, and in particular the general rules for diagram construction, are based on Wick's theorem and are therefore independent of a particular approach. Those who find the explanations unclear should refer to older, classical books.

In order to make the discussion less abstract, I will use a particular example, the so-called O(N)-symmetric vector model, described by the following action:

$$S = \int_0^\beta d\tau \int d^D x \left[\frac{1}{2} (\nabla \Phi)^2 + \frac{m^2}{2} \Phi^2 + \frac{g}{8} \Phi^2 \Phi^2 \right] \tag{5.1}$$

where $\Phi = (\Phi_1, \ldots, \Phi_N)$ is a vector field with N components, $g > 0$ and the ∇-operator includes both Euclidean time and space derivatives.

I choose the O(N)-symmetric vector model mainly because this relatively simple theory possesses an inexhaustible richness of physical properties. For this reason it is frequently used for educational purposes; a detailed discussion can be found, for example, in the book by Zinn-Justin (see the select bibliography). In studying this model we shall consider such important concepts in QFT as spontaneous symmetry breaking, renormalizability and scaling, and many others. The model is not simply a mathematical toy, however, but describes real systems. As a classical model it is a Ginzburg–Landau functional describing the vector order parameter in the vicinity of the second-order phase transition. In this case $m^2 \sim (T - T_c)/T_c$, where T_c is the critical temperature. Recall that the two transitions most frequently discussed in condensed matter physics, the superfluid and superconducting phase transitions, are described by the model (5.1) with $N = 2$; ferromagnetic phase transitions are described by this model with $N = 3$. The quantum version of the O(N) model describes, for example, anharmonic optical phonons or the Higgs bosons in high energy physics.

Perturbation expansion can be developed both in real and in momentum space. Since we have formulated Wick's theorem in momentum space, let us do it in the momentum representation. After the Fourier transformation of the fields the action (5.1) acquires the following form:

$$S = S_0 + S_1$$

$$S_0 = \frac{1}{2\beta V} \sum_p \Phi(-p)(p^2 + m^2)\Phi(p) \tag{5.2}$$

$$S_1 = \frac{g}{8(\beta V)^3} \sum_{p,q,k} [\Phi(-q + p/2)\Phi(q + p/2)][\Phi(-k - p/2)\Phi(k - p/2)] \tag{5.3}$$

Assuming that $m^2 > 0$ we shall expand the correlation functions in powers of the coupling constant g. In this case it is difficult to continue the discussion in the general form analysing the generating functional. Therefore I shall discuss only three particular correlation functions: the partition function itself, the two-point and the four-point correlation functions.

The expansion for the partition function is given by:

$$Z[0] = \int D\Phi e^{-S} = \int D\Phi e^{-S_0} \left(1 - gS_1 + \frac{g^2}{2!}S_1^2 + \cdots\right)$$

$$= Z_0 \left(1 + \sum_{n=1}^{\infty} \frac{g^n(-1)^n}{n!} \langle S_1^n \rangle_0\right) \tag{5.4}$$

The quantity $\langle S_1^n \rangle_0$ is the $4n$-point correlation function

$$\langle \Phi_{\alpha_1}(1) \cdots \Phi_{\alpha_{4n}}(4n) \rangle_0$$

which we can calculate using Wick's theorem expressing it as a sum of products of two-point correlation functions. For reasons which will become clear later, the expansion for Z is not the simplest one. It turns out, however, that it is simpler to study the two-point correlation function

$$G_a(p) \equiv \langle\langle \Phi_a(-p)\Phi_a(p)\rangle\rangle = \frac{\int D\Phi\, \Phi_a(-p)\Phi_a(p)e^{-S}}{\int D\Phi e^{-S}}$$

$$= \frac{\int D\Phi\, \Phi_a(-p)\Phi_a(p)[1 - gS_1 + \frac{1}{2!}g^2S_1^2 + \cdots]e^{-S_0}}{\int D\Phi[1 - gS_1 + \frac{1}{2!}g^2S_1^2 + \cdots]e^{-S_0}} \tag{5.5}$$

As we shall see below, this function forms a 'building block' for the free energy expansion.

According to (5.5), in the first order in g the Green's function is given by

$$G_a^{(1)}(p) = -\frac{g}{8(\beta V)^3} \sum_{q,k',k} [\langle \Phi_a(-p)\Phi_a(p)\Phi_b(-k + q/2)\Phi_b(k + q/2)\Phi_c(-k' - q/2)$$

$$\times \Phi_c(k' - q/2)\rangle - \langle \Phi_a(-p)\Phi_a(p)\rangle\langle \Phi_b(-k + q/2)\Phi_b(k + q/2)$$

$$\times \Phi_c(-k' - q/2)\Phi_c(k' - q/2)\rangle] \tag{5.6}$$

The second term comes from the expansion of the denominator.

Let us apply Wick's theorem to this average. As you know, we have to rewrite this as a sum of two-point averages, making all sorts of pairings inside the original average. We shall denote each pairing pattern with a Feynman diagram. Because of the summation over

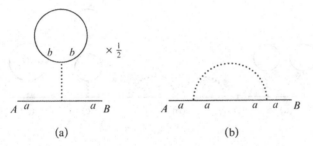

Figure 5.1.

internal momenta, some pairings yield the same expressions and therefore certain diagrams will have numerical prefactors. For expression (5.6) we have the following pairings:

$$\langle \Phi(A)\Phi(1)\Phi(1)g(1-1')\Phi(1')\Phi(1')\Phi(B) \rangle$$

$$- \langle \Phi(A)\Phi(B) \rangle \langle \Phi(1)\Phi(1)g(1-1')\Phi(1')\Phi(1') \rangle$$

$$= \langle \overbrace{\Phi(A)\Phi(1)}^{4}\ \overbrace{\Phi(1)g(1-1')\Phi(1')}^{1}\ \overbrace{\Phi(1')\Phi(B)}^{2} \rangle$$

$$+ \langle \overbrace{\Phi(A)\Phi(1)}^{4}\ \underbrace{\Phi(1)g(1-1')\ \overbrace{\Phi(1')\Phi(1')}\ \Phi(B)}_{1} \rangle \tag{5.7}$$

where the number of ways by which each connection can be chosen is shown with a brace. The quantity $g(1-1')$ represents the interaction with its delta-functions in the isotopic and momentum space.

The resulting diagrams are shown in Fig. 5.1 where solid lines denote the bare Green's function

$$\langle\langle \Phi_a(p')\Phi_a(p) \rangle\rangle = \delta(p'+p)G_0(p) \quad G_0(p) = (p^2+m^2)^{-1}$$

and the dotted line is a graphical representation of the bare interaction vertex. Sometimes a vertex is denoted by a dot, but I prefer to represent it as a line, to make it easier to keep track of the indices. The corresponding analytical expression is

$$G^{(1)}(p) = G_0(p)\Sigma_1(p)G_0(p) \tag{5.8}$$

$$\Sigma_1 = -g(N/2+1)T\sum_n \int \frac{\mathrm{d}^D k}{(2\pi)^D}\left(\frac{1}{\omega_n^2 + \mathbf{k}^2 + m^2}\right) \tag{5.9}$$

Exercise

Check that in the second order in g we have the diagrams shown in Fig. 5.2 and write down analytical expressions for them (some are given below in the text).

It can be shown that the diagram expansion of the $2n$-point correlation function obeys the following rules: (i) in each order one has to draw all possible connected graphs consisting of thin lines connected by dotted vertex lines; disconnected diagrams cancel by the expansion

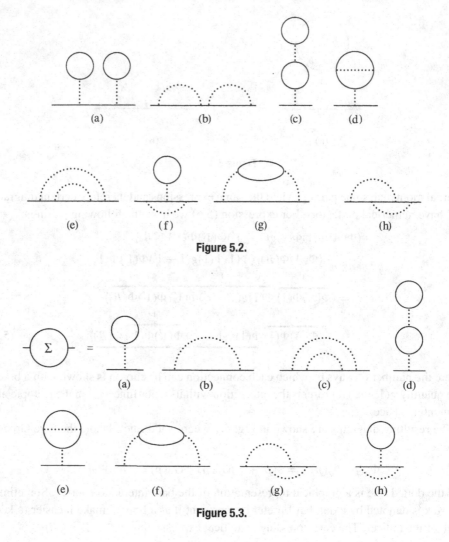

Figure 5.2.

Figure 5.3.

of the denominator; (ii) there are no numerical coefficients related to the order of interaction n; (iii) dotted lines carry $-g$ everywhere except for diagrams with internal 'bubbles', as in Fig. 5.1(a), Fig. 5.2(d), (f), and (g) (one bubble), and Fig. 5.2(c) (two bubbles), where it is $-g/2$.

Using these rules one can derive several important theorems for the diagram expansion. First, we can classify all diagrams with two external lines according to the principle introduced by Dyson. Namely, we can distinguish between diagrams which can and cannot be cut over a single line. A sum of diagrams which cannot be cut over a single line is called the self energy part Σ.

Some diagrams for the self energy part are represented in Fig. 5.3. Then the complete single-particle Green's function is a geometric progression (see Fig. 5.4):

$$G_0(p) + G_0(p)\Sigma(p)G_0(p) + G_0(p)\Sigma(p)G_0(p)\Sigma(p)G_0(p) + \cdots$$

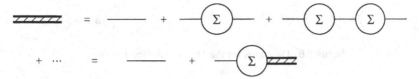

Figure 5.4. The Dyson equation.

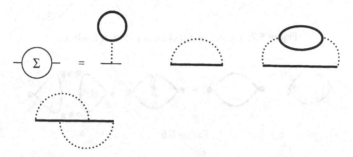

Figure 5.5.

Analytically this progression can be represented as the Dyson equation:

$$G(p) = G_0(p) + G_0(p)\Sigma(p)G(p)$$
$$G^{-1}(p) = G_0^{-1}(p) - \Sigma(p)$$

(5.10)

It is almost never possible to sum the entire diagram series and thus find Σ exactly. In general only a partial summation is possible. Therefore one has to find out which diagrams give the largest contribution and sum them first. To find such diagrams is always a matter of skill. Some examples will be given in subsequent chapters. The entire diagram expansion can be reformulated in terms of the exact Green's functions, i.e. Feynman diagrams can be redrawn with fat lines instead of thin ones. The rules for this expansion are almost the same as those cited above, with one difference: it is not allowable to insert self energy corrections into lines. Indeed, this would lead to over-counting because all possible insertions have already been done. Therefore this expansion contains only the so-called skeleton diagrams (Fig. 5.5).

This sort of expansion can be productive if we manage somehow to find a general form of the Green's function. Then the Dyson equation becomes not a functional equation for Σ, but an algebraic one. One example of this will be given at the end of Chapter 7.

Diagrams of the type depicted in Fig. 5.3(c), (d), (e), and (g) contain insertions of Σ into internal lines and therefore do not appear in this expansion. The expansion with fat lines converts the Dyson equation into a nonlinear equation for G. Following the same principle, one can define the Dyson equation for the interaction and introduce a renormalized interaction, a fat dotted line $D(q)$. Figure 5.6 represents the Dyson equation, where the bubble $\Pi(q)$ is a sum of all diagrams which cannot be cut along a single dotted line. Some

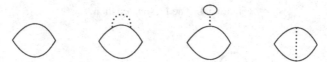

Figure 5.6. Diagrams for the Dyson equation for $D(q)$.

Figure 5.7. Diagrams for the polarization bubble.

Figure 5.8.

diagrams for $\Pi(q)$ are represented in Fig. 5.7.

$$D(q) = -g - \frac{1}{2}g\Pi(q)D(q)$$
$$D(q) = -[g^{-1} + 1/2\Pi(q)]^{-1} \qquad (5.11)$$

With both dotted and solid lines being fat, one can leave in the diagram expansion only the so-called 'skeleton' diagrams (see Fig. 5.8).

Sometimes it is more suitable to use Feynman diagrams in real space. The difference in formulation of the diagram expansion is minor. We have the same rules, but the integrals look different. Namely, we have to introduce Fourier transformed Green's functions:

$$G(\tau - \tau'; \mathbf{r} - \mathbf{r}') = T \sum_n \int \frac{d^D k}{(2\pi)^D} e^{-i\omega_n(\tau-\tau') - i\mathbf{k}(\mathbf{r}-\mathbf{r}')} G(\omega_n, \mathbf{k}) \qquad (5.12)$$

Now there are no momenta conservation laws in vertices and instead one has to integrate over all internal coordinates. For example, the analytical expressions for the 'tadpole' and the 'shell' diagrams of Fig. 5.1 are, respectively,

$$-\frac{1}{2} \int dx_1 dx_2 G_{aa}(x_1, x_1) g(x_1 - x_2) G_{bb}(x_2, x_2) \qquad (5.13)$$

$$-\int dx_1 dx_2 G_{aa}(x_1, x_2) g(x_1 - x_2) G_{aa}(x_2, x_1) \qquad (5.14)$$

where $g(x - y) = g\delta(x - y)$ and $x = (\tau, \mathbf{r})$. The Dyson equation in real space is

$$G(x, y) = G_0(x, y) + \int dx' dx'' G_0(x, x')\Sigma(x', x'')G(x'', y) \qquad (5.15)$$

The reason I have not discussed an expansion for the free energy is that in this case one cannot make the lines fat because the diagrams with different numbers of interactions

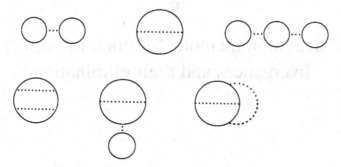

Figure 5.9. The diagrams with repeating bubbles.

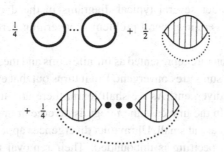

Figure 5.10. The diagrams for equation (5.17).

contain different numerical coefficients. This difficulty can be avoided, however, if instead of the free energy F we consider its derivative:

$$g\frac{1}{V}\frac{\partial F}{\partial g} = \frac{g}{8(\beta V)^3}\sum_{p,q,k}\langle(\Phi(-q+p/2)\Phi(q+p/2))(\Phi(-k-p/2)\Phi(k-p/2))\rangle$$

(5.16)

Thus the problem of the calculation of F is reduced to the problem of the calculation of the four-point correlation function. The diagram expansion for the latter does not contain numerical coefficients. Expanding the right-hand side of (5.16) in powers of g leads to the diagrams shown in Fig. 5.9 (up to the second order in g).

These are summed to give the following expression (see Fig. 5.10):

$$\frac{1}{V}\frac{\partial F}{\partial g} = \frac{1}{8}(\mathrm{Tr}G)^2 + \frac{1}{4}T\sum_n\int\frac{d^Dk}{(2\pi)^D}\left[\frac{\Pi(\omega_n,k)}{1+\frac{1}{2}g\Pi(\omega_n,k)}\right]$$

(5.17)

6

Calculation methods for diagram series: divergences and their elimination

In this chapter we consider several typical diagrams in the diagram expansion of the theory (5.1) and discuss certain general problems concerning perturbation expansions in QFT.

The correlation functions are represented as infinite sums and the first thing one should do is to check whether these sums are convergent. But it turns out that even individual members of these sums are often divergent! As we shall learn, there are divergences of two sorts: ultraviolet and infrared. In the first case the integrals diverge at large frequencies and momenta, and in the second case at small. Ultraviolet divergences appear in those field theories where the bare particle spectrum is unbounded. Their removal presents a severe ideological problem in high energy physics, where the unbounded spectrum $\epsilon^2 = p^2 c^2 + (mc^2)^2$ follows from the Lorentz invariance. In models of condensed matter physics, ultraviolet divergences present not a problem but a nuisance. Their presence indicates that the continuum description is incomplete, i.e. the behaviour of long-wavelength excitations depends on shorter length scales. It is certainly a nuisance, because the description becomes nonuniversal. Usually it is much easier to write down a continuous field theory; for this purpose one can employ quite general arguments, ones based on symmetry requirements, for example. It is really a disappointment when this beautiful castle built from pure ideas crumbles, and it turns out that a realistic description requires a careful study of processes occurring on the lattice scale. There are such nasty models in condensed matter physics, but there are also others, where the lattice does not play such an important role. In some cases ultraviolet divergences can be removed by a rearrangement of the perturbation expansion. I give some examples of this below. A more complete discussion of this topic can be found in the book by Zinn-Justin.

Infrared divergences are more interesting. Their appearance is always an indication of an incorrectly chosen reference ground state. For example, if electrons in a metal attract, it is wrong to approximate them as free particles; the real ground state is a superconducting condensate of electron pairs. This ground state is orthogonal to the ground state of the noninteracting electron gas and therefore is unreachable by a perturbation expansion. There is no universal recipe for how to choose a correct ground state. Sometimes it is relatively easy (as for ordinary superconductors), sometimes it is not easy at all. In fact, this book is all about the possible consequences of infrared divergences. Leaving this subject to more detailed discussion, I now concentrate on ultraviolet divergences.

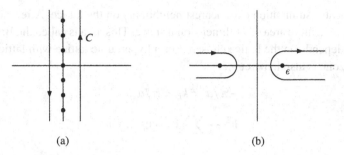

(a)   (b)

Figure 6.1.

Let us consider the diagram expansion for the self energy in the Φ^4 theory. In our discussion I shall combine two tasks: I shall discuss the divergences and explain how to calculate the integrals. In the first order in g we have two diagrams represented in Fig. 5.1 and the analytical expression is given by (5.9).

There is a standard trick for calculating sums over discrete frequencies, which we have already discussed briefly in Chapter 2. Namely, we should rewrite the sum as an integral in the complex plane over a countour surrounding the points $z = 2i\pi T n$ (see Fig. 6.1(a)):

$$T \sum_n \frac{1}{\omega_n^2 + \epsilon^2} = -\frac{1}{4i\pi} \int_C dz \coth\left(\frac{z}{2T}\right) \frac{1}{z^2 - \epsilon^2} \tag{6.1}$$

This is possible because the value of the integral on the right-hand side is given by the sum of the residues in the poles of the integrand.

These poles inside the contour are poles of the function $\coth(z/2T)$; in the vicinity of a pole we have

$$\coth\left(\frac{z}{2T}\right) \approx \frac{2T}{z - 2i\pi T n}$$

The integrand has an infinite number of poles inside the contour and only two poles outside at the points $z = \pm\epsilon$. Now we can bend the countour onto these poles (see Fig. 6.1(b)) and get the following answer:

$$\Sigma_1 = -\frac{1}{4}g(N/2 + 1) \int \frac{d^D k}{(2\pi)^D} \frac{1}{\sqrt{k^2 + m^2}} \coth\left(\frac{\sqrt{k^2 + m^2}}{2T}\right) \tag{6.2}$$

This expression diverges at large $|k|$ for any $D > 0$, i.e. it displays an ultraviolet divergence, the very type of divergence which we are discussing now. There are various formal methods to remove divergences; they are called 'regularization procedures'. The simplest one is to put our theory on a lattice as was described in Chapter 1. In this case instead of continuous derivatives we get discrete ones:

$$(\nabla \Phi)^2 \rightarrow \frac{1}{|e|^2} \sum_{\langle e \rangle} [\Phi(\mathbf{x}) - \Phi(\mathbf{x} + \mathbf{e})]^2$$

$$\frac{1}{2}k^2 \rightarrow \frac{1}{|e|^2} \sum_{\langle e \rangle} [1 - \cos(\mathbf{k}\mathbf{e})] \tag{6.3}$$

where $\langle \mathbf{e} \rangle$ indicates summation over nearest neighbours on the lattice. After the discretization, $\mathbf{k}$ belongs to a finite area of D-dimensional space. This area is called the Brillouin zone and its shape depends on the lattice chosen. For a hypercubic lattice with lattice spacing a the Brillouin zone is also hypercubic:

$$-\pi/a < k_i < \pi/a$$

$$\mathbf{k}^2 \rightarrow \sum_{i=1}^{D} [1 - \cos(k_i a)]$$

Thus the integrals over the $\mathbf{k}$-space are cut off on large wave vectors $|\mathbf{k}| \approx 1/a \equiv \Lambda$ and converge. The described regularization procedure is natural for statistical models or models of condensed matter physics which are really defined on a lattice, but for high energy physics one can accept it only as a formal method to calculate integrals if the results do not depend on the lattice spacing. The latter is not the case for the diagram we discuss:

$$\Sigma_1 \propto \begin{cases} g\Lambda^{D-1} & D > 1 \\ g\ln(\Lambda/\max\{T, m\}) & D = 1 \end{cases}$$

which is lattice dependent.

The physical quantities depending on regularization procedures are called nonuniversal. There is a wider class of theories where the perturbation expansion can be reorganized in such a way that nonuniversal quantities appear as a finite number of parameters, the vector Φ^4 theory being a good example. Such theories are called renormalizable. Let us consider the Green's function:

$$G(p) = [p^2 + m^2 - \Sigma(p)]^{-1}$$

The nonuniversal diagram Σ_1 does not depend on external momenta; this suggests the idea of subtracting it from the self energy and including it in a redefinition of m^2:

$$G(p) = \{p^2 + [m^2 - \Sigma(0)] - [\Sigma(p) - \Sigma(0)]\}^{-1}$$

Now we can consider $m^2 - \Sigma(0)$ as a new independent parameter $\tilde{m}^2$ related to the old one in a nonuniversal way. Check that the diagrams of Fig. 5.1 do not appear for the following action:

$$S = \int_0^{1/T} d\tau d^D x \left[\frac{1}{2}(\nabla \Phi_a)^2 + \frac{\tilde{m}^2}{2}(\Phi_a)^2 \right.$$
$$\left. + \frac{g}{8}(\Phi_a \Phi_a \Phi_b \Phi_b - 2\Phi_a \Phi_a \langle \Phi_b \Phi_b \rangle - 2\Phi_a \langle \Phi_a \Phi_b \rangle \Phi_b) \right] \tag{6.4}$$

where $\langle \cdots \rangle$ stands for the exact average. The only possible trouble with this procedure is that $m^2(T)$ is temperature dependent. If this dependence is also nonuniversal, it introduces into our theory an essential uncontrollable element. Therefore let us check the universality of the temperature dependence. We subtract from Σ_1 its value at $T = 0$:

$$\Sigma_1(T) - \Sigma_1(0) = -\frac{g}{4}(N/2 + 1) \int \frac{d^D k}{(2\pi)^D} \frac{1}{\sqrt{k^2 + \tilde{m}^2}} \left[\coth\left(\frac{\sqrt{k^2 + \tilde{m}^2}}{2T}\right) - 1 \right]$$

This integral converges at $|\mathbf{k}| \approx T$ and therefore does not depend on Λ as $\Lambda \to \infty$. Thus the temperature dependence of $\tilde{m}^2(T)$ is universal which is good news. We have to carry on, however, and make sure that $\Sigma(p) - \Sigma(0)$ is a universal quantity. Let us consider the next diagram depicted in Fig. 5.3(c). According to our rule we have to subtract from this diagram its value at $p = 0$ which contributes to $\tilde{m}^2(T)$. We therefore get

$$
\Sigma_2(p) = \frac{Tg^2}{2(2\pi)^D} \sum_n \int d^D q \Pi(q)[G(p-q) - G(-q)]
$$
$$
\Pi(q) = \frac{T}{(2\pi)^D} \sum_n \int d^D k G(-k+q)G(k)
$$

(6.5)

The integral for $\Pi(q)$ converges at $D < 3$ because

$$
G(-p+q)G(q) \propto 1/q^4
$$

at $q \to \infty$.

This analysis can be repeated for higher diagrams with the same result: nothing depends on the lattice at $D < 3$. The case $D = 3$ is marginal in the sense that all diagrams depend logarithmically on Λ. This slow logarithmic dependence is not too dangerous since the diagrams really depend on Λ only in order of magnitude, which is easier to estimate than the exact value. The majority of model theories of QFT have ultraviolet logarithmic divergences. Therefore this situation is very important. There is an entire philosophy behind such theories. According to this philosophy, the models with such behaviour describe only a low energy sector of reality, but they describe it thoroughly. The upper energy cut-off Λ is an element of the theory; it cannot be calculated by its means, being established by some processes on a deeper level. Within quantum electrodynamics (QED), for instance, one cannot calculate the electron mass m_e or account for the quantization of electric charge. The electron mass depends on the ultraviolet cut-off Λ. Within its limits, however, QED is quite self-consistent and e and m_e are the only parameters taken for granted.

Thus we have understood, at least vaguely, how to deal with divergences at large momenta. As I have said, integrals in perturbation series can also diverge at small momenta. In our theory this can happen only if $\tilde{m}^2 \leq 0$. As we have seen, the parameter $\tilde{m}^2(T)$ depends on temperature. Therefore we can get into such a dangerous region by a temperature change. The case of strictly negative $\tilde{m}^2 < 0$ will be considered in later chapters, the case $\tilde{m}^2 \to +0$ being considered in Chapter 7.

Let us continue our discussion of integrals. I have already explained one trick, but there are several more. In dealing with perturbation expansions we frequently perform partial summations of diagram series. For example, let us imagine that in our $O(N)$-symmetric vector model N is a large parameter. Let us calculate the self energy part. Diagrams containing maximal numbers of summations over j for a given order g^n should be summed first. Such diagrams for the interaction are presented in Fig. 5.8. Then the leading term for the self energy part is given by the skeleton 'rainbow' diagram, Fig. 5.5(lower). Tadpole diagrams of the sort depicted in Fig. 5.5(upper) renormalize the $\tilde{m}^2$ and can be dismissed. The diagrams like Fig. 5.5(lower) are essential because they change the momentum dependence of the self energy part.

$$D(q) = \frac{1}{g^{-1} + \Pi(q)/2}$$

$$\Pi(q) = \frac{T}{(2\pi)^D} \sum_n \int d^D k G(-k + q)G(k)$$

(6.6)

Inserting this interaction into the self energy part we get

$$\Sigma(p) = -T \sum_n \int \frac{d^D q}{(2\pi)^D} D(p - q)G_0(q)$$

(6.7)

The latter expression can be rewritten in a simplified way. To do this we can use general theorems about Green's functions. According to these theorems, which are explained in Chapter 1, any thermodynamic Green's function can be written in terms of its retarded function (see (1.29)). Using this representation for $D(q)$ we get:

$$\Sigma(\Omega, Q) = -\int \frac{d^D k}{(2\pi)^D} \left\{ \frac{iT}{\pi} \sum_n \int dy \frac{\Im m D^{(R)}(y, Q - k)}{(i\omega_n - y)[-(i\Omega - i\omega_n)^2 + k^2 + \tilde{m}^2]} \right\}$$

(6.8)

Now it is much easier to perform the summation over frequencies (you can say that we do not know $D^{(R)}$, but just be patient). Using the standard method explained above, we get

$$T \sum_n \frac{1}{(i\omega_n - y)[-(i\Omega - i\omega_n)^2 + k^2 + m^2]} = \frac{1}{4\epsilon(k)} \left(\frac{1}{i\Omega - y + \epsilon(k)} \{\coth(y/2T) \right.$$

$$\left. - \coth[\epsilon(p)/2T]\} - \frac{1}{i\Omega - y - \epsilon(k)} \{\coth(y/2T) + \coth[\epsilon(p)/2T]\} \right)$$

(6.9)

where $\epsilon(k) = \sqrt{k^2 + \tilde{m}^2}$.

The expression (6.8) can be simplified further if we take into account the fact that for practical purposes it is sufficient to have $\Im m \Sigma^{(R)}(\Omega)$. With this quantity at hand one is always able to restore $\Sigma(i\Omega_n)$ itself. Therefore we can make an analytic continuation $i\Omega_n \to \Omega + i0$ and use the identity:

$$\Im m \left(\frac{1}{\Omega - y + i0} \right) = -\pi \delta(\Omega - y)$$

(6.10)

Then substituting (6.9) into (6.8) and taking its imaginary part we get

$$\Im m \Sigma^{(R)}(\Omega, Q) = -\int \frac{d^D k}{(2\pi)^D} \int dy \frac{\Im m D^{(R)}(y, Q - k)}{4\epsilon(k)} \{\delta(\Omega - y - \epsilon)[\coth(y/2T)$$

$$+ \coth(\epsilon/2T)] - \delta(\Omega - y + \epsilon)[\coth(y/2T) - \coth(\epsilon/2T)]\}$$

(6.11)

The integral over y is easily calculated, yielding

$$\Im m \Sigma^{(R)}(\Omega, Q) = -\int \frac{d^D k}{(2\pi)^D} \frac{1}{4\epsilon(k)} \{\Im m D^{(R)}(\Omega - \epsilon, Q - k)[\coth(\epsilon/2T) + \coth(\Omega - \epsilon/2T)]$$

$$- \Im m D^{(R)}(\Omega + \epsilon, Q - k)[\coth(\epsilon + \Omega/2T) - \coth(\epsilon/2T)]\}$$

(6.12)

This expression simplifies further in the limit $T \to 0$ where

$$[\coth(\epsilon/2T) - \coth(\epsilon + \Omega/2T)] \to 2\theta(-\epsilon - \Omega)$$

As a result we get

$$\Im m \Sigma^{(R)}(\Omega, Q; T = 0) = -\int \frac{d^D k}{2(2\pi)^D \epsilon(k)}$$

$$\times [\Im m D^{(R)}(\Omega - \epsilon, Q - k; T = 0)\theta(\Omega - \epsilon) - (\Omega \to -\Omega)] \qquad (6.13)$$

In the present case, when our theory is Lorentz invariant at $T = 0$, it is possible to carry the calculation even further in its general form. Due to the Lorentz invariance the Green's function itself and the self energy in particular depend only on $s^2 = \Omega^2 - Q^2$. Therefore we do not need to calculate the integral (6.13) for finite Q; it is sufficient to do it for $Q = 0$ and then substitute Ω for s. The advantage is that we can perform integration over angles:

$$\Im m \Sigma^{(R)}(s; T = 0) = -\alpha_D \int \frac{d\epsilon}{\epsilon}(\epsilon^2 - \tilde{m}^2)^{D/2-1}$$

$$\times \{\Im m D^{(R)}(s - \epsilon, \sqrt{\epsilon^2 - \tilde{m}^2}; T = 0)\theta(\Omega - \epsilon) - (\Omega \to -\Omega)\} \qquad (6.14)$$

where

$$\alpha_D = \frac{S_{D-1}}{4(2\pi)^D}$$

and S_D is the volume of the D-dimensional sphere of unit radius.

Now we should calculate the renormalized interaction Γ. We shall do this at $T = 0$ only. In fact, it is sufficient to calculate the imaginary part of the polarization loop $\Pi^{(R)}$. Indeed, we have

$$\Im m D^{(R)}(\omega, k) = \Im m \left[\frac{1}{g^{-1} + \Pi^{(R)}(\omega, k)/2}\right] = -\frac{2\Im m \Pi^{(R)}}{(2g^{-1} + \Re e \Pi^{(R)})^2 + (\Im m \Pi^{(R)})^2}$$

$$\qquad (6.15)$$

and the real part of Π can be restored from its imaginary part. It turns out that it is more convenient to calculate the polarization loop in real space. Indeed, in real space we should not integrate over the momenta:

$$\Pi(r - r') = NG^2(r - r') \qquad (6.16)$$

Since $G(r)$ is different for different D, the further calculations depend on the space dimensionality D. Please check the following expressions:

$$G(r) = \begin{cases} K_0(\tilde{m}|r|)/2\pi & D = 1 \\ \exp(-\tilde{m}|r|)/4\pi|r| & D = 2 \\ (2\pi^2)^{-1}[K_2(\tilde{m}|r|) - K_0(\tilde{m}|r|)] & D = 3 \end{cases} \qquad (6.17)$$

where

$$K_\gamma(x) = \int_0^\infty dy \cosh(\gamma y)e^{-x\cosh y}$$

The further calculations I do only for $D = 2$ where they are especially easy. Substituting (6.17) with $D = 2$ into (6.16) and performing the Fourier transformation we get:

$$\Pi_2(s) = \frac{N}{4\pi s} \tan^{-1}\left(\frac{s}{2\tilde{m}}\right) \tag{6.18}$$

In order to obtain the retarded function, we have to perform the analytic continuation of the above expression

$$\omega = iy$$

Here I use the following identity:

$$\tan^{-1} x = \frac{1}{2i} \ln\left(\frac{x-i}{x+i}\right)$$

The result is

$$\Im m\, \Pi^{(R)}(y, k) = \frac{N}{8\sqrt{y^2 - k^2}}\theta(y^2 - k^2 - 4\tilde{m}^2)$$

$$\Re e\, \Pi^{(R)}(y, k) = \frac{N}{8\pi\sqrt{y^2 - k^2}} \ln\left(\left|\frac{\sqrt{y^2 - k^2} + 2\tilde{m}}{\sqrt{y^2 - k^2} - 2\tilde{m}}\right|\right) \tag{6.19}$$

Substituting it into (6.15) and then (6.15) into (6.14), we get the following horrific expression for the imaginary part of Σ at $\Omega > 0$:

$$\Im m\, \Sigma^{(R)}(\Omega, Q; T = 0)$$

$$= -\alpha_D \int_{\tilde{m}}^{\infty} \frac{d\epsilon}{\epsilon} \Im m\, D^{(R)}(s - \epsilon, \sqrt{\epsilon^2 - \tilde{m}^2}; T = 0)\theta(s^2 - 2\epsilon s - 3\tilde{m}^2) \tag{6.20}$$

where Γ is given by (6.15) and (6.19) and the theta-function comes from (6.19). Resolving the inequality

$$s^2 - 2\epsilon s - 3m^2 > 0$$

we get

$$s > \epsilon + \sqrt{\epsilon^2 + \tilde{m}^2} \tag{6.21}$$

and since $\epsilon > \tilde{m}$ we get that the imaginary part of the self energy does not vanish only if $\Omega^2 - k^2 > 9\tilde{m}^2$. It is clear that particles scatter on pair excitations and in order to create such a complex, one has to exceed the $3\tilde{m}$ energy threshold. Finally we have:

$$\Im m\, \Sigma^{(R)}(s; T = 0) = -\alpha_D \int_{\tilde{m}}^{(s/2 - 3\tilde{m}^2/2s)} \frac{d\epsilon}{\epsilon} \Im m\, D^{(R)}(s - \epsilon, \sqrt{\epsilon^2 - \tilde{m}^2}; T = 0) \tag{6.22}$$

This integral has finite limits and therefore is always convergent. It is not too difficult to calculate it in the vicinity of the threshold where $s = 3\tilde{m} + x$, $x \ll \tilde{m}$. In this area we have

$\epsilon = \tilde{m} + \epsilon'$ with $0 < \epsilon' < 2x/3$ and

$$\Im m \Pi \approx \frac{N}{16\tilde{m}}$$

$$\Re e \Pi \approx \frac{N}{16\pi\tilde{m}} \ln\left(\frac{4\tilde{m}}{2x - 3\epsilon'}\right)$$

(6.23)

Substituting this into the integral (6.22) we get with logarithmic accuracy:

$$\Im m \Sigma^{(R)}(s = 3\tilde{m} + x; T = 0) = -\frac{Nx/48\pi\tilde{m}^2}{[2g^{-1} + N\ln(\tilde{m}/x)/16\pi\tilde{m}]^2 + (N/16\tilde{m})^2}$$

(6.24)

The goal of all these tedious calculations is to demonstrate typical procedures and methods. Using analytical properties of the Green's functions, we have managed to get an analytical expression for the sum of an infinite series of diagrams and thus obtain a nontrivial result for a physically observable quantity $\Sigma^{(R)}$ (6.24).

7

Renormalization group procedures

The topic of this chapter is one of the most beautiful and profound concepts in QFT: the concept of criticality, renormalization and scaling. This concept is relevant for systems which have no natural intrinsic scale. In the theory we have been discussing so far, the $O(N)$-symmetric vector field model, there is such a scale: it is the inverse mass m^{-1}. The correlation functions decay exponentially at distances (or time scales) larger then m^{-1}. Meanwhile there are extrinsic scales both large and small in the theory: the large scales are $1/T$ and the size of the system L, and the small scale is a, the lattice size. If we have an intrinsic scale we can apply it to all external agents acting on the system and make qualitative estimates of their strength. However, imagine that $m \rightarrow 0$; it seems that the system becomes self-similar on all scales.

Let me give an example of what I mean by self-similar. This example is not taken from physics, but from social institutions. Imagine such a venerable system as an army. An army has a hierarchical structure represented as a collection of units of different size. All this structure is united under the principles of seniority of ranks and subordination of lower ranks to the senior. These principles form a pattern for relations between army personnel inside units of any size. This pattern is very similar for relations between a soldier and a sergeant inside the smallest unit and for relations between a lieutenant of this unit and a captain of the platoon, and between this captain and his colonel, etc. Therefore one can say that the army hierarchical structure is self-similar. Self-similarity is not a property of every hierarchical structure, however. For example, if we consider a class society, then we can discover that patterns of relations in different classes are quite different. There are clear scales in the latter case, but the army has no such scale. Of course, army relations are similar on all levels, but are still not quite the same!

This is also very important; relations change continuously in such a way that one can derive the relations on a higher level from ones on a lower level by adjusting them to the rank or by *rescaling* them.

I hope that you have been able to catch the spirit of the subject I am going to discuss. And now let us descend from the heavens of military matters to this sinful earth of QFT and consider how the ideas of self-similarity work there.

Let us summon again our faithful $O(N)$-symmetric vector field model and consider a limit $\tilde{m}^2(T) \rightarrow 0$. What happens in this case with our diagram expansion? I shall demonstrate that at $\tilde{m}^2(T) \rightarrow 0$: (i) some diagrams become divergent at small momenta; since this happens

Figure 7.1.

Figure 7.2.

Figure 7.3.

only at $D \leq 4$, $D = 4$ is called the 'upper critical' dimension of this theory; (ii) the degree of this divergence depends drastically on whether we are at $T = 0$ or not.

The reason behind statement (ii) is that at finite T the integrals over frequencies are converted into sums

$$\int \frac{\mathrm{d}q_0}{2\pi} \to T \sum_s$$

and the discrete frequencies in *bosonic* theories include $\omega_s = 0$. Therefore if in $D = 3$ the polarization loop

$$\Pi(q) = T \sum_s \int \frac{\mathrm{d}^3 p}{(2\pi)^D} \frac{1}{[\omega_s^2 + \mathbf{p}^2][(\Omega - \omega_s)^2 + (\mathbf{p} - \mathbf{q})^2]}$$

diverges logarithmically at $T = 0$ at small momenta, then at finite T the same loop diverges much more strongly, namely as T/q (q is an external momentum). The additional divergence comes exclusively from the term with $\omega_s = 0$. Therefore if the external momentum $|q| \ll T$ one can neglect contributions from frequencies other than $\omega_s = 0$ and treat our field as time independent!

This change can be made in a general way on the level of the effective action. Treating the Φ-field as time independent, we integrate over τ to get

$$\int_0^{1/T} \mathrm{d}\tau \int \mathrm{d}^D x L[\Phi] = \frac{1}{T} \int \mathrm{d}^D x L[\Phi(\mathbf{x})] \tag{7.1}$$

Thus in the vicinity of $T = T_c$ where the 'mass' $\tilde{m}$ vanishes, bosonic QFTs become effectively classical theories provided $T_c > 0$!

Let me make the last preparatory remark. How do we get $\tilde{m}(T) \to 0$? At $T = 0$ we can do this 'by hand' by treating it as a parameter we have the liberty to change. At finite T the temperature-dependent part of $\tilde{m}^2(T)$ is not an independent quantity. Its magnitude changes with T and can reach zero at some particular $T = T_c$. For example, in our theory in the first order in g we have

$$\tilde{m}^2(T) - \tilde{m}^2(0) \equiv -\Sigma_1(T) + \Sigma_1(0)$$

$$= g(N + 1) \int \frac{\mathrm{d}^D k}{(2\pi)^D} \frac{1}{\sqrt{k^2 + \tilde{m}^2(T)}} \left\{ \coth \left[\frac{\sqrt{k^2 + \tilde{m}^2(T)}}{2T} \right] - 1 \right\}$$

If $\tilde{m}^2(T) \to 0$ (that is what we need!) then the right-hand side is of the order of $T^{(D-1)}$, i.e.

$$\tilde{m}^2(T) = \tilde{m}^2(0) + \alpha T^{(D-1)}$$

Now if $\tilde{m}^2(0) < 0$ we see that $\tilde{m}^2(T)$ is positive above $T_c = [-\tilde{m}^2(0)/\alpha]^{1/(D-1)}$ and vanishes at $T = T_c$. That is, the situation $\tilde{m}^2(T) \to 0$ is achieved by a change of temperature. It could well be that at T_c we have a phase transition. Indeed, we see that the system cannot just go through the point T_c into the region of negative $\tilde{m}^2(T)$ because in this region our perturbation expansion based on the positivity of

$$\frac{1}{p^2 + \tilde{m}^2(T) - \Sigma(p) + \Sigma(0)}$$

becomes meaningless (recall how we have estimated all Gaussian integrals!). So something really profound must happen at T_c: a phase transition. Therefore the problem we are interested in can be stated as: *what happens near a phase transition?* We have seen already that quantum mechanics is irrelevant here, that in the vicinity of T_c we have a completely classical problem.

Now let us consider our theory in $D = 4$ (finite temperatures) or in $D = 3$ ($T = 0$) at $\tilde{m}^2(T) \to 0$. In both these cases the diagram expansion contains logarithmically divergent diagrams, but there are no stronger singularities! This slow and boring function, $\ln x$, is like a marvellous gift from God for QFT. The essential point is that if we have logarithmically divergent integrals we can estimate them with logarithmic accuracy, i.e. not distinguishing between $\ln x$ and $\ln ax$ with a being a constant of the order of one. Using this property one can roughly divide the momentum space into regions of different momenta

$$\Lambda > |p| > \Lambda_1 > |p| > \Lambda_2 > \cdots > |p| > 0$$

The principle behind this separation is that

$$\Lambda \gg \Lambda_1 \gg \Lambda_2 \gg \cdots \gg 0$$

but

$$\ln(\Lambda/\Lambda_1) \approx \ln(\Lambda_1/\Lambda_2) \approx \cdots$$

In this approach it is even customary to treat

$$\ln(\Lambda_{n-1}/\Lambda_n) - \ln(\Lambda_n/\Lambda_{n+1}) \equiv d\xi$$

as an infinitesimal quantity!

Introducing these strange definitions we can evaluate our path integral step by step (with logarithmic accuracy!). As a first step we shall integrate over $\Phi(p)$ fields with $\Lambda > |p| > \Lambda_1$ and get some effective action for the fields with $\Lambda_1 > |p|$. As the next step we integrate over the region $\Lambda_1 > |p| > \Lambda_2$ and get the effective action for $\Lambda_2 > |p|$, etc. The idea is that the effective action for the nth step will be the same as the bare one, but with another coupling constant $g(\ln \Lambda_n)$. Thus we separate our measure

$$\prod_p d\Phi_a(p) = \prod_{\Lambda > |p| > \Lambda_1} d\Phi_a(p) \prod_{\Lambda_1 > |p| > \Lambda_2} d\Phi_a(p) \cdots \prod_{\Lambda_{n-1} > |p| > \Lambda_n} d\Phi_a(p) \cdots \quad (7.2)$$

and for the first step we separate the fields as follows:

$$\Phi_a(x) = \sum_{\Lambda > |p| > \Lambda_1} \Phi_a(p)e^{ipx} + \sum_{\Lambda_1 > |p|} \Phi_a(p)e^{ipx} \equiv \phi_a(x) + \Phi_a^{cl}(x) \quad (7.3)$$

From the point of view of $\phi_a(x)$ the field $\Phi_a^{cl}(x)$ is a very slow background.

Substituting this separation into the expression for the action we get:

$$S(\Phi) = S(\Phi_{cl}) + \int d^D x \phi(x) \frac{\delta S}{\delta \Phi_{cl}(x)} + \frac{1}{2} \int d^D x d^D y \phi(x) \frac{\delta^2 S}{\delta \Phi_{cl}(x) \delta \Phi_{cl}(y)} \phi(y) + \cdots$$

(7.4)

The term linear in ϕ vanishes because ϕ and Φ_{cl} do not contain common Fourier harmonics. Let us denote $\hat{A} = \delta^2 S / \delta \Phi_{cl}(x) \delta \Phi_{cl}(y)$ and integrate over ϕ in the path integral:

$$Z[\eta] = e^{-S(\Phi_{cl})} \int \Pi d\phi(p) \exp\left(-\frac{1}{2}\phi \hat{A} \phi\right) = (\det \hat{A})^{-1/2} e^{-S(\Phi_{cl})} \equiv e^{-S_1(\Phi_{cl})} \quad (7.5)$$

$$S_1(\Phi_{cl}) = S(\Phi_{cl}) + \frac{1}{2} \text{Tr} \ln \hat{A} \tag{7.6}$$

Thus we get a renormalized effective action S_1.

Let us perform this procedure for our theory. We have

$$S(\phi + \Phi_{cl}) - S(\Phi_{cl}) = \frac{1}{2}(\partial_\mu \phi_a)^2 + \frac{m^2}{2}(\phi_a^2) + \frac{g}{2}[\phi_a^2 \Phi_{cl}^2 + (\phi \Phi_{cl})^2] + \mathcal{O}(\phi^3) \quad (7.7)$$

The term quadratic in ϕ is

$$\frac{1}{2}\phi \hat{A} \phi = \frac{1}{2} \sum_p \phi_a(-p)[\delta_{ab}(p^2 + m^2) + g B_{ab}]\phi_b(p)$$

(7.8)

$$B_{ab} = (\delta_{ab}\Phi^2 + 2\Phi_a \Phi_b)$$

So we have

$$\frac{1}{2}\text{Tr} \ln \hat{A} - \frac{1}{2} \ln(p^2 + m^2)$$

$$= \frac{1}{2} \ln(p^2 + m^2) + \frac{1}{2}\text{Tr} \ln \left(I + \frac{g\hat{B}}{p^2 + m^2}\right)$$

$$\approx \frac{g}{2} \int \frac{d^4 p}{(2\pi)^4} \frac{1}{p^2 + m^2} \text{Tr} \hat{B} - \frac{g^2}{4} \int \frac{d^4 p}{(2\pi)^4} \frac{1}{(p^2 + m^2)^2} \text{Tr} \hat{B}^2 \tag{7.9}$$

Taking into account the following relations

$$\text{Tr} \hat{B} = (N + 2)(\Phi)^2 \qquad \text{Tr} \hat{B}^2 = (N + 8)(\Phi)^2(\Phi)^2$$

we get from (7.9) the new action:

$$S_1(\Phi) = \frac{1}{2}(\partial_\mu \Phi)^2 + \frac{m_1^2}{2}(\Phi)^2 + \frac{g_1}{4}(\Phi)^2(\Phi)^2 \tag{7.10}$$

$$g_1 = g_0 - g_0^2(N + 8) \int_{\Lambda_1}^{\Lambda} \frac{d^4 p}{(2\pi)^4} \frac{1}{p^4} \tag{7.11}$$

$$m_1^2 = m^2 + g_0(N + 2) \int_{\Lambda_1}^{\Lambda} \frac{d^4 p}{(2\pi)^4} \frac{1}{p^2} \tag{7.12}$$

where I have substituted g_0 for the bare coupling g.

Integrating over the momenta we get

$$\int_{\Lambda_1}^{\Lambda} \frac{d^4 p}{(2\pi)^4} \frac{1}{p^4} \approx \frac{1}{8\pi^2} \ln(\Lambda/\Lambda_1) \tag{7.13}$$

provided $\Lambda_1 \gg m$ and

$$\int_{\Lambda_1}^{\Lambda} \frac{d^4 p}{(2\pi)^4} \frac{1}{p^2} \approx \frac{1}{8\pi^2} (\Lambda^2 - \Lambda_1^2) \tag{7.14}$$

As I have mentioned above, in the renormalization group theory the variable

$$\xi = \ln(\Lambda_n/\Lambda_{n+1})$$

is treated as infinitesimal. The idea behind this is to represent renormalization transformations as a continuous process.

Let us see how this idea works for the coupling constant g (I shall discuss renormalization of m^2 later). Let us rewrite (7.11) as a differential equation:

$$\frac{dg}{d\ln(\Lambda/k)} = -g^2 \frac{N+8}{8\pi^2} \tag{7.15}$$

$$g(k = \Lambda) = g_0 \qquad \ln(\Lambda/k)|_{\max} = \ln(\Lambda/m^*)$$

One can notice that there is an ambiguity here. Indeed, I have defined this equation in such a way that its right-hand side depends only on the renormalized coupling constant $g(k)$ and does not contain bare parameters. This is the *universality hypothesis*; such a definition is consistent with the idea of renormalization. Equation (7.15) is a particular case of the general renormalization group equation called the Gell-Mann–Low equation:

$$\frac{dg}{d\ln(\Lambda/k)} = \beta(g) \tag{7.16}$$

$$g(k = \Lambda) = g_0$$

where $\beta(g)$ is a Gell-Mann–Low function of a given theory. It is usually calculated perturbatively and the perturbation theory gives $\beta(g)$ in the form of an expansion in powers of g. In the present case we have calculated $\beta(g)$ in the second order in g. Integrating (7.16) we get

$$F(g) - F(g_0) = \int_{g_0}^{g} \frac{dg'}{\beta(g')} = \ln \Lambda - \ln k \tag{7.17}$$

One can rewrite this equation in a form where the bare parameters are united in one single parameter k_0:

$$F(g) = \ln(k_0/k) \qquad k_0 = \Lambda e^{-F(g_0)} \tag{7.18}$$

If $k_0 < \Lambda$, this usually marks a crossover scale where a crossover from one regime to another occurs. In the case of (7.15) we have

$$\frac{1}{g} = \frac{1}{g_0} + \frac{N+8}{8\pi^2} \ln(\Lambda/k) \tag{7.19}$$

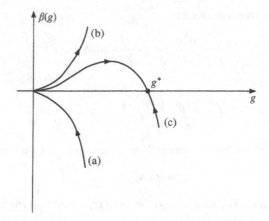

Figure 7.4.

We see that at $g_0 > 0$ the renormalized coupling constant decreases at small momenta. It justifies our neglect of the higher order terms in the expansion of $\beta(g)$. If the bare coupling is negative, $g_0 < 0$, the renormalized coupling becomes infinitely strong at

$$k < k_0 = \Lambda \exp\left[-\frac{8\pi^2}{(N+8)|g_0|}\right]$$

which means that one cannot use the perturbation theory at $k < k_0$. The reason for this failure of the perturbation theory is obvious in this case: at $g_0 < 0$ the integrand in the path integral has a positive exponent and the integral diverges. Strictly speaking, the entire theory is ill defined in this case.

Let us return now to m^2. We see that the perturbation correction for this quantity diverges at large momenta. This means that the behaviour of m^2 is nonuniversal and is determined in the ultraviolet region. In fact, for long-wavelength processes we have to substitute the bare m^2 for a renormalized value m^{*2}. This new m^{*2} is nonuniversal and we can only hope that it is small. Otherwise the renormalization procedure will not continue for long since all divergences are cut on the scale m^*. Thus the scale m^* plays the role of the infrared cut-off.

Let us return to the Gell-Mann–Low equation (7.16). Its solutions can be of three types. The first type is called *zero-charge* behaviour: the coupling constant decreases for small momenta (large distances). If $m^* = 0$ the coupling constant simply vanishes for small momenta and the theory in this limit becomes a free theory. As we have just discussed, such behaviour takes place in the $D = 3$ quantum ($D = 4$ classical) O(N)-symmetric vector model. A well known physical example is the renormalization of electric charge in $(3 + 1)$-dimensional quantum electrodynamics: an individual electric charge is screened at large distances by vacuum fluctuations of electron–positron pairs. The zero-charge type β-function has a qualitative form shown in Fig. 7.4(a). The second type of behaviour is called *asymptotic freedom* (see Fig. 7.4(b)): the coupling constant grows at large distances. This means that excitations with small momenta interact more and more strongly. In the theory we have just discussed such a thing occurs due to a trivial reason: at $g_0 < 0$ the theory is ill defined. It is not the only source of asymptotic freedom, however. 'All happy families look the same, all unhappy ones are unhappy in their own way' (Leo Tolstoy, *Anna Karenina*). Asymptotically free theories, being the most interesting objects of thought in QFT, like the unhappy families in Tolstoy's novel, all have their own reason to be so. Some such theories will be discussed in later chapters, where I shall describe various nonperturbative methods which have been suggested for such theories. Finally, the third type of behaviour occurs when the β-function has a zero at some finite value of $g = g^*$ (see Fig. 7.4(c)). Then, if $m^* = 0$, the scaling terminates at g^*. Suppose that around g^* we have the following expansion:

$$\beta(g) = -\alpha(g - g^*) + \mathcal{O}[(g - g^*)^2] \qquad (7.20)$$

Integrating the Gell-Mann–Low equation (7.16) with this β-function we get

$$|g(k) - g^*| = C(k/\Lambda)^\alpha \qquad (7.21)$$

where C is some constant. This type of behaviour is called *critical point*. Usually a power law behaviour of the coupling constant leads to power laws for physical quantities. If a critical point occurs at a finite temperature $T = T_c$, the scaling usually terminates on the scale $m^* \sim (T - T_c)^\nu$. Therefore in the vicinity of a critical point physical quantities depend on powers of $(T - T_c)$. Critical points are points of second-order phase transitions. We shall deal with such theories later in this chapter and also in subsequent chapters, particularly in Part IV.

Now let us consider one example of critical point behaviour, namely a point of the second-order phase transition in the same classical $O(N)$-symmetric model, but with $D = 3$. We shall assume that $N \gg 1$ and consider an approach alternative to the renormalization group. The condition $N \gg 1$ will allow us to select the most essential diagrams and sum them up. It turns out that in this case one can figure out a general functional form of the two-point Green's function $G(k)$. This makes it possible to work with Feynman diagrams containing the exact Green's functions – fat lines – and treat the expansion as a self-consistent equation for the unknown parameters of $G(k)$. Let me explain this important point. The most general form for $G(k)$ is given by the Dyson equation:

$$G(k) = [k^2 - \Sigma(k)]^{-1}$$

Since we are at the transition point, $\Sigma(0) = 0$; also since the theory is Lorentz invariant, the self energy part is a function of $\mathbf{k}^2 \equiv k^2$. Let us now assume that at small k the self energy part decreases slower than k^2. That is, there is a scale k_0 below which the k-dependence of the Green's function is determined entirely by interactions and the bare Green's function is forgotten. This assumption expresses the spirit of universality: properties at criticality do not depend on details of interactions at short distances. We assume that at $k < k_0$ the self energy part and the Green's function have the following form:

$$\Sigma(k) = -Z^{-1}k^{2-\eta}$$
$$G(k) \approx -[\Sigma(k)]^{-1} = \frac{Z}{k^{2-\eta}} \qquad (7.22)$$

We choose a power law dependence because at criticality, where there is no scale, this is the only functional form which does not contain a scale.

Our purpose now is to check that our assumption is self-consistent and then find Z and η. As I have said, we shall use the Feynman diagrams with renormalized (fat) lines. Figure 5.5 represents several such diagrams for the self energy part and Fig. 5.8 shows several diagrams for the bubble.

As we know from Chapter 5, the renormalized interaction $D(q)$ is

$$D(q) = -\left[g^{-1} + \frac{1}{2}\Pi(q)\right]^{-1} \tag{7.23}$$

where $\Pi(q)$ is a sum of diagrams which cannot be cut along a single dotted line (see Fig. 5.8). Here we also assume that the bare interaction is irrelevant, i.e. that $\Pi(q)$ diverges at small q. It is easy to check that all the diagrams in Fig. 5.8 are proportional to $Z^2 q^{-1+2\eta}$:

$$D(q) \approx \frac{2}{\Pi(q)} = 2B^{-1}(N, \eta)Z^{-2}q^{1-2\eta} \tag{7.24}$$

where B is some numerical coefficient dependent only on N and η. Substituting this result into the diagrams for the self energy part we find each diagram is proportional to $k^{2-\eta}$. Indeed, all the Z cancel, i.e. our assumption has a chance to be self-consistent. Why 'has a chance to ...' and not 'is self-consistent'? Because we have not yet found η and Z.

So far we have not used the property $N \gg 1$; we shall need it now to find η in the leading order in $1/N$. The first diagram on the right-hand side of Fig. 5.8 contains the maximal power of N; therefore in the leading order in N we have

$$\Pi(q) = \frac{NZ^2}{(2\pi)^3} \int \frac{d^3k}{k^{2-\eta}|k-q|^{2-\eta}} = q^{-1+2\eta}\frac{NZ^2}{(2\pi)^3} \int \frac{d^3k}{k^{2-\eta}|k-n|^{2-\eta}} \tag{7.25}$$

where $\mathbf{n}$ is the unit vector in the direction of $\mathbf{q}$. To get the last integral I made a substitution $\mathbf{k} = q\mathbf{k'}$. Thus in the leading order in N we get

$$\Pi(q) = \frac{Z^2}{q^{1-2\eta}}B(N, \eta) \quad B(N, \eta) = \frac{N}{(2\pi)^3} \int \frac{d^3k}{k^{2-\eta}|k-n|^{2-\eta}} \tag{7.26}$$

The latter integral is calculated in the Appendix with the result

$$(2\pi)^2 B(N, \eta) = \frac{1}{2}\pi^2 N + \mathcal{O}(1) \tag{7.27}$$

Substituting this into the expression for the self energy given by the diagram in Fig. 5.5(b), we get

$$\Sigma(k) - \Sigma(0) = -2\frac{k^{2-\eta}}{BZ(2\pi)^3} \int d^3p\, p^{-2+\eta}(|\mathbf{p}-\mathbf{n}|^{1-2\eta} - p^{1-2\eta}) = -\frac{A(N, \eta)}{B(N, \eta)}Z^{-1}k^{2-\eta} \tag{7.28}$$

where

$$(2\pi)^2 A(N, \eta) = \frac{4}{3\eta} + \mathcal{O}(1) \tag{7.29}$$

Clearly, the self-consistency condition in this order of N is

$$1 = A(N, \eta)/B(N, \eta) = \frac{8}{3\pi^2 N\eta} + \mathcal{O}(N^{-2}) \tag{7.30}$$

from which we find

$$\eta = \frac{8}{3\pi^2 N} + \mathcal{O}(N^{-2}) \tag{7.31}$$

Exercise

Prove the result (7.29).

Calculate the next correction to η coming from the diagrams shown in Fig. 5.5(b),(d) and Fig. 5.8(c),(d).

Appendix

$$
\begin{aligned}
(2\pi)^2 N^{-1} B(N, \eta) &= \frac{1}{(2\pi)} \int \frac{d^3 k}{k^{2-\eta} |\mathbf{k} - \mathbf{n}|^{2-\eta}} \\
&= \int_0^\infty dk \int_0^\pi d\cos\theta \frac{k^\eta}{(k^2 - 2k\cos\theta + 1)^{1-\eta/2}} \\
&= \eta^{-1} \int_0^\infty dk k^{-1+\eta}[(k+1)^\eta - |1 - k|^\eta] \\
&= \int_0^1 dx(x^{-1+\eta} + x^{-1-\eta})[(1+x)^\eta - (1-x)^\eta] \tag{7.32}
\end{aligned}
$$

The last integral is a table one; the final result is given above.

8

O(N)-symmetric vector model below the transition point

Let us discuss what happens with the O(N)-symmetric model at $\tau = -m^2 < 0$, when its action has the following form:

$$S = \int d\tau d^D x \left[\frac{1}{2}(\partial_\mu \Phi \partial_\mu \Phi) - \frac{m^2}{2}\Phi^2 + \frac{g}{4}(\Phi^2)(\Phi^2) \right] \tag{8.1}$$

Obviously, since the action at $g = 0$ is not definitely positive, we cannot expand correlation functions in powers of g. This is a case when interacting and noninteracting systems have orthogonal ground states. Therefore we need another procedure.

As a preliminary step, however, let us discuss the most primitive case, namely $D = 0$. In this case we can switch back from the path integral to the quantum mechanical description, the quantum Hamiltonian for the action (8.1) being given by

$$\hat{H} = -\frac{1}{2}\frac{\partial^2}{\partial \Phi_a^2} - \frac{m^2}{2}\Phi^2 + \frac{g}{4}(\Phi^2)(\Phi^2) \tag{8.2}$$

This Hamiltonian describes a particle of unit mass in the N-dimensional potential well $V(\Phi)$, where

$$V(\Phi) = -\frac{m^2}{2}\Phi^2 + \frac{g}{4}(\Phi^2)(\Phi^2) = -\frac{m^4}{4g} + \frac{g}{4}\left(\Phi^2 - \frac{m^2}{g}\right)^2 \tag{8.3}$$

This potential favours configurations of Φ with a fixed modulus:

$$\Phi^2 = \frac{m^2}{g}$$

Therefore one can expect that for certain potentials the energy levels corresponding to radial degrees of freedom lie essentially above the rotational levels. This suggestion can be formally justified as follows. Let us choose the spherical coordinate system:

$$\Phi_a = M n_a \qquad \mathbf{n}^2 = 1 \tag{8.4}$$

Then the Hamiltonian (8.2) is

$$\hat{H} = -\frac{1}{2M^{N-1}}\partial_M [M^{N-1}\partial_M] + \frac{g}{4}\left(M^2 - \frac{m^2}{g}\right)^2 - \frac{1}{2M^2}\Delta'_N \tag{8.5}$$

where Δ'_N is the angular part of the Laplace operator. The angular part becomes a perturbation provided

$$\frac{1}{2M_0^2} \equiv \left\langle 0 \left| \frac{1}{2M^2} \right| 0 \right\rangle \ll E_1 - E_0$$

where $|0\rangle = \psi_0(M)$ is the ground state function of the radial part of the Hamiltonian and E_1, E_0 are its first two lowest levels. The fine structure of the radial ground state is described by the angular part of the Hamiltonian averaged over $\psi_0(M)$:

$$\hat{H}_{eff} = -\frac{1}{2M_0^2}\Delta'_N \tag{8.6}$$

The Hamiltonian $\hat{H}_{eff}$ is a Hamiltonian of the O(N)-symmetric rotator. It describes the lowest energy levels of the system. The radial motion participates in spectrum formation only through the parameter M_0; for all other purposes one can forget about fluctuations of the modulus. We can say that these high energy fluctuations decouple from the low energy ones.

Before coming back to the quantum field theory with $D > 0$, I make the following remark. Notice that the angular wave function of the ground state is independent of angles, i.e. there is no preferential direction in the isotopical space. Of course, one cannot expect otherwise for a quantum mechanical system; in quantum mechanics a ground state always has the same symmetry as the Hamiltonian. I make this remark only because for higher dimensionalities we shall see something very different.

Now let us generalize the procedure of separation of variables for $D > 0$. Choosing the spherical system of coordinates in the $\mathbf{\Phi}$-space:

$$\Phi_a(x) = M(x)n_a(x) \qquad \mathbf{n}^2 = 1 \tag{8.7}$$

and substituting (8.7) into the action and the measure of integration we get the following expression for the partition function:[1]

$$Z = \int M^{(N-1)}(x)DM(x) \int D\mathbf{n}(x)\delta(\mathbf{n}^2(x) - 1)\exp\{-S[M] - S[\mathbf{n}, M]\} \tag{8.8}$$

$$S[M] = \int d\tau d^D x \left\{ \frac{1}{2}(\partial_\mu M \partial_\mu M) + \frac{g}{4}\left(M^2 - \frac{m^2}{g}\right)^2 \right\} \tag{8.9}$$

$$S[\mathbf{n}, M] = \int d\tau d^D x \frac{M^2}{2}(\partial_\mu \mathbf{n}\partial_\mu \mathbf{n}) \tag{8.10}$$

Let us show that, as in the quantum mechanical example, (i) the modulus field fluctuates weakly around its average value, and (ii) the interaction between fluctuations of the direction (transverse fluctuations) and the modulus (radial fluctuations) does not lead to singular corrections and therefore one can find a region of parameters where it is small.

---

[1] The cross terms in the gradient term vanish due to the following property of the unit vector: $2\mathbf{n}\partial\mathbf{n} = \partial\mathbf{n}^2 = 0$.

Let us make a shift of variable in our path integral:

$$M(x) = M_0 + l(x)$$

where M_0 is determined from the condition

$$\left.\frac{\delta \ln Z}{\delta l(x)}\right|_{\delta l = 0} = 0 \qquad (8.11)$$

which is equivalent to

$$M_0^2 = \frac{1}{g}\left[m^2 - \langle(\partial_\mu \mathbf{n})^2\rangle + \frac{(N-1)}{a^D M_0^2}\right] \qquad (8.12)$$

Then the partition function transforms as follows:

$$Z = \int Dl(x) \int D\mathbf{n}(x)\delta(\mathbf{n}^2(x) - 1)\exp[-S_1 - S_2 - S_3] \qquad (8.13)$$

$$S_1 = \int d\tau d^D x \left(\frac{1}{2}\partial_\mu l \partial_\mu l + m^{*2} l^2\right) \qquad (8.14)$$

$$S_2 = \frac{M_0^2}{2}\int d\tau d^D x (\partial_\mu \mathbf{n} \partial_\mu \mathbf{n}) \qquad (8.15)$$

$$S_3 = \int d\tau d^D x \{(M_0 l + l^2/2) : (\partial_\mu \mathbf{n} \partial_\mu \mathbf{n}) :$$
$$- (N-1)a^{-D}\left[\ln(1 + l/M_0) - l/M_0 - l^2/2M_0^2\right] + g M_0 l^3 + g l^4/4\} \qquad (8.16)$$

where dots indicate that the average of the operator is subtracted. The log term in the action comes from the $\prod_x M^{(N-1)}(x)$ term in the measure. The coefficient m^{*2} is defined as

$$m^{*2} = \frac{\delta^2 \ln Z}{\delta l(x)^2}$$

The condition (8.11) guarantees that terms linear in l do not appear; all quadratic terms are by definition included in the $m^{*2} l^2$ term and therefore the S_3 term in the action contains, in fact, only terms of higher order in l. Let us neglect for a moment the interaction between fluctuations of the modulus and the direction fields, described by the S_3 term, and consider the two actions (8.14) and (8.15) as independent. The first of them describes a model of a free massive scalar field whose excitations have the spectrum

$$\epsilon^2(k) = k^2 + m^{*2}$$

The second theory is nonlinear due to the geometrical constraint $\mathbf{n}^2 = 1$. Quantization of this model for $D = 0$ gives the model (8.6). If we forget for a moment about this nonlinearity, we can conjecture that the excitations of the model (8.15) are gapless particles, i.e.

$$\epsilon^2(k) = k^2$$

This result is in strong contradiction to the quantum mechanical example described above. Therefore we can conclude that there are situations where the nonlinearity is essential and gives rise to a mass gap. Let us assume, however, that this mass (if it appears) is always much smaller than the mass for radial excitations m^*. Then it is possible to average over the radial modes and get a self-consistent description of the angular fluctuations in terms of the director field $\mathbf{n}$ only. Can the interaction S_3 be a principal obstacle? No, because the corresponding perturbation expansion contains derivatives of $\mathbf{n}$ and therefore is infrared finite. There are severe ultraviolet divergences, but they are curable. Thus one can always find a region of parameters m^2 and g and a (the lattice spacing), where corrections from S_3 lead only to a renormalization of M_0 without changing the form of the effective action itself. Since this renormalization depends on the lattice it does not make any sense to discuss it here.

Thus for the low energy excitations we have the effective action in the form of a *nonlinear sigma* model:

$$S = \frac{M^2}{2} \int d\tau d^D x (\partial_\mu \mathbf{n} \partial_\mu \mathbf{n}) \qquad \mathbf{n}^2 = 1 \tag{8.17}$$

As has already been mentioned, this theory describes excitations with energies smaller than the energies of the radial modes. Therefore its ultraviolet cut-off is equal to the mass of radial excitations: $\Lambda = m^*$. M^2 is a renormalized stiffness and in general does not coincide with M_0^2. Let us suppose that the field $\mathbf{n}$ fluctuates weakly around some preferential direction given by some constant vector $\mathbf{e}$:

$$\mathbf{n} = \mathbf{e}\sqrt{1 - \mathbf{m}^2} + \mathbf{m} \qquad \langle \mathbf{m}^2(x) \rangle \ll 1 \qquad \mathbf{e}\mathbf{m} = 0 \tag{8.18}$$

If we are able to demostrate that fluctuations are really weak, then it will mean something very remarkable. Namely, the preferential direction appears in the wave functions, being absent in the action itself. This is called spontaneous symmetry breaking. It is unknown in quantum mechanics, and exists only in theories with an infinite number of degrees of freedom, i.e. in field theories. All second-order phase transitions occur via symmetry breaking. A state with a broken symmetry is characterized by an additional quantity – in the given case it is $\mathbf{e}$ – which is called an order parameter. The current discussion leads to the suggestion that for $\tau < 0$ the ground state of the O(N)-symmetric vector model has a broken O(N) symmetry characterized by the vector order parameter $\mathbf{e} \sim \langle \mathbf{n}(x) \rangle$. As we shall see later, this symmetry breaking occurs only in higher dimensionalities. As many of the readers probably know, symmetry breaking is accompanied by the appearance of massless particles: Goldstone bosons. These particles describe those fluctuations of the order parameter which do not violate its symmetry. They appear only if the broken symmetry is continuous. In this case, since the energy cost of uniform rotations of the order parameter is zero, the dispersion of fluctuations of the order parameter vanishes at zero wave vector. This means that such excitations have zero spectral gap, i.e. they are massless. In the present case the Goldstone particles are described by the effective action (8.17). This effective action provides a good description of low energy modes sufficiently below the transition, in the area where fluctuations of the modulus of the order parameter are small enough.

Let us now take a look at details of the symmetry breaking. We need to check that the assumption $\langle \mathbf{m}^2 \rangle \ll 1$ is self-consistent. Substituting (8.18) into the action (8.17) we get:

$$S_m = \frac{M^2}{2} \int d\tau d^D x \left[\partial_\mu \mathbf{m} \partial_\mu \mathbf{m} + \frac{(\mathbf{m}\partial_\mu \mathbf{m})(\mathbf{m}\partial_\mu \mathbf{m})}{1 - \mathbf{m}^2} \right] \tag{8.19}$$

The measure of integration is also affected:

$$\mathbf{Dn}\delta(\mathbf{n}^2 - 1) = \prod_{x,\tau} \frac{\prod_a dm_a(x, \tau)}{\sqrt{1 - \mathbf{m}^2(x, \tau)}} \tag{8.20}$$

One can make the measure trivial by transferring the factor $[1 - \mathbf{m}^2(x, \tau)]^{-1/2}$ into the action:

$$S = S_m + \frac{1}{2a^{(D+1)}} \int d\tau d^D x \, \ln[1 - \mathbf{m}^2(x, \tau)] \tag{8.21}$$

Let us suppose that the nonlinear terms of the action (8.21) are small and can be considered as a perturbation. Then in the zeroth order we have

$$\langle m^a(-p)m^b(p) \rangle = M^{-2} \frac{1}{\omega_n^2 + k^2} \delta_{ab} \tag{8.22}$$

Therefore we have

$$\langle \mathbf{m}^2 \rangle = \frac{(N-1)}{M^2} T \sum_n \int \frac{d^D k}{(2\pi)^D} \frac{1}{\omega_n^2 + k^2} = \frac{(N-1)}{2M^2} \int \frac{d^D k}{(2\pi)^D} \coth(|k|/2T) \frac{1}{|k|} \tag{8.23}$$

At finite T this integral converges at $D > 2$ and at $T = 0$ it also converges at $D = 2$.

Let us first consider the case $D > 2$, $T \neq 0$. In this case one can satisfy the requirement $\langle \mathbf{m}^2 \rangle \ll 1$ provided M^2 is large enough:

$$M^2 \gg (N - 1)/a^{(D-1)} \tag{8.24}$$

In this case one can check that the diagram expansion for the model (8.21) does not contain infrared divergences. Thus the smallness of transverse fluctuations persists in all orders. The nonlinear terms in (8.21) give corrections $\sim M^{-2}$ to the Green's function (8.22). The important point is that the perturbations do not change the singular character of the Green's function (8.22) at small ω and k; the excitation spectrum remains gapless. Physically this property follows from the fact that the broken symmetry is continuous. The energy of the system does not depend on the direction of the order parameter; on the other side fluctuations of $\mathbf{m}$ with $\mathbf{k} \to 0$ are uniform rotations of the order parameter $\mathbf{e}$, and therefore must have a zero energy (the Goldstone theorem). Technically this physical property reveals itself as a cancellation of certain diagrams, and here the change of the measure plays an important role.

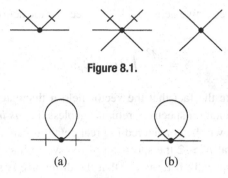

Figure 8.1.

(a)   (b)

Figure 8.2.

The action (8.21) has the vertices shown in Fig. 8.1. There are two sorts of corrections to the self energy: those which contain differentiations on the external tails and are thus proportional to $\mathbf{k}^2$, and those which do not.

Examples of the diagrams of the first and the second type are given in Figs. 8.2(a) and (b) respectively.

Exercise

For the first two orders check that diagrams of the second type cancel due to the presence of terms coming from the expansion of the measure.

Now let us return to the complicated cases $D = 1$ and $D = 2$. For $D = 1$ the integral (8.23) diverges logarithmically at $T = 0$ and as T/k at finite T. Therefore the following is valid:

Theorem I. A *continuous* symmetry cannot be spontaneously broken in one dimension.[2]

For $D = 2$ ($T \neq 0$) the integral (8.23) is also logarithmically divergent. This means that:

Theorem II. A continuous symmetry cannot be spontaneously broken in two dimensions at finite temperatures.

At zero temperatures the phenomenon of spontaneous symmetry breaking can occur in two dimensions. For this reason, two-dimensional magnets can have spontaneous magnetization (or staggered magnetization) at $T = 0$.

In the following chapters I shall describe the $\tau < 0$ state in both these cases. Now I only want to remark that this description is different for $N = 2$ and $N > 2$. Indeed, let us consider the case $N = 2$. The constraint $\mathbf{n}^2 = 1$ can be easily resolved in the explicit form:

$$\mathbf{n} = (\cos \alpha, \sin \alpha) \qquad (8.25)$$

[2] Strictly speaking, this statement is valid only in the absence of long distance forces.

Substituting this into the effective action (8.17) we see that the action for α is the free field action:

$$S = \frac{M^2}{2} \int d\tau d^D x \partial_\mu \alpha \partial_\mu \alpha \tag{8.26}$$

Thus in this case, despite the fact that the vector field $\mathbf{n}$ fluctuates wildly and $\langle \mathbf{n}(x) \rangle \propto \langle \exp[i\alpha(x)] \rangle = 0$, the excitation spectrum remains gapless, i.e. *as in the state with broken symmetry*. This situation will be considered in great detail in Part IV. In the next chapters I shall demonstrate that at $N > 2$ the spectrum of transverse fluctuations acquires a gap. Why then is $N = 2$ so special? It turns out that the difference is purely topological: the group O(2) is an Abelian group: it consists of only commuting elements – rotations of two-dimensional Euclidean space. The groups O(N) with $N > 3$ consist of noncommuting rotations of N-dimensional Euclidean spaces. Therefore the following statement is valid:

Theorem III. Two-dimensional systems with Abelian and non-Abelian symmetries behave differently.

Exercise

Apply the procedures of this chapter to a nonrelativistic analogue of the model (8.1), the model of repulsive bosons in two dimensions:

$$S = \int d\tau d^2 x \left[\Phi^+ \partial_\tau \Phi + \frac{1}{2} \nabla \Phi^+ \nabla \Phi - \mu \Phi^+ \Phi + \frac{1}{2} g (\Phi^+ \Phi)^2 \right] \tag{8.27}$$

At $\mu > 0$ the bosons condense. For $D > 2$ this condensation corresponds to the appearance of the nonzero average $\langle \Phi \rangle$. What symmetry is broken? For $D = 2$ this average is zero, but the correlation function $\langle \Phi(x) \Phi^+(y) \rangle$ decays as a power law. Consider the following steps. (a) Rewrite the path integral in the radial and angular variables. Which of the following is the right choice of variables?

$$\Phi = r e^{i\alpha}$$

or

$$\Phi = \sqrt{\rho} e^{i\alpha}$$

Why? (b) Repeating the procedures outlined above, separate slow and fast variables and obtain the following expression:

$$S = \int d\tau d^2 x \left[i\rho \partial_\tau \alpha + \frac{1}{8\rho} (\nabla \rho)^2 + \frac{1}{2} \rho (\nabla \alpha)^2 + \frac{1}{2} g (\rho - \rho_0)^2 \right] \tag{8.28}$$

When μ is large enough, the radial component of Φ fluctuates weakly around its nonzero average value $\sqrt{\rho_0}$. Obtain the effective action (8.26) for α and an effective action for $\tilde{\rho} = \rho - \rho_0$. Neglecting higher terms in the expansion in $\tilde{\rho}$, find the pair correlation functions of $\tilde{\rho}$ and α. The answer is

$$\langle\langle\tilde{\rho}(-\omega_n,-k)\tilde{\rho}(\omega_n,k)\rangle\rangle = \frac{\rho_0 k^2}{\omega_n^2 + \frac{1}{4}k^2(k^2+4\rho_0 g)}$$

$$\langle\langle\alpha(-\omega_n,-k)\tilde{\rho}(\omega_n,k)\rangle\rangle = \frac{\omega_n}{\omega_n^2 + \frac{1}{4}k^2(k^2+4\rho_0 g)} \qquad (8.29)$$

$$\langle\langle\alpha(-\omega_n,-k)\alpha(\omega_n,k)\rangle\rangle = \frac{g+k^2/4\rho_0}{\omega_n^2 + \frac{1}{4}k^2(k^2+4\rho_0 g)}$$

Derive from these expressions the excitation spectrum. (c) Obtain an explicit expression for the pair correlation function of the phase field $\langle\langle\alpha(\tau,x)\alpha(0,0)\rangle\rangle$ in real space. What term in the sum over frequencies gives the most singular contribution? What effect does it have on the behaviour of $\langle\Phi(x)\Phi^+(0)\rangle$? (If the last part is too difficult for you, consult Chapter 22.)

9

Nonlinear sigma models in two dimensions: renormalization group and $1/N$-expansion

Let us consider what happens with the Goldstone modes of the $O(N)$-symmetric vector model in one and two dimensions. We have established already that in this case the $O(N)$ symmetry remains unbroken. It does not invalidate, however, the nonlinear sigma model (8.17), provided the relevant energy scale for transverse fluctuations is much smaller than the gap for longitudinal fluctuations. We shall see now whether this assumption is self-consistent.

I shall discuss the $D = 1$ case first. As theories with strong interaction, nonlinear sigma models in $(1 + 1)$ dimensions cannot be treated perturbatively. Indeed, the renormalization group calculations for the $O(N)$ nonlinear sigma model give the following Gell-Mann–Low equations for the renormalized stiffness:[1]

$$\frac{d\tilde{M}^2}{d\ln(\Lambda/|k|)} = -\frac{(N-2)}{2\pi} + \mathcal{O}(\tilde{M}^{-2}) \tag{9.1}$$

whose solution

$$\tilde{M}^2 = M^2 - \frac{(N-2)}{2\pi}\ln(\Lambda/|k|) \tag{9.2}$$

vanishes at

$$|k_c| = \Lambda \exp[-2\pi M^2/(N-2)] \tag{9.3}$$

which indicates that at $|k| < |k_c|$ one can no longer use the perturbation theory.

To find out what happens at small momenta one should employ nonperturbative methods. This will be done later in this chapter, but I now want to show that the case $D = 2$, $T \neq 0$ can be described within the same scheme. Namely, I claim that leading singularities in the $D = 2$ nonlinear sigma model are the same as for the $D = 1$ nonlinear sigma model with the bare stiffness

$$M^2(D = 1) = \frac{M^2(D = 2)}{T} \tag{9.4}$$

and ultraviolet cut-off $\Lambda \sim T$. Expressed in other words, the claim is that only time-independent configurations of the **n**-field are important. Indeed, the $(2 + 1)$-dimensional

---

[1] The corresponding derivation can be found in textbooks, for example the books by Amit, Fradkin and Zinn-Justin.

action on such configurations reduces to the two-dimensional action with the new stiffness (9.4). As we know from (9.3), the singularities in the corresponding perturbation series come from the area

$$|k| \sim |k_c| \sim \Lambda_{2D} \exp[-2\pi M^2 (D = 2)/T(N-2)] \ll T \qquad (9.5)$$

The fact that $|k| \ll T$ means that these singularities do not appear in diagrams containing nonzero frequencies $\omega_n = 2\pi Tn > 2\pi T$. As an illustration of this statement, consider the following integral:

$$T \sum_n \int \frac{d^2k}{(2\pi)^2} \frac{1}{\omega_n^2 + k^2} = T \int \frac{d^2k}{(2\pi)^2} \frac{1}{k^2} + T \sum_{n \neq 0} \int \frac{d^2k}{(2\pi)^2} \frac{1}{\omega_n^2 + k^2} \qquad (9.6)$$

The divergence comes from the term containing zero frequency. Neglecting quantum fluctuations we restrict our description to the area $|k| < T$. For this reason the ultraviolet cut-off for the theory of static fluctuations is $\Lambda_{2D} \sim T$. This discussion will be repeated in part in Chapter 17, where I discuss applications of the O(3) nonlinear sigma model to the theory of magnetism.

Thus the O(N) nonlinear sigma model

$$S = \frac{M_D^2}{2} \int d^2x \partial_\mu \mathbf{n} \partial_\mu \mathbf{n} \qquad (9.7)$$

$$= \frac{M^2}{2} \int d^2x \left[\partial_\mu \mathbf{m} \partial_\mu \mathbf{m} + \frac{(\mathbf{m}\partial_\mu \mathbf{m})(\mathbf{m}\partial_\mu \mathbf{m})}{1 - \mathbf{m}^2} \right] \qquad (9.8)$$

describes long-wavelength fluctuations of the vector order parameter $\mathbf{n}$ in one and two spatial dimensions. This model is a particular example of a wider class of nonlinear sigma models defined by the action

$$S = \frac{1}{2} \int d^2x G_{ab}(X) \partial_\mu X^a \partial_\mu X^b \qquad (9.9)$$

with the measure given by

$$D\mu[X] = \prod_x \sqrt{G[X(x)]}$$

Comparing this expression with (9.8), we find that for the O(N) nonlinear sigma model

$$G_{ab} = M_D^2 \left(\delta_{ab} + \frac{X^a X^b}{1 - \mathbf{X}^2} \right) \qquad (9.10)$$

All these models have a clear geometrical interpretation: the O(N) nonlinear sigma model describes transverse fluctuations of vectors belonging to an N-dimensional sphere, and the general sigma model describes fluctuations of fields belonging to a hyperspace with the metric G_{ab}. This hyperspace can appear as an internal space of some order parameter; then at low temperatures, when longitudinal fluctuations are small (recall the discussion in the previous chapter), the model describes fluctuations of this order parameter in directions transverse to this hyperspace. For the O(N) nonlinear sigma model, (9.10) describes a

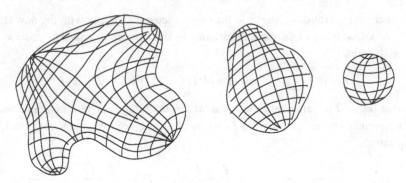

Figure 9.1.

metric of an N-dimensional sphere with radius M_D. The general formulation is particularly illuminating because it allows us to appreciate the geometrical aspects of the problem. Let us take a look at the Gell-Mann–Low equations for nonlinear sigma models derived by Friedan (1980, 1985):[2]

$$\frac{dG_{ab}}{d\ln(\Lambda/k)} = -\frac{1}{2\pi}R_{ab} - \frac{1}{8\pi^2}\left(R_{a\beta\gamma\eta}R_b^{\beta\gamma\eta}\right) + \cdots \tag{9.11}$$

where $R_{a\beta\gamma\eta}$ is the Riemann tensor of the hyperspace and $R_{ab} = R^\beta_{a\beta b}$ is the Ricchi tensor (see the discussion in Chapter 4). The latter is related to the metric tensor and the connection coefficients:

$$R_{ab} = \frac{1}{\sqrt{G}}\partial_c\left(\Gamma^c_{ab}\sqrt{G}\right) - \partial_a\partial_b(\ln\sqrt{G}) - \Gamma^r_{ac}\Gamma^c_{br}$$

$$\Gamma^a_{bc} = \frac{1}{2}G^{ak}(\partial_c G_{bk} + \partial_b G_{kc} - \partial_k G_{bc}) \tag{9.12}$$

Equation (9.11) suggests a very beautiful picture: under the renormalization process, changes of the hyperspace geometry are determined solely by the geometry itself (see Fig. 9.1).

Here we have a chance to study a change of not just a single coupling constant, but of the entire function G_{ab}. Of course, in a primitive case of the $O(N)$ nonlinear sigma model the hyperspace described by the metric G_{ab} is just a sphere and so remains under the renormalization. There are models, however, where the metric is less trivial (see, for example, Fateev et al. (1993)).

From (9.11) one can find general conditions for strong coupling. Namely, let us suppose that our hyperspace has a finite volume and study how this volume changes under the renormalization. Using the identity

$$d\sqrt{G} = \frac{1}{2}\sqrt{G}G^{ab}dG_{ab} \tag{9.13}$$

---

[2] See also (14.85) in the 1993 edition of the book by Zinn-Justin.

we get from (9.11)

$$\frac{d\Omega}{d\ln(\Lambda/k)} \equiv \frac{d}{d\ln(\Lambda/k)}\left(\int d^N X\sqrt{G}\right) = -\frac{1}{4\pi}\int d^N X\sqrt{G}R + \mathcal{O}(R^2) \quad (9.14)$$

where $R = G^{ab}R_{ab}$ is the scalar curvature and N is the dimension of the hyperspace. We see that if the scalar curvature is positive the volume decreases under renormalization which leads to the eventual collapse of the hyperspace. As a rule, non-Abelian groups have closed group manifolds with positive scalar curvature. Therefore the corresponding sigma models always scale to strong coupling. If the hyperspace is two-dimensional (as for the O(3) sigma model, for example), (9.14) simplifies even further. In two dimensions we have the Gauss–Bonnet theorem:

$$\int d^2 X\sqrt{G}R = 8\pi(1-h) \quad (9.15)$$

where h is the number of handles of a given closed manifold. For a sphere $h = 0$ and we have from (9.14)

$$\frac{d\Omega}{d\ln(\Lambda/k)} = -2 + \mathcal{O}(R^2) \quad (9.16)$$

For a torus $h = 1$ and the right-hand side of (9.14) is zero (in the one-loop approximation). So we can hope that sigma models on tori are critical. No wonder: a torus can be represented as a product of two circles, i.e. its symmetry group is U(1) × U(1) and does not contain noncommuting operations. We shall return to this problem at the end of Chapter 23, where the sigma model on a torus will be considered in more detail. Sigma models on two-dimensional manifolds with many handles $h > 1$ may be critical or exhibit a zero charge behaviour; as far as I know, this problem has not yet been studied.

As I have already said, the perturbation theory cannot predict what happens in the strong coupling regime; there are different methods for solving strong coupling problems. Sometimes it is possible to do this exactly: the exact solution for the O(3) sigma model was found by Zamolodchikov and Zamolodchikov (1979) and by Wiegmann (1985), and for the O(4) sigma model by Polyakov and Wiegmann (1983). Though these solutions are very beautiful and instructive. I shall discuss only their results. For the derivation I resort to a simpler and more general method, the $1/N$-expansion. The expansion is based on the fact that at $N \to \infty$ the O(N) sigma model is a model of free massive particles. At finite N these particles begin to interact, but since the spectrum is already massive there are no infrared divergences in the diagram expansion and the interaction can be treated perturbatively. The results obtained by the $1/N$-expansion remain qualitatively correct down to $N = 3$. The exact solution shows that the excitations are interacting massive particles and their multiplet is N-fold degenerate.[3] The latter is really remarkable: at large energies we have $N - 1$ transverse modes and at low energies there are N degenerate massive branches. It seems

---

[3] Recall that in integrable systems interaction does not create new particles; interacting excitations just change their phase.

Figure 9.2. A 'Mexican hat' potential.

that the longitudinal degree of freedom is restored! Indeed, it can be shown (see Nicopoulos and Tsvelik, 1991) that the effective action for the O(3) nonlinear sigma model in the strong coupling limit is

$$S = \int d^2x \left[\frac{1}{2}\partial_\mu \Phi \partial_\mu \Phi + \frac{m^2}{2}(\Phi)^2 + U(\Phi)^2(\Phi)^2 \right] \tag{9.17}$$

where Φ is a *three*-dimensional vector. This means that at low energies the constraint $\mathbf{n}^2 = 1$ is relaxed. As we shall see in a moment, the same picture arises in the $1/N$-expansion. Now recall that the nonlinear sigma model originates from the vector model at $\tau < 0$ where the effective potential is like a deep circular gutter (it is also called a 'Mexican hat', see Fig. 9.2). The field moves in this gutter which effectively diminishes its dimensionality. The fluctuations, however, renormalize this Mexican hat potential removing its central part.

Therefore the system returns effectively to the region $\tau > 0$, which corresponds to the fact that in $D \leq 2$, continuous symmetry remains unbroken.

$1/N$-expansion

Let us rewrite the partition function for the $\mathbf{n}$-field using the identity for the functional delta-function:

$$
\begin{aligned}
Z &= \int D\mathbf{n}(x)\delta[\mathbf{n}^2(x) - 1] \exp\left[-\frac{M^2}{2} \int d\tau dx (\partial_\mu \mathbf{n}\partial_\mu \mathbf{n}) \right] \\
&= \int D\mathbf{n}(x)D\lambda(x) \exp\left(-\int d\tau dx \left\{ \frac{M^2}{2}(\partial_\mu \mathbf{n}\partial_\mu \mathbf{n}) + i\lambda(x)[\mathbf{n}^2(x) - 1] \right\} \right)
\end{aligned} \tag{9.18}
$$

Now let us redefine the variables

$$\lambda = -\frac{i}{2}m^2 M^2 + \frac{u}{\sqrt{N}}M^2 \qquad \mathbf{n} = M^{-1}\mathbf{l}$$

which is equivalent to the shift of the contour of integration from the real axis as shown in Fig. 9.3.

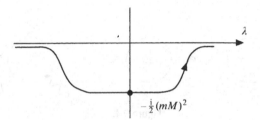

$-\frac{i}{2}(mM)^2$

Figure 9.3.

After integrating over **l** we get:

$$Z = \int Du(x)\exp\{-S_{\text{eff}}[u]\} \tag{9.19}$$

$$S_{\text{eff}}[u] = \int d\tau dx \left\{ M^2\left(-i\frac{u}{N^{-1/2}} - \frac{1}{2}m^2\right) + \frac{N}{2}\text{Tr}\ln\left[-\partial^2 + m^2 + \frac{2iu}{N^{-1/2}}\right]\right\} \tag{9.20}$$

The exponent of this integral contains N and for $N \gg 1$ one can calculate the integral using the saddle point approximation. The quantity $-m^2 M^2/2$ is the point where the new contour of integration crosses the imaginary axis. This point is determined self-consistently by the saddle point condition:

$$\frac{\partial}{\partial u}\ln Z(m^2, u)|_{u=0} = 0 \tag{9.21}$$

In the leading order in $1/N$ we can calculate the free energy, neglecting the fluctuations of u, i.e. just setting $u = 0$. Then we have

$$\ln Z = V\left[\frac{1}{2}M^2 m^2 - \frac{N}{2(2\pi)^2}\int d^2p\ln(p^2 + m^2)\right] \tag{9.22}$$

where V is the volume occupied by the system in Euclidean space-time. The saddle point condition (9.21) gives

$$M^2 = N\int\frac{d^2p}{(2\pi)^2}G(p) \approx \frac{N}{2\pi}\ln(\Lambda/m) \tag{9.23}$$

The solution of this equation is

$$m = \Lambda\exp[-2\pi M^2/N] \tag{9.24}$$

The condition (9.21) is, in fact, equivalent to the condition

$$\langle \mathbf{n}^2(x)\rangle = 1 \tag{9.25}$$

Thus in our approximation the local constraint is substituted by the global one. The validity of such a substitution is ensured by the large N.

It follows from (9.24) that excitations of the **n**-field are massive particles. Their mass coincides with the energy scale (9.3) where the strong coupling regime establishes and the perturbation theory ceases to work.[4] If $M^2/N > 1$, the mass particle is much smaller than

---

[4] In the limit of large N we cannot distinguish between N and $N - 2$.

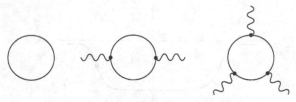

Figure 9.4.

the cut-off – the mass of the longitudinal modes. Only in this case is the nonlinear sigma model applicable.

As I have stated above, the approximation we are discussing is for large N. The result we have just obtained is exact at $N \to \infty$, M^2/N finite. Let us now discuss $1/N$ corrections. Let us consider the effective action (9.20); it contains the determinant of the sort discussed in Chapter 4. Expanding the determinant in powers of $u/\sqrt{N}$ we get the diagrams represented in Fig. 9.4, where the solid lines correspond to the propagator

$$G(p) = \frac{1}{p^2 + m^2}$$

and the wavy lines to $iu(x)/\sqrt{N}$.

The term of this expansion linear in u cancels; the quadratic terms give the zeroth order effective action for the fluctuations:

$$S_{\text{eff}} = \frac{1}{2} \sum_p u(-p)\Pi(p)u(p) + \mathcal{O}\left(\frac{1}{\sqrt{N}}u^3\right)$$

(9.26)

$$\Pi(p) = \int \frac{d^2q}{(2\pi)^2} G(p/2 + q)G(p/2 - q)$$

Therefore, in the leading order in $1/N$, the correlation function of the u-fields is given by

$$\langle u(-p)u(p)\rangle = [\Pi(p)]^{-1}$$

(9.27)

Formally $\langle u^2 \rangle$ does not contain large N. However, the correlation function $\langle u(-p)u(p)\rangle$ is very singular at large $|p| \gg m$:

$$\langle u(-p)u(p)\rangle \to \frac{2\pi p^2}{\ln(p^2/m^2)}$$

(9.28)

These divergences can be made less severe if we relax the δ-function constraint on the length of **n**. Recalling the original theory (8.2) (i.e. the O(N)-symmetric vector model) we can rewrite the δ-function in (9.18) as follows:

$$\delta(\mathbf{n}^2 - 1) \to \exp\left[-\frac{\alpha}{2}(\mathbf{n}^2 - 1)^2\right] = \int D\lambda \exp\left[-i\lambda(\mathbf{n}^2 - 1) - \frac{1}{2\alpha}\lambda^2\right]$$

(9.29)

where $\alpha = gM_0^4/2$. Then the propagator of u-fields (9.27) becomes finite at large p:

$$\langle u(-p)u(p)\rangle = \frac{1}{\Pi(p) + \alpha^{-1}}$$

(9.30)

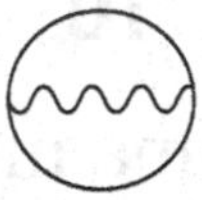

Figure 9.5.

which improves convergence of the diagrams at large momenta. However, not all divergences can be removed this way. We see that the integral for $\langle u^2 \rangle$, for example, is still divergent and gives a value proportional to the cut-off:

$$\langle u^2 \rangle \sim \int \frac{\mathrm{d}^2 p}{\alpha^{-1} + \pi/[2p^2 \ln(p^2/m^2)]} \propto \alpha \Lambda^2$$

The ultraviolet divergences renormalize the saddle point value of m. The first correction to the saddle point is given by the diagram shown in Fig. 9.5 where the wavy line represents the propagator of u-fields (9.27) times $1/N$.

Adding this diagram to the self energy part of $G(p)$ and substituting the latter into (9.23) we get the improved saddle point equation:

$$M^2 = \frac{N}{(2\pi)^2} \int \mathrm{d}^2 p G(p) - \frac{1}{(2\pi)^2} \int \mathrm{d}^2 p G^2(p) G(p+q) \Pi^{-1}(q) \qquad (9.31)$$

After carrying out all necessary integrations we obtain the improved mass gap:

$$m = \Lambda M^{2/N} \exp(-2\pi M^2/N) \qquad (9.32)$$

The renormalization group calculations give the same form, but with $(N-2)$ instead of N.

I shall not go into further details of the $1/N$-expansion; it is thoroughly analysed in the books by Polyakov and by Sachdev in particular (see references in the Select bibliography).

References

Fateev, V. A., Onofri, E. and Zamolodchikov, Al. B. (1993). *Nucl. Phys. B*, **406**, 521.
Friedan, D. (1980). *Phys. Rev. Lett.*, **45**, 1057; *Ann. Phys.*, **163**, 318 (1985).
Nicopoulos, V. N. and Tsvelik, A. M. (1991). *Phys. Rev. B*, **44**, 9385.
Polyakov, A. M. and Wiegmann, P. B. (1983). *Phys. Lett. B*, **131**, 121.
Zamolodchikov, A. and Zamolodchikov, Al. (1979). *Ann. Phys.*, **120**, 253.
Wiegmann, P. B. (1985). *Phys. Lett. B*, **152**, 209.

10

O(3) nonlinear sigma model in the strong coupling limit

So far I have developed a path integral approach to QFT. However, common sense tells us that it is wiser to view things not just from one point, but to develop different descriptions of the same thing. Therefore I shall now consider the Hamiltonian quantization of the $O(N)$-symmetric vector model. For the Hamiltonian quantization it is suitable to consider real time, so in this chapter I abandon the thermodynamic time τ and use the real time t instead.

Let us imagine that our system is a material consisting of granulas. Each granula has a linear size a and contains many lattice sites. Therefore on each granula one can introduce an averaged variable

$$\Phi(x_n) = a^{-D/2} \int_{|x-x_n|<a} d^D x \, \Phi(x)$$

and then rewrite approximately the initial action as the effective action for these averaged variables:

$$S = \int dt \sum_x \left\{ \frac{1}{2}(\partial_t \Phi)^2 - \frac{J}{2} \sum_e [\Phi(x+e) - \Phi(x)]^2 - \frac{ga^{-D}}{4} (\Phi^2 - a^D M_0^2)^2 \right\} \quad (10.1)$$

where e are elementary vectors of the lattice of granulas and x are coordinates of centres of the granulas. $Ja^2 \approx v^2$, where v is the velocity of bare excitations. In previous calculations I have always considered $v = 1$, but here it is better to express it explicitly.

Let us consider the case $J = 0$ first. As we shall see below, this approximation provided a good starting point for the case of small $a^D M_0^2$, when the radial fluctuations of the Φ-field are weak but the transverse ones are large. For $J = 0$ all lattice sites can be considered independently. Each site is described as a quantum mechanical particle of mass $m = 1$ moving in the N-dimensional space in the potential

$$U(\Phi_1, \ldots, \Phi_N) = \frac{ga^{-D}}{4} (\Phi^2 - a^D M_0^2)^2$$

In the case when this potential is very deep we can neglect the radial motion and put

$$\Phi(t, x) = a^{D/2} M_0 n(t, x) \quad (10.2)$$

Since the characteristic energy distance between the levels of $U(\Phi)$ is of the order of $g^{1/2} M_0$ and, as will be clear later, the characteristic energy scale for the angular fluctuations is $\sim a^{-D} M_0^{-2}$, this approach is justified for

$$g^{1/2} M_0^3 \gg a^{-D} \tag{10.3}$$

The remaining angular motion on each site is described by the following Lagrangian:

$$L = \frac{a^D M_0^2}{2} (\partial_t \mathbf{n})^2 \tag{10.4}$$

Below I shall consider only the case $N = 3$ because the corresponding quantization procedure is described in textbooks on elementary quantum mechanics. The generalization for $N > 3$ is straightforward. For $N = 3$ the problem under consideration is the problem of a quantum rotator whose quantization leads to the following Hamiltonian:

$$\hat{H} = \frac{[\hat{\mathbf{l}}]^2}{2a^D M_0^2} \tag{10.5}$$

where the components of the angular momentum operator satisfy the following commutation relations[1]

$$[\hat{l}_a, \hat{l}_b] = i\epsilon_{abc} \hat{l}_c \tag{10.6}$$

between themselves and have similar commutation relations with the coordinates

$$[\hat{l}_a, \hat{n}_b] = i\epsilon_{abc} \hat{n}_c \tag{10.7}$$

The coordinates commute:

$$[\hat{n}_a, \hat{n}_b] = 0 \tag{10.8}$$

Now we can write the quantum Hamiltonian for the problem with $J \neq 0$:

$$\hat{H} = \sum_x \left\{ \frac{[\hat{\mathbf{l}}(x)]^2}{2a^D M_0^2} + \frac{J a^D M_0^2}{2} \sum_e [\mathbf{n}(x+e) - \mathbf{n}(x)]^2 \right\} \tag{10.9}$$

The Hamiltonian is fully defined by the commutation relations (10.6) with the additional remark that all operators belonging to different sites commute.

The obtained model appears to be similar to models of lattice magnets. As will become clear later the similarity does exist. Let us consider the case when

$$J M_0^4 \ll a^{-2D}$$

where one can treat the coupling between different sites as a perturbation. Then on each site we have a Hilbert space defined by the eigenfunctions of the angular momentum. In

---

[1] For general N the operators $\hat{l}_a$ are generators of the o(N) algebra whose commutation relations contain f_{abc} (the structure constant of the o(N) algebra) instead of ϵ_{abc}.

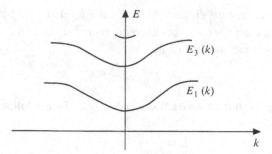

Figure 10.1. A qualitative picture of the spectrum in the strong coupling limit.

the zeroth order in J the energy and the wave functions of the many-body Hamiltonian are given by

$$E = \frac{1}{2a^D M_0^2} \sum_i l_i(l_i + 1)$$ (10.10)

$$\Psi = |l_1, m_1; l_2, m_2; \ldots; l_M, m_M\rangle$$

where M is the total number of sites.

The ground state wave function corresponds to $l_i = 0$. This state is fully isotropic and

$$\langle 0|n^a(x)|0\rangle = 0$$

The first excited state is a threefold degenerate state with angular momentum $l = 1$:

$$\Psi(1, m) = \frac{1}{\sqrt{M}} \sum_x e^{ikx}|1, m\rangle_x$$ (10.11)

Its energy in the first order of perturbation theory is equal to (see Fig. 10.1)

$$E(k) = \frac{1}{a^D M_0^2} + \frac{J a^D M_0^2}{2}|\langle 0|n^a|1, m\rangle|^2 \sum_e [1 - \cos(\mathbf{k}\mathbf{e})]$$ (10.12)

The elementary calculation gives

$$|\langle 0|n^a|1, m\rangle|^2 = |\langle 0|n^3|1, 0\rangle|^2 = 2/3$$

The spectrum (10.12) is separated from the ground state by the gap $1/a^D M_0^2$. Since the matrix element $\langle 0|n^a|l, m\rangle \neq 0$ only for odd l, the next propagating excitation has $l = 3$. Its spectral gap is equal to

$$E_3(0) = \frac{6}{a^D M_0^2}$$

Therefore at $J a^{2D} M_0^4 \ll 1$ two-particle states with $l = 1$ are always lower than one-particle states with $l = 3$. The spectrum is represented in Fig. 10.1.

Recall the requirements imposed during this derivation.

(i) The radial fluctuations are weak: $g^{1/2} M_0 \gg 1/a^D M_0^2$.
(ii) The intersite exchange is small: $J a^D M_0^2 \approx v^2 a^{D-2} M_0^2 \ll 1/a^D M_0^2$, which means that the excitation bandwidth is smaller than the spectral gap.

For theories where the renormalization process leads to a strong coupling it is not necessary to satisfy these two requirements on the level of the bare action. Instead, they can
be satisfied on some intermediate scale after the appropriate renormalization. In particular,
one can apply the strong coupling approach to models with a disordered ground state. Such
models have a finite correlation length and in a rough approximation we can represent the
system as an array of granulas of the size of the correlation length. In this approximation
the order parameter is constant within each granula and granulas interact as shown in
(10.1). Such a 'coarse grained' description gives a qualitatively correct picture for the
$(1 + 1)$-dimensional O(3) nonlinear sigma model, where the exact solution is available. The
latter model scales to the strong coupling regime and the correlation length is always finite.
The exact solution shows that the spectrum of the O(3) nonlinear sigma model consists of
triplets of massive bosons in agreement with the strong coupling approximation. There is
a disagreement, however, about excitations with higher angular momenta: according to the
exact solution these modes do not exist. This is because the interaction between 'granulas'
is not weak in this case and the requirement (ii) above is not satisfied. Indeed, when the
bandwidth of the $l = 1$ particles exceeds the gap between the $l = 1$ and $l = 2$ energy
levels, particles with larger angular momenta become unstable with respect to decay into
fundamental particles with $l = 1$.

The calculations described in this chapter can be applied directly to real experimental
systems. Granulated materials are not mathematical abstractions, but phenomena of nature.
When granulas are large enough to allow suppression of the radial fluctuations, we have an
exciting opportunity to study the collective quantum dynamics of the order parameter. The
best studied granulated materials are granulated superconductors. They are described by
the same model (10.1), but with the symmetry O(2). The vector Φ in this case describes the
real and the imaginary parts of the superconducting order parameter $\Phi = (\Re e\Delta, \Im m\Delta) =
|\Delta|(\cos\phi, \sin\phi)$. The hopping term describes the Josephson coupling between granulas.
The model with O(3) symmetry is relevant for granulated antiferromagnets. Here the vector
Φ represents the staggered magnetization averaged over a granula. Recent years have seen
a great deal of progress in the preparation of microscopic magnetic devices. The quantum
processes of magnetization tunnelling are now the subject of experimental studies. A good
review on this subject is given by Chudnovsky (1993).

Reference

Chudnovsky, E. M. (1993). Macroscopic quantum tunneling of the magnetic moment. Proc. 37th Ann.
    Conf. on Magnetism and Magnetic Materials, *J. Appl. Phys.*, **73**, 6697.

II

Fermions

11

Path integral and Wick's theorem for fermions

In this chapter we consider path integral representations for fermionic theories. Fermionic theories are even more typical of QFT than bosonic ones. The reason why their discussion has been postponed is purely pedagogical: the introduction of fermionic path integrals requires certain new concepts whose acceptance is psychologically difficult. It is believed that a thorough acquaintance with the bosonic path integral helps one to 'digest' its fermionic analogue.

Let us recall some facts about fermionic many-body systems. The principal one is that a fermionic many-body wave function is antisymmetric with respect to permutation of coordinates. This means that a wave function of N fermions

$$\psi(x_1, \sigma_1; \ldots ; x_N, \sigma_N) \tag{11.1}$$

acquires the factor $(-1)^P$ under a permutation of coordinates $\xi_i = (x_i, \sigma_i)$, where P is the parity of this permutation.

As an example of a fermionic theory we consider a model of nonrelativistic fermions interacting via a pair potential $U(x - y)$. In this case the many-body wave function (11.1) is an eigenfunction of the following Hamiltonian:

$$\hat{H} = \sum_i -\frac{1}{2m}(\Delta_i) + \frac{1}{2}\sum_{i \neq j} U(x_i - x_j) \tag{11.2}$$

Both the Hamiltonian and the wave function depend explicitly on N (the number of particles), which is very inconvenient for QFT. To overcome this inconvenience we use the procedure of the second quantization. In this procedure we introduce the operators $\hat{a}^+(x, \sigma)$, $\hat{a}(x, \sigma)$ (creation and annihilation operators) with the following anticommutation relations:

$$\{\hat{a}^+(x, \sigma), \hat{a}(y, \sigma')\} = \delta(x - y)\delta_{\sigma,\sigma'} \tag{11.3}$$

$$\{\hat{a}(x, \sigma), \hat{a}(y, \sigma')\} = 0 \tag{11.4}$$

$$\{\hat{a}^+(x, \sigma), \hat{a}^+(y, \sigma')\} = 0 \tag{11.5}$$

Then instead of the wave function (11.1) depending on coordinates of particles and their number, we introduce the coordinate-independent wave functions:

$$\Psi_N(E) = \sum_{\sigma_i} \int \prod_i d^D x_i \psi(x_1, \sigma_1; \ldots; x_N, \sigma_N) \hat{a}^+(x_1, \sigma_1), \ldots; \hat{a}^+(x_N, \sigma_N)|0\rangle \quad (11.6)$$

$$\Psi_N^*(E) = \sum_{\sigma_i} \int \prod_i d^D x_i \langle 0|\hat{a}(x_1, \sigma_1) \cdots \hat{a}(x_N, \sigma_N) \psi^*(x_1, \sigma_1; \ldots; x_N, \sigma_N) \quad (11.7)$$

where $|0\rangle$ is the state annihilated by all operators $\hat{a}(x, \sigma)$ (the fermionic vacuum):

$$\hat{a}(x, \sigma)|0\rangle = 0$$

$$\langle 0|\hat{a}^+(x, \sigma) = 0$$

The Hamiltonian acting in the Hilbert space of Ψ-functions is expressed in terms of the creation and the annihilation operators:

$$\hat{H} = -\int d^D x \hat{a}_\sigma^+(x) \frac{1}{2m} \Delta \hat{a}_\sigma(x) + \frac{1}{2} \int d^D x \int d^D y \hat{a}_\sigma^+(x) \hat{a}_\sigma(x) U(x - y) \hat{a}_{\sigma'}^+(y) \hat{a}_{\sigma'}(y)$$

$$(11.8)$$

(I assume summation over repeated indices.)

It is often convenient to work in momentum representation. The new operators $\hat{a}_\sigma(\mathbf{p})$ are Fourier harmonics of the operators $\hat{a}^+(x)$, $\hat{a}(x)$:

$$\hat{a}_\sigma(\mathbf{p}) = \int d^D x \hat{a}_\sigma(\mathbf{x}) e^{i\mathbf{p}\mathbf{x}/\hbar} \quad (11.9)$$

In momentum representation one has the advantage of dealing with a lattice of discrete $\mathbf{p}$ rather than with a continuous space of $\mathbf{x}$. The transformation (11.9) preserves the anticommutation relations:

$$\{\hat{a}^+(\mathbf{p}, \sigma), \hat{a}(\mathbf{q}, \sigma')\} = \delta_{\mathbf{p},\mathbf{q}} \delta_{\sigma,\sigma'} \quad (11.10)$$

$$\{\hat{a}(\mathbf{p}, \sigma), \hat{a}(\mathbf{q}, \sigma')\} = 0 \quad (11.11)$$

$$\{\hat{a}^+(\mathbf{p}, \sigma), \hat{a}^+(\mathbf{q}, \sigma')\} = 0 \quad (11.12)$$

It is also convenient to subtract from the Hamiltonian the quantity $\mu \hat{N}$, where μ is the chemical potential. Substituting (11.9) into (11.8) we get the operator $\hat{H} - \mu \hat{N}$ in momentum representation:

$$\hat{H} - \mu \hat{N} = \sum_p \hat{a}_\sigma^+(p)[E(p) - \mu]\hat{a}_\sigma(p)$$

$$+ \frac{1}{2} \sum_{p,q,k} \hat{a}_\sigma^+(p - q/2)\hat{a}_\sigma(p + q/2)U(q)\hat{a}_{\sigma'}^+(k + q/2)\hat{a}_{\sigma'}(k - q/2) \quad (11.13)$$

where in the present case

$$E(p) = p^2/2m$$

but below we shall consider various generalizations and therefore do not specify the form of the spectrum.

This is exactly the type of formulation suitable for QFT because $\hat{a}_\sigma^+(x)$, $\hat{a}_\sigma(x)$ are *operator fields* defined at every point in space and there is no explicit dependence of the Hamiltonian on the number of particles.

I am not going to dwell any more on matters of second quantization. One remark is, however, worth making. As I have said, the Hamiltonian (11.8) expressed in terms of creation and annihilation operators acts in the space of the objects Ψ. These objects characterize the quantum states of many-body systems and belong to a metric space with a scalar product

$$(\Psi_N^*(E), \Psi_M(E'))$$

$$= \delta_{N,M} \sum_{\sigma_i} \int \prod_i d^D x_i \psi_E^*(x_1, \sigma_1; \ldots; x_N, \sigma_N) \psi_{E'}(x_1, \sigma_1; \ldots; x_N, \sigma_N) \quad (11.14)$$

thus certainly having all the properties of regular wave functions. However, in contrast to wave functions in coordinate representation, they are not numbers! This curious property leads us to the gateway of the fermionic path integral.

Our objective is to define the fermionic path integral in such a way that we may reproduce all the results obtained in the operator representation. As a minimum, we must reproduce the expression for the two-point correlation function:

$$\int_0^\beta d\tau \int dr^D e^{i\omega_n \tau + i\mathbf{pr}} \langle\langle a(\tau, \mathbf{r}) a^+(0, 0) \rangle\rangle = \frac{1}{i\omega_n - \epsilon(\mathbf{p})} \quad (11.15)$$

where $\epsilon(p) = E(p) - \mu$.

From our previous experience with bosonic path integrals we know that in order to define a path integral one needs to know the classical Lagrangian. Can we find such a Lagrangian for the fermions? Let us try the following one:

$$L = \sum_p i\hbar a_\sigma^*(p) \partial_t a_\sigma(p) - H[a_\sigma^*(p), a_\sigma(p)]$$

where a^*, a are now not operators, but numbers. In fact, this expression is $\hbar$-independent being expressed in the canonical coordinates $Q = a_\sigma(p)$ and

$$P_\sigma(p) = \frac{\partial L}{\partial \dot{Q}_\sigma(p)} = i\hbar a_\sigma^*(p) \quad (11.16)$$

According to the quantization rules we declare P and Q operators with the famous commutation relations

$$[\hat{Q}, \hat{P}] = i\hbar$$

Stop! There is a contradiction here with the previous discussion: we have the anticommutation relations instead! In order to overcome this difficulty, one should treat $a_\sigma(p)$, $a_\sigma^*(p)$ as Grassmann numbers.

Grassmann numbers $\psi(p)$, $\psi^*(p)$ are defined as objects satisfying the following conditions:

$$\psi(p)\psi(q) + \psi(q)\psi(p) = 0$$
$$\psi^*(p)\psi(q) + \psi(q)\psi^*(p) = 0 \qquad (11.17)$$
$$\psi^*(p)\psi^*(q) + \psi^*(q)\psi^*(p) = 0$$

Do not confuse Grassmann numbers with fermionic creation and annihilation operators! According to this definition we have

$$\psi(p)^2 = \psi^*(p)^2 = 0$$

Grassmann numbers compose an algebra in the same way as the ordinary numbers do. They satisfy all algebraic axioms and the pair operation defined on their field is anticommutator.

We can define functionals on the *field* of Grassmann numbers:

$$F(\psi^*, \psi) = \sum_{a_i, b_i = 0, 1} C(a_1, b_1, \ldots, a_N, b_N)(\psi_1)^{a_1} \ldots (\psi_N)^{a_N}(\psi_N^*)^{b_N} \cdots (\psi_1^*)^{b_1}$$

$$F^*(\psi^*, \psi) = \sum_{a_i, b_i = 0, 1} C^*(a_1, b_1, \ldots, a_N, b_N)(\psi_1)^{b_1} \cdots (\psi_N)^{b_N}(\psi_1^*)^{a_1} \cdots (\psi_N^*)^{a_N}$$

$$(11.18)$$

where C are some complex coefficients (these equations look very similar to (11.6) and (11.7), but they are not quite the same thing!). We introduce the integrals:

$$\int F(\psi^*, \psi) \mathrm{d}\psi^* \mathrm{d}\psi = \int F(\psi^*, \psi) \mathrm{d}\psi_1^* \cdots \mathrm{d}\psi_N^* \mathrm{d}\psi_1 \cdots \mathrm{d}\psi_N \qquad (11.19)$$

The definition is made precise by setting

$$\int \mathrm{d}\psi(p) = \int \mathrm{d}\psi^*(p) = 0 \quad \int \psi(p)\mathrm{d}\psi(p) = \int \psi^*(p)\mathrm{d}\psi^*(p) = 1 \quad (11.20)$$

The 'infinitesimal' Grassmann numbers $\mathrm{d}\psi$, $\mathrm{d}\psi^*$ anticommute with each other and with the 'finite' Grassmann numbers. Substituting the definitions (11.20) into (11.19) we get

$$\int F(\psi^*, \psi) \mathrm{d}\psi^* \mathrm{d}\psi = C(1, 1; 1, 1; \ldots; 1, 1) \qquad (11.21)$$

Only now we are able to define the 'classical' Lagrangian whose quantization gives the Hamiltonian (11.13) and then define the path integral. The classical Lagrangian is defined in terms of Grassmann numbers:

$$L = i\hbar \sum_p \psi_\sigma^*(p)\partial_t \psi_\sigma(p) - H[\psi_\sigma^*(p), \psi_\sigma(p)] \qquad (11.22)$$

In Matsubara time representation one has to substitute $t = i\tau$ and treat ψ^* not as a complex conjugate of ψ, but as an independent variable. Then the generating functional and the partition function are defined in the same spirit as for bosonic theories. The only difference is that now we integrate over Grassmann variables:

$$Z[\eta(x), \eta^*(x)] = \int D\psi^* D\psi \exp\left[-\int_0^{1/T} d\tau L(\psi^*, \psi) + \int \eta^*\psi + \psi^*\eta\right] \quad (11.23)$$

($\eta^*(x), \eta(x)$ are Grassmann numbers). To obtain the correct results it is necessary to impose on ψ, ψ^* the antiperiodicity conditions in the Matsubara time:

$$\psi(\mathbf{x}, 0) = -\psi(\mathbf{x}, 1/T) \qquad \psi^*(\mathbf{x}, 0) = -\psi^*(\mathbf{x}, 1/T) \quad (11.24)$$

As usual, in order to develop a diagram expansion, one needs to learn how to calculate Gaussian integrals. Due to the anticommuting properties of Grassmann numbers it is very easy to integrate them: for each variable there is only one non-zero integral (11.20). In particular, applying the definition of the Grassmann integral (11.20) to the Gaussian integrals we get:

$$\int D\psi^*(i)D\psi(i)\exp[-\psi^*(i)A_{ij}\psi(j)] = \det A \quad (11.25)$$

$$\frac{\int D\psi^*(i)D\psi(i)\exp[-\psi^*(i)A_{ij}\psi(j) + \eta^*(i)\psi(i) + \psi^*(i)\eta(i)]}{\int D\psi^*(i)D\psi(i)\exp[-\psi^*(i)A_{ij}\psi(j)]} = \exp\left[\eta^*(i)A_{ij}^{-1}\eta(j)\right]$$
$$(11.26)$$

With these definitions we can calculate the pair correlation function for free fermions. Substituting the quadratic part of the Hamiltonian (11.13) into (11.22) we get the following thermodynamic action:

$$S = \int d\tau \left\{\sum_p \psi_\sigma^*(p, \tau)[\partial_\tau - \epsilon(p)]\psi_\sigma(p, \tau)\right\} \quad (11.27)$$

In order to reduce the path integral to the canonical form, we have to diagonalize this action. This is achieved by the transformation of variables:

$$\psi(p, \tau) = T \sum_{n=-\infty}^{\infty} \exp[i\pi T(2n + 1)\tau]\psi_n(p)$$
$$(11.28)$$
$$\psi^*(p, \tau) = T \sum_{n=-\infty}^{\infty} \exp[-i\pi T(2n + 1)\tau]\psi_n^*(p)$$

After this transformation the generating functional (11.23) acquires a form especially convenient for calculations:

$$Z[\eta_n(p), \eta_n^*(p)]$$

$$= \int D\psi^* D\psi \exp\left(-\sum_{n,p}\{\psi_{\sigma,n}^*(p)[i\omega_n - \epsilon(p)]\psi_{\sigma,n}(p) - \eta_n^*(p)\psi_n(p) - \psi_n^*(p)\eta_n(p)\}\right)$$

$$(11.29)$$

Figure 11.1.

$(\omega_n = \pi T(2n + 1))$. This integral is calculated by the shift of variables which removes the terms linear in ψ, ψ^*:

$$\psi(p) \to \psi(p) - [i\omega - \epsilon(p)]^{-1}\eta(p)$$
$$\psi^*(p) \to \psi^*(p) - [i\omega - \epsilon(p)]^{-1}\eta^*(p) \tag{11.30}$$

The result is

$$\frac{Z[\eta_n(p), \eta_n^*(p)]}{Z[0, 0]} = \exp\left[\sum_{p,n} \frac{\eta_n^*(p)\eta_n(p)}{i\omega_n - \epsilon(p)}\right] = \prod_{n,p}\left[1 + \frac{\eta_n^*(p)\eta_n(p)}{i\omega_n - \epsilon(p)}\right] \tag{11.31}$$

from which it is easy to derive the familiar expression for the pair correlation function (11.15):[1]

$$\langle\langle\psi(\omega_n, \mathbf{p})\psi^*(\omega_n, \mathbf{p})\rangle\rangle = Z^{-1}[0, 0]\frac{\partial^2 Z[\eta, \eta^*]}{\partial\eta_n(p)\partial\eta_n^*(p)} = [i\omega_n - \epsilon(p)]^{-1} \tag{11.32}$$

Wick's theorem for fermionic fields is derived in the same way as for bosonic scalar fields. As for bosons, any multi-point correlation function decouples into a sum of products of the pair correlation functions. There are a few differences, however. The first is that due to the anticommuting nature of fermionic variables, each diagram with a loop acquires a $(-1)^F$ factor, where F is the number of loops. The second difference is that the fermionic variables which we have discussed in this chapter are complex. Therefore fermionic Green's functions have arrows which point toward ψ^*. For this reason diagrams with bubbles do not carry factors $1/2$. Later in Chapter 19 we shall discuss real, or Majorana, fermions where there are no arrows. These fermions are analogous to the bosonic scalar field in that their diagram expansion contains factors of $1/2$.

I shall not discuss fermionic diagram expansion in detail because it does not depend on the derivation procedure. Whether one uses a path integral approach or works with time ordered exponents, the pictures remain the same. This allows me not to dwell one more time on the material explained in hundreds of other books. Thus, without going into any further detail, I can formulate the diagram technique for the model (11.13). The diagram technique can be formulated in frequency momentum space or in real space-time. It contains the elements shown in Fig. 11.1, where the solid line is the two-point Green's function (11.15) and the wavy line is the matrix element of the interaction $U(q)$ ($\delta(\tau_1 - \tau_2)U(x_1 - x_2)$ in real space-time).

---

[1] The order of differentiation is important; the η anticommute!

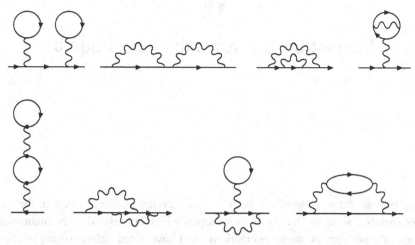

Figure 11.2.

The general rules which are used to calculate the correction of order n for the pair correlation function are as follows.

1. Form all connected, topologically nonequivalent diagrams with $2n$ vertices and two external points, where two solid lines and one wavy line meet at each vertex.
2. If operating in real space-time, integrate over all the vertex coordinates and sum over all internal spin variables. If operating in frequency momentum space, associate with each vertex the factor

$$\delta_{n_1+n_2-n_3-n_4}\delta(\mathbf{k}_1 + \mathbf{k}_2 - \mathbf{k}_3 - \mathbf{k}_4)$$

sum over all internal frequencies $T\sum_n$ and integrate over all internal momenta.
3. Multiply the resulting expression by $(-1)^F$, where F is the number of closed loops.
4. If there are any Green's functions G_0 whose time arguments are the same, interpret them as

$$\lim_{\tau\to+0} G_0(-\tau, x_{12})$$

In the second order in U we have the diagrams for the single-fermion Green's function G shown in Fig. 11.2. Compare these diagrams with those presented in Fig. 5.2.

12

Interacting electrons: the Fermi liquid

It is quite beyond the purpose of this book to give a comprehensive review of the modern theory of metals. I shall touch only on those topics which are related to the mainstream of this course: the problem of strong interactions. As I have noted before, strong interactions can appear as a result of renormalization, and from the aesthetic point of view this is, perhaps, the most interesting case. As we shall see in Part IV, such renormalizations are particularly strong in one-dimensional systems, where practically any interaction leads to quite dramatic effects. Therefore a one-dimensional electron gas neither undergoes a phase transition (this is forbidden due to the low dimensionality), nor behaves like a system of free electrons. On the contrary, in higher dimensions at low temperatures a system of electrons either undergoes a phase transition (superconductivity, magnetic ordering, etc.), or behaves like a free electron gas.[1]

When temperatures as low as several degrees became experimentally available, physicists discovered that an enormous amount of experimental data on normal metals can be described by the model where one neglects electron–electron interactions. This apparent miracle was explained by Landau who demonstrated that the interaction pattern drastically simplifies close to the Fermi surface. Provided the system does not undergo a symmetry breaking phase transition, all interactions except forward scattering effectively vanish on the Fermi surface and low-lying excitations carry quantum numbers of electrons. Therefore they are called *quasi-particles*. The meaning of the word *quasi* will become clear in a moment.

Suppose the bare Hamiltonian of interacting fermions is

$$\hat{H} - \mu \hat{N} = \sum_p \hat{a}_\sigma^+(p) E(p) \hat{a}_\sigma(p) + \frac{1}{2V} \sum_{p,q,k} \hat{a}_\sigma^+(p-q) \hat{a}_{\sigma'}^+(k) U(q) \hat{a}_{\sigma'}(k-q) \hat{a}_\sigma(p) \tag{12.1}$$

where V is the volume of the system. Then according to the Landau theory, for a wide class of bare interactions the low energy limit of the theory is described by the effective Hamiltonian

$$\hat{H}_L \equiv [\hat{H} - \mu \hat{N}]_{\text{eff}} = \sum_{p,\sigma} \epsilon(\mathbf{p}) n_\sigma(\mathbf{p}) - \frac{1}{2V} \sum_{p,k,\sigma} n_\sigma(\mathbf{p}) F(\mathbf{p} - \mathbf{k}) n_\sigma(\mathbf{k}) \tag{12.2}$$

---

[1] The situation becomes considerably more complicated if disorder is involved.

where $n_\sigma(\mathbf{p}) = \hat{a}_\sigma^+(p)\hat{a}_\sigma(p)$ and the functions $\epsilon(\mathbf{p})$ and $F(\mathbf{p})$ are determined by the parameters of the bare Hamiltonian (12.1). Note that the interaction term in the Hamiltonian (12.2) has less momentum summations than the original one. The most important property of the Hamiltonian (12.2) is that it conserves individual occupation numbers:

$$[\hat{n}_\sigma(\mathbf{p}), \hat{H}_L] = 0 \qquad (12.3)$$

Therefore the corresponding operators can be replaced by numbers. Then each point in the spin momentum space can be characterized by an occupation number $n_\sigma(\mathbf{p})$ and the Hamiltonian (12.2) becomes a function (or rather a *functional*) of these numbers.

I will not enter into a detailed discussion of the Landau Fermi liquid theory; there are quite a few books which do it much better than I would ever be able to. The best traditional description is given in the book by Abrikosov, Gorkov and Dzyaloshinsky. The modern approach is very well presented in the review article by Shankar (1994). Nevertheless, since the concept of Fermi liquid is so important in condensed matter theory, I will briefly discuss the most essential points.

Using (12.2) it is easy to establish that thermodynamic properties of Fermi liquids at low temperatures resemble those of a free electron gas: the specific heat is linear in temperature, $C_v = \gamma T$, and the paramagnetic susceptibility is almost temperature independent. I would strongly recommend to every reader to follow the corresponding derivation in one of the textbooks or even to do it as an exercise.

Since the Hamiltonian (12.2) becomes exact only at the Fermi surface, the quasi-particles have an infinite lifetime only at $\epsilon(p) = 0$, otherwise the occupation number does not commute with the Hamiltonian. As a consequence of this nonconservation at nonzero temperatures, when the density of excitations is finite, one observes a finite electrical resistivity. As we shall see in a moment, the quasi-particle decay rate is quadratic in energy and temperature:

$$1/\tau(\epsilon, T) = a\epsilon^2 + bT^2 \qquad (12.4)$$

where a, b are constants. This leads to the resistivity which at low temperatures is quadratic in T: $R = R_0 + aT^2$.[2] At $D > 1$ one can always recognize a Fermi liquid by these properties.

Even though many metals undergo phase transitions at low temperatures, the Fermi liquid theory usually provides a good description for the range of temperatures $T_c < T \ll \epsilon_F$. In other words the transitions can be described as originating from *weak*, though relevant, corrections to the Landau Hamiltonian. One such relevant correction corresponds to superconducting pairing. So according to this standard scenario a correlated electron system first becomes almost a Landau Fermi liquid, though its effective Hamiltonian still contains relevant terms which eventually destabilize Fermi liquid behaviour. Consequently, the corresponding instabilities can be analysed as weak ones. It is quite amazing that this

---

[2] Strictly speaking, the resistivity is finite only in the presence of a lattice and the above estimate is valid only provided some additional conditions are met. Here I refer the reader to the book by A. A. Abrikosov.

scenario works even in those metallic systems where the bare Coulomb interaction between electrons is quite strong. The best examples of that kind are the so-called *heavy fermion* compounds like $CeCu_6$, $CeAl_3$ where the effective mass of the quasi-particles becomes of the order of 500–1000 bare electron masses. There are systems, however, whose behaviour is outside the basin of attraction of the Fermi liquid theory fixed point. One well established example is the doped copper oxides, layered metals like $La_{2-x}Sr_xCuO_2$, YBaCuO, etc. All these materials in their metallic phase are close to a metal–insulator transition and are rather poor conductors. Paradoxically, at relatively high temperatures[3] these poor conductors become superconductors. This is particularly strange because the Coulomb repulsion in these materials is strong. However, the behaviour of these materials at temperatures higher than T_c is even more puzzling than the superconductivity itself. For example, contrary to the predictions of the Fermi liquid theory, the resistivity at $T > T_c$ is linear in temperature: $R \sim T$ which suggests a very strong scattering of elementary excitations.

To understand why the quasi-particle description is so robust, let us estimate their decay rate. The decay originates from the part of the interaction discarded at the transition to the Landau Hamiltonian (12.2). The eigenfunctions of the Hamiltonian (12.2) (not eigenvalues!) coincide with the eigenfunctions of the Hamiltonian of noninteracting electrons with the same Fermi surface. The ground state is represented by the filled Fermi surface (FS):

$$\text{ground state} = |FS\rangle \tag{12.5}$$

For simplicity we shall consider a spherical FS. Due to the Pauli principle the state with filled FS is annihilated by creation operators with $|\mathbf{p}| < p_F$:

$$\hat{a}^+_{\mathbf{p},\sigma}|FS\rangle = 0 \qquad |\mathbf{p}| < p_F \tag{12.6}$$

However, acting on the ground state by a creation operator with $|\mathbf{p}| > p_F$, or by an annihilation operator with $|\mathbf{p}| < p_F$ (removing a particle from the FS), we create new states. Since the energy of these states is greater than the energy of the ground state, they are called 'excited' states:

$$\text{single-particle excited states} = \begin{cases} \hat{a}^+_{\mathbf{p},\sigma}|FS\rangle & |\mathbf{p}| > p_F \\ \hat{a}_{\mathbf{p},\sigma}|FS\rangle & |\mathbf{p}| < p_F \end{cases} \tag{12.7}$$

These states are eigenstates of (12.2) with energies $\epsilon(p)$ (the single-particle state) and $-\epsilon(p)$ (the single-hole state) respectively. Acting on $|FS\rangle$ by creation and annihilation operators one can create yet new states. These states are only approximate eigenstates of the original Hamiltonian.

The following factors impose restrictions on the decay rate of the quasi-particle states. First of all, since both the effective Hamiltonian (12.2) and the original one (12.1) commute

---

[3] High in comparison with other superconductors: $T_c \sim 30$–$110\,\mathrm{K}$.

with the total charge

$$Q = N_{\text{particles}} - N_{\text{holes}} \tag{12.8}$$

the interaction may have matrix elements only between states with the same charge. Therefore looking at the interaction

$$U = V^{-1} \sum_{\mathbf{q},\mathbf{p},\mathbf{p}'} U_q \hat{a}^+_{\mathbf{p}+\mathbf{q},\sigma_1} \hat{a}^+_{\mathbf{p}'-\mathbf{q},\sigma_2} \hat{a}_{\mathbf{p}',\sigma_2'} \hat{a}_{\mathbf{p},\sigma_1'} \delta_{\sigma_1+\sigma_2,\sigma_1'+\sigma_2'} \tag{12.9}$$

we conclude that in the leading order of perturbation theory a state $\hat{a}^+_{\mathbf{p},\sigma} |FS\rangle$ may decay only into the states with two particles and one hole:

$$\hat{a}^+_{\mathbf{q}_1,\sigma_1} \hat{a}^+_{\mathbf{q}_2,\sigma_2} \hat{a}_{-\mathbf{p}+\mathbf{q}_1+\mathbf{q}_2,\sigma'} |FS\rangle$$

$$|\mathbf{q}_1|, |\mathbf{q}_2| > p_F \qquad |-\mathbf{p}+\mathbf{q}_1+\mathbf{q}_2| < p_F \tag{12.10}$$

To estimate a decay rate of the state $\hat{a}^+_{\mathbf{p},\sigma} |FS\rangle$ into multi-particle states (there are many of them!) (12.10) we use the Fermi Golden Rule formula:

$$W = \frac{2\pi}{\hbar} |\langle \text{in}| \hat{V} |\text{out}\rangle|^2 \delta(E_{\text{in}} - E_{\text{out}}) \tag{12.11}$$

In our case

$$E_{\text{in}} = \epsilon(\mathbf{p})$$

and

$$E_{\text{out}} = \epsilon(\mathbf{q}_1) + \epsilon(\mathbf{q}_2) - \epsilon(-\mathbf{p}+\mathbf{q}_1+\mathbf{q}_2)$$

It follows from the definition of state (12.10) that $\epsilon(\mathbf{q}_{1,2}) > 0$ and $\epsilon(-\mathbf{p}+\mathbf{q}_1+\mathbf{q}_2) < 0$. Thus the total decay rate is

$$W = 2\frac{2\pi}{\hbar} \int \frac{d^3 q_1}{(2\pi)^3} \int \frac{d^3 q_2}{(2\pi)^3} U^2(\mathbf{p}-\mathbf{q}_1)\theta[\epsilon(\mathbf{q}_1)]\theta[\epsilon(\mathbf{q}_2)]\theta[-\epsilon(-\mathbf{p}+\mathbf{q}_1+\mathbf{q}_2)]$$
$$\times \delta[\epsilon(\mathbf{p}) + \epsilon(-\mathbf{p}+\mathbf{q}_1+\mathbf{q}_2) - \epsilon(\mathbf{q}_1) - \epsilon(\mathbf{q}_2)] \tag{12.12}$$

(the factor two comes from the summation over spin indices). I am going to demonstrate now that this decay rate becomes smaller and smaller when the energy of the initial state $\epsilon(p)$ approaches zero (that is the momentum of the particle becomes closer to the Fermi surface). This justifies the use of perturbation theory and also means that the closer we are to the Fermi surface the better the quasi-particle description is.

To simplify matters I estimate integral (12.12) for a case when the Fourier matrix element $U(\mathbf{k})$ is k-independent, that is U is a local interaction.

As is evident from Fig. 12.1, when $\epsilon_p \equiv \epsilon$ is much smaller than the Fermi energy, the integrals in $\mathbf{q}_{1,2}$ are restricted to a thin layer near the Fermi surface.

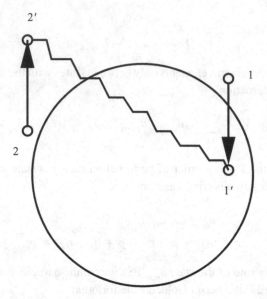

Figure 12.1. A picture of quasi-particle decay.

Let us use this fact to our advantage and write down these integrals as integrals over energy and angles:

$$\int \frac{d^3q}{(2\pi)^3} = \int \rho(\epsilon)d\epsilon \int \frac{d\hat{n}}{4\pi}$$

At the Fermi surface one can replace the density of states by a constant and write

$$\int \frac{d^3q}{(2\pi)^3} = \rho(\epsilon_F) \int d\epsilon \int \frac{d\hat{n}}{4\pi}$$

Substituting this into (12.12) we get

$$W(\epsilon) = \frac{4\pi}{\hbar}[U\rho(\epsilon_F)]^2 \int_0^\epsilon d\epsilon_1 \int_0^\epsilon d\epsilon_2 \int \frac{d\hat{n}_1 d\hat{n}_2}{(4\pi)^2} \theta[-\epsilon(-\mathbf{p}+\mathbf{q}_1+\mathbf{q}_2)]$$
$$\times \delta[\epsilon + \epsilon(-\mathbf{p}+\mathbf{q}_1+\mathbf{q}_2) - \epsilon_1 - \epsilon_2] \tag{12.13}$$

At $|\epsilon|/\epsilon_F \ll 1$ we can simply neglect ϵ, $\epsilon_{1,2}$ in the delta-function and write down

$$\epsilon(-\mathbf{p}+\mathbf{q}_1+\mathbf{q}_2) = \frac{1}{2m}[-p_F^2 + \mathbf{p}^2 - 2\mathbf{p}(\mathbf{q}_1+\mathbf{q}_2) + \mathbf{q}_1{}^2 + \mathbf{q}_2{}^2 + 2\mathbf{q}_1\mathbf{q}_2]$$

$$\approx \frac{p_F^2}{m}[1 + \mathbf{n}_1\mathbf{n}_2 - \mathbf{n}\mathbf{n}_2 - \mathbf{n}_1\mathbf{n}] \tag{12.14}$$

where we have replaced $\mathbf{q}_{1,2} = p_F\mathbf{n}_{1,2}$ where $\mathbf{n}_a$ are unit vectors. Then the delta-function becomes

$$\delta[\epsilon(-\mathbf{p}+\mathbf{q}_1+\mathbf{q}_2)] = \frac{m}{p_F^2}\delta[1 + \mathbf{n}_1\mathbf{n}_2 - \mathbf{n}\mathbf{n}_2 - \mathbf{n}_1\mathbf{n}] \tag{12.15}$$

Now we can integrate over energies in (12.13):

$$W(\epsilon) = \frac{2\pi}{\hbar}\frac{\epsilon^2}{\epsilon_F}[U\rho(\epsilon_F)]^2 I \tag{12.16}$$

$$I = \int \frac{d\hat{n}_1 d\hat{n}_2}{(4\pi)^2}\delta[1 + \mathbf{n}_1\mathbf{n}_2 - \mathbf{n}\mathbf{n}_2 - \mathbf{n}_1\mathbf{n}] \tag{12.17}$$

It is shown in the Appendix that the angular integral $I = 1$.

Expression (12.16) has a fundamental importance for the theory of metals and we have to discuss it in more detail. The decay rate is inversely proportional to the lifetime of the particle:

$$\tau = 2\pi/W \sim \epsilon^{-2}$$

However, the period of the de Broglie oscillations for the particle is $T \sim \epsilon^{-1}$. Therefore quasi-particle states close to the Fermi surface are well defined plane waves: they manage to oscillate many times before decay. Hence one can really think about them as particles, well, quasi-particles, because they do not live for ever.

Appendix: calculation of the integral (12.17)

Let us choose the z-axis along $\mathbf{n}$. Then we have

$$\mathbf{n}\mathbf{n}_1 = \cos\theta_1 \qquad \mathbf{n}\mathbf{n}_2 = \cos\theta_2$$
$$\mathbf{n}_1\mathbf{n}_2 = \cos\theta_1\cos\theta_2 + \sin\theta_1\sin\theta_2\cos(\phi_1 - \phi_2)$$

and

$$I = \int \frac{d\phi_1 d\phi_2 d\cos\theta_1 d\cos\theta_2}{(4\pi)^2}\delta[1 - \cos\theta_1 - \cos\theta_2 + \cos\theta_1\cos\theta_2 + \sin\theta_1\sin\theta_2\cos(\phi_1 - \phi_2)]$$
$$= \int \frac{d\alpha d\beta d\cos\theta_1 d\cos\theta_2}{(4\pi)^2}\delta(1 - \cos\theta_1 - \cos\theta_2 + \cos\theta_1\cos\theta_2 + \sin\theta_1\sin\theta_2\cos\alpha)$$

where $\beta = (\phi_1 + \phi_2)/2, \alpha = (\phi_1 - \phi_2)$. Now

$$\int d\alpha\delta(B + A\cos\alpha) = \int \frac{d(A\cos\alpha)}{A\sin\alpha}\delta(B + A\cos\alpha) = \frac{1}{A\sin\alpha} = (A^2 - B^2)^{-1/2}$$

In our case $A = \sin\theta_1\sin\theta_2, B = 1 + \cos\theta_1\cos\theta_2 - \cos\theta_1 - \cos\theta_2$. So the remaining integral gives

$$\frac{1}{2\pi}\int \frac{d\cos\theta_1 d\cos\theta_2}{[(\sin\theta_1\sin\theta_2)^2 - (1 + \cos\theta_1\cos\theta_2 - \cos\theta_1 - \cos\theta_2)^2]^{1/2}}$$
$$= \frac{1}{2\pi}\int_{-1}^{1}dx\int_{-1}^{1}dy[(1 - x^2)(1 - y^2) - (1 + xy - x - y)^2]^{-1/2}$$
$$= \frac{1}{2\pi\sqrt{2}}\int_{-1}^{1}dx\int_{-1}^{1}dy[(x + y)(1 - x)(1 - y)]^{-1/2}$$

Changing the variables $1 - x = x'$, $1 - y = y'$ we get

$$I = \frac{1}{2\pi\sqrt{2}} \int_0^2 dx \int_0^{2-x} dy [xy(2 - x - y)]^{-1/2}$$

Obviously, the integration must be restricted to the area $2 > x + y$, so after the change of variables $x = z_1^2$, $y = z_2^2$ we have

$$I = \frac{\sqrt{2}}{\pi} \int_0^{\sqrt{2}} dz_1 \int_0^{\sqrt{2-z_1^2}} dz_2 (2 - z_1^2 - z_2^2)^{-1/2} = \frac{\sqrt{2}}{\pi} \int_0^{\sqrt{2}} dz_1 \int_0^1 \frac{d\tau}{\sqrt{1 - \tau^2}} = 1$$

where the last substitution was $z_2 = \sqrt{2 - z_1^2}\tau$.

Reference

Shankar, R. (1994). *Rev. Mod. Phys.*, **66**, 129.

13
Electrodynamics in metals

The possiblity of a violation of Fermi liquid theory remains one of the most intriguing problems in condensed matter theory. One loophole in the Landau reasoning is easy to find – in the derivation of decay rate given in the previous chapter I assumed that the effective interaction U is weakly momentum dependent. If this is not the case and $U(k)$ is singular at $|k| \to 0$, the decay rate will not be that small at the Fermi surface. Strong momentum dependence corresponds to long distance interactions; hence the presence of such interactions makes the Landau theory inapplicable. The problem is where to get such interactions. It appears that the electrostatic (Coulomb) interaction will do the job. This is a vain hope, however, because the Coulomb force is long distance only in vacuum and in insulators; in metals conduction electrons screen the Coulomb interaction and make it effectively short range. The interaction which remains long distance in metals is a much weaker interaction of electronic currents via the exchange of transverse photons. The fact that this interaction may lead to violations of the Fermi liquid theory was discovered by Holstein *et al.* (1973), but the real interest in this fact arose after the papers by Reizer (1989, 1991). Since photons are everywhere, this interaction exists in any metal. However, since the current–current interaction is proportional to the square of the ratio of the Fermi velocity to the speed of light, the corresponding bare coupling constant is small. As we shall see, deviations from the Fermi liquid theory may become noticeable only at very low temperatures and only provided the metal is very pure.

There have been many attempts to prove the existence of nonphoton mediated current–current interactions in strongly correlated electronic systems. It is suggested that such interactions are effectively generated by the strong short-range Coulomb repulsion and therefore their dimensionless coupling constant is significant. Such interaction also emerges in the vicinity of a ferromagnetic phase transition.

In order to illustrate this interesting possibility of the breakdown of the Fermi liquid theory, I shall discuss the simplest model where this breakdown certainly occurs.[1] Namely, following Reizer, we consider a model of conduction electrons interacting with an electromagnetic field. In order to feel more confident about the results, let us start our discussion with the more familiar semiclassical picture of electrodynamics in metals.

---

[1] At very low temperatures, but it is the principle we are concerned about!

Electrodynamics in metals: the semiclassical approach

The behaviour of an electromagnetic field in metals is determined by the Maxwell equations and the linear response functions of the charge and current densities:[2]

$$\rho(\omega, \mathbf{q}) = e^2 \Pi(\omega, \mathbf{q})\phi(\omega, \mathbf{q})$$
$$\mathbf{j}(\omega, \mathbf{q}) = \sigma(\omega, \mathbf{q})\mathbf{E}(\omega, \mathbf{q})$$

(13.1)

For the sake of simplicity we shall neglect anisotropy and treat the conductivity as a scalar rather than a tensor. Calculation of the response functions is a problem of a microscopic theory; some general statements can be made, however, without calculations. As we know from statistical mechanics, the following property holds for electrons in thermodynamic equilibrium, i.e. in a very slowly varying potential:

$$e\phi(x) + \mu(x) = \text{const}$$

From this property it follows that

$$-\Pi(\omega = 0, \mathbf{q} \to 0) = \frac{\partial N}{\partial \mu} = \rho(\epsilon_F)$$

(13.2)

where N is the number of particles per unit volume and $\rho(\epsilon_F)$ is the density of states at the Fermi level.

Let us substitute the expressions (13.1) into the Maxwell equations:

$$\nabla \times \mathbf{H} = \sigma * \mathbf{E} - \frac{\epsilon}{c}\frac{\partial \mathbf{E}}{\partial t} + 4\pi \nabla \times \mathbf{m}$$
$$\nabla \cdot \mathbf{E} = 4\pi \Pi * \phi + 4\pi \rho_0(x, t)$$
$$\nabla \cdot \mathbf{H} = 0$$
$$\nabla \times \mathbf{E} = \frac{\mu}{c}\frac{\partial \mathbf{H}}{\partial t}$$

(13.3)

where μ and ϵ are the magnetic permeability and the dielectric constant, and the terms with ρ_0 and $\mathbf{m}$ represent external sources for the scalar potential and magnetic field, respectively. The star means a convolution; it substitutes for the conventional multiplication because the response functions have a dispersion. Now let us express the field strengths in terms of the potentials

$$\mathbf{E} = -\nabla\phi + \frac{1}{c}\frac{\partial \mathbf{A}}{\partial t}$$
$$\mu \mathbf{H} = \nabla \times \mathbf{A}$$

(13.4)

and choose the following gauge:

$$\nabla \cdot \mathbf{A} = 0$$

(13.5)

In this gauge the equations for the scalar potential and magnetic field become independent:

$$-\Delta\phi = 4\pi e^2 \Pi * \phi + 4\pi \rho_0$$

(13.6)

$$-\Delta \mathbf{H} + \frac{\mu\epsilon}{c^2}\frac{\partial^2 \mathbf{H}}{\partial t^2} - \frac{4\pi\mu}{c^2}\sigma * \frac{\partial \mathbf{H}}{\partial t} = 4\pi \nabla \times \nabla \times \mathbf{m}$$

(13.7)

---

[2] This representation assumes a certain choice of gauge, namely the one given by (13.5) below.

Rewriting these equations in Fourier components we get:

$$[\mathbf{q}^2 - 4\pi e^2 \Pi(\omega, \mathbf{q})]\phi(\omega, \mathbf{q}) = 4\pi \rho_0(\omega, \mathbf{q}) \tag{13.8}$$

$$[q^2 - \mu\epsilon\omega^2/c^2 + 4i\pi\mu\omega\sigma(\omega, \mathbf{q})/c^2]H_a(\omega, \mathbf{q}) = 4\pi(q^2\delta_{ab} - q_a q_b)m_b \tag{13.9}$$

The solutions define the Green's functions of the scalar potential

$$\langle\phi(-\omega, -\mathbf{q})\phi(\omega, \mathbf{q})\rangle \equiv \frac{\partial\phi(\omega, \mathbf{q})}{\partial\rho_0(\omega, \mathbf{q})} = \frac{4\pi}{q^2 - 4\pi e^2 \Pi(\omega, \mathbf{q})} \tag{13.10}$$

and the magnetic field

$$\langle H_a(-\omega, -q)H_b(\omega, q)\rangle \equiv \frac{\partial H_a(\omega, \mathbf{q})}{\partial m_b(\omega, \mathbf{q})} = \frac{(\delta_{ab} - q_a q_b/q^2)}{1 - \mu\epsilon\omega^2/c^2 q^2 + 4i\pi\mu\omega\sigma(\omega, \mathbf{q})/c^2 q^2} \tag{13.11}$$

Later we shall obtain the same Green's functions from the microscopic theory.

Looking at (13.10) and (13.11) we can recognize how differently the different fields propagate. The static scalar potential is screened; from (13.10) and (13.2) we derive that

$$\langle\phi(0, -\mathbf{q})\phi(0, \mathbf{q})\rangle = \frac{4\pi}{q^2 + 4\pi e^2 \rho(\epsilon_F)} \tag{13.12}$$

In real space this expression gives the famous Debye potential:

$$V(r) = \frac{1}{|r|}\exp(-|r|/\xi) \quad \xi^{-2} = 4\pi e^2 \rho(\epsilon_F)$$

We shall see later that excitations of the scalar potential (plasma waves) have a spectral gap and therefore cannot directly influence electrons in the vicinity of the Fermi surface.

The magnetic field is also screened, but in a quite different way. As follows from (13.11), the penetration length of magnetic fields grows at small frequencies:

$$\xi_H \sim \frac{c}{\sqrt{\omega\mu\sigma(0, q \sim \xi^{-1})}} \tag{13.13}$$

For impure metals where the conductivity for small wave vectors is finite we have

$$\xi_H \sim \omega^{-1/2}$$

(the skin effect); for pure metals where $\sigma(0, q) \sim q^{-1}$, we have

$$\xi_H \sim \omega^{-1/3}$$

(the anomalous skin effect). Thus one can expect a strong interaction between low energy electrons and the fluctuating magnetic field. We shall see below that this is indeed the case.

Microscopic description

Let us now consider a model for nonrelativistic fermions in a metal interacting with an electromagnetic field. This model is described by the following Lagrangian:

$$L = \frac{1}{8\pi} \int d^D x \left(\mathbf{E}^2 + \mathbf{H}^2 \right) + \int d^D x \left[\psi_\sigma^* (\partial_\tau + ie\phi + \epsilon_F) \psi_\sigma - \psi_\sigma^* \hat{\epsilon} \left(-i\nabla - \frac{e}{c} \mathbf{A} \right) \psi_\sigma \right]$$

(13.14)

where $\epsilon(p)$ is the electron's dispersion law and $\hat{\epsilon}$ is the operator obtained from $\epsilon(p)$ by a substitution $\mathbf{p} \to -i\nabla - (e/c)\mathbf{A}$.[3]

Here I have to make an important remark. First, we can consider the model (13.14) as an abstract theory in an arbitrary space dimension D. This makes sense because, as I have said, there are arguments that such 'electrodynamics' describes low energy excitations in strongly correlated systems. Second, we can take the model (13.14) in a literal sense, that is as a model of electrons interacting with real photons. In this case photons are always three-dimensional and electrons may not be, if the electron dispersion law does not depend on certain components of momentum. For example, in layered metals $\epsilon(p)$ depends only on two components and does not depend (or depends weakly) on the component perpendicular to the layers. I shall call the first the 'abstract' and the second the 'realistic' electrodynamical model and consider both cases.

The Lagrangian (13.14) is invariant with respect to the gauge transformations:

$$\phi \to \phi + \partial_\tau \chi$$
$$\mathbf{A} \to \mathbf{A} - c\nabla\chi$$
$$\psi \to e^{ie\chi} \psi$$

In what follows we shall treat the interaction of electrons with the electromagnetic field as a perturbation expanding correlation functions in powers of e. In order to define this diagram expansion we need to fix the gauge, otherwise the partition function would diverge. Let us choose the same gauge as before, (13.5); the advantage is that the correlation functions of the scalar and vector potentials decouple: $\langle\langle\phi(x)\mathbf{A}(y)\rangle\rangle = 0$. The gauge condition (13.5) should be interpreted as a condition imposed on the correlation functions; that is any correlation function of $A^a(x)$-fields must satisfy the condition

$$\sum_{a=1}^{D} \frac{\partial}{\partial x_a} \langle\langle A^a(x) A^{b_1}(x_1) \cdots A^{b_N}(x_N)\rangle\rangle = 0$$

(13.15)

or, in momentum space,

$$q_a^j \langle\langle A^a(q) A^{a_2}(q_2) \cdots A^{a_N}(q_N)\rangle\rangle = 0$$

(13.16)

Let us integrate over fermions and consider the effective action for the electromagnetic

---

[3] It is assumed that electrons carry charge '$-e$'.

Figure 13.1. The bare propagators of model (13.21).

Figure 13.2. The interaction vertices of model (13.21).

field. This Euclidean effective action can be written in symbolic form:

$$S_{\text{eff}} = \frac{1}{8\pi} \int d\tau d^D x [\mathbf{E}^2 + \mathbf{H}^2] - 2\text{Tr}\ln\left\{(\partial_\tau + ie\phi + \epsilon_F) - \hat{\epsilon}\left(-i\nabla - \frac{e}{c}\mathbf{A}\right)\right\} \quad (13.17)$$

(the factor 2 is due to the spin degeneracy). This form will remain symbolic, until we learn how to calculate the fermionic determinant. For simplicity, let us consider a quadratic dispersion law $\epsilon(p) = p^2/2m$. Let us rewrite the Lagrangian (13.14) in the form where the quadratic terms are separated from the interaction:

$$L = L_f + L_{\text{el}} + L_{\text{int}} \quad (13.18)$$

$$L_f = \frac{1}{8\pi} \int d^D x [(\nabla\phi)^2 + c^{-2}(\partial_\tau \mathbf{A})^2 + (\nabla \times \mathbf{A})^2] \quad (13.19)$$

$$L_{\text{el}} = \int d^D x \left[\psi_\sigma^*(\partial_\tau + \epsilon_F)\psi_\sigma - \frac{1}{2m}\nabla\psi_\sigma^* \nabla\psi_\sigma\right] \quad (13.20)$$

$$L_{\text{int}} = \int d^D x \left\{ie\phi\psi_\sigma^*\psi_\sigma - \frac{e\mathbf{A}}{2mc}\left[i\psi_\sigma^*\nabla\psi_\sigma - i(\nabla\psi_\sigma^*)\psi_\sigma - \frac{e}{c}\mathbf{A}\psi_\sigma^*\psi_\sigma\right]\right\} \quad (13.21)$$

From (13.19) and (13.20) we find the following correlation functions for the noninteracting fields (see Fig. 13.1):

$$G_0(\omega_n, p) \equiv \langle\langle\psi(\omega_n, p)\psi^*(\omega_n, p)\rangle\rangle = \frac{1}{i\omega_n - p^2/2m + \epsilon_F}$$

$$D_0^{00}(\omega_n, p) \equiv \langle\langle\phi(-\omega_n, q)\phi(\omega_n, q)\rangle\rangle = \frac{4\pi}{q^2} \quad (13.22)$$

$$D_0^{ab}(\omega_n, p) \equiv \langle\langle A^a(-\omega_n, q)A^b(\omega_n, q)\rangle\rangle = \frac{4\pi(\delta_{ab} - q^a q^b/q^2)}{\omega_n^2/c^2 + q^2}$$

(notice that D^{ab} satisfies the gauge fixing condition (13.16)). Here the straight line denotes the bare electronic Green's function G_0, and the wavy and zigzag lines correspond to D_0^{00} and D_0^{ab}, respectively. From Eq. (13.21) it follows that our theory has the vertices shown in Fig. 13.2, where the vector vertex represented by the triangle is equal to

$$\gamma^a = \frac{e}{mc}(p^a + q^a/2).$$

Let us now repeat the derivation of the spectrum of the electromagnetic field given by the conditions (13.10) and (13.11) on the microscopic level. We have the following Dyson

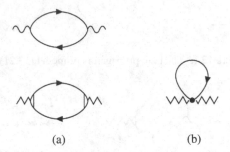

(a)   (b)

Figure 13.3. Examples of diagrams giving contributions to the polarization bubbles.

equations for the complete two-point correlation functions D^{00} and D^{ab}:

$$D^{00}(\omega_n, q) = D_0^{00}(\omega_n, q) + D_0^{00}(\omega_n, q)e^2\Pi(\omega_n, q)D^{00}(\omega_n, q) \qquad (13.23)$$

$$D^{ab}(\omega_n, q) = D_0^{ab}(\omega_n, q) + D_0^{ac}(\omega_n, q)e^2\Pi^{cd}(\omega_n, q)D^{db}(\omega_n, q) \qquad (13.24)$$

where the self energy parts Π, Π^{ab} (the polarization operators) are diagrams which cannot be cut along a single wavy or zigzag line. Examples of such diagrams are given in Fig. 13.3.

The solution of (13.23) reproduces (13.10) derived from Maxwell's equations.[4] In the zeroth order in e^2 we have

$$\Pi(\Omega_n, q) = 2T\sum_m \int \frac{d^D q}{(2\pi)^D} G_0(\Omega_n + \omega_m, q + p)G_0(\omega_m, p)$$

$$= 2\int \frac{d^D q}{(2\pi)^D} \frac{n(p) - n(p+q)}{i\Omega_n - \epsilon(p+q) + \epsilon(p)} \qquad (13.25)$$

where $\epsilon(p) = p^2/2m - \epsilon_F$ and

$$n(p) = \frac{1}{e^{\epsilon(p)/T} + 1}$$

In what follows we restrict ourselves to the case of a degenerate electron gas where $T \ll \mu$. As usual we are interested in retarded Green's functions and it is more convenient to carry out the analytic continuation $i\Omega_n = \Omega + i\delta$ before integrals are calculated. For the polarization bubble (13.25) we have at small momenta:

$$\Pi^{(R)}(\Omega, q) = 2\int \frac{d^D q}{(2\pi)^D} \frac{n(p) - n(p+q)}{\Omega + i\delta - \epsilon(p+q) + \epsilon(p)} \approx -2\int \frac{d^D q}{(2\pi)^D} \frac{\partial n}{\partial \epsilon} \frac{qv}{\Omega + i\delta - qv} \qquad (13.26)$$

where $v = \partial\epsilon/\partial p = p/m$. In the zeroth order in T/μ the Fermi distribution becomes a step function and $\partial n/\partial \epsilon = -\delta(\epsilon)$. The integral in the momentum space becomes an integral over the Fermi surface. This fact is important to remember because it is widely used in the theory of metals. The value of the integral (13.26) depends essentially on the dimensionality D.

---

[4] Notice, however, that $D^{00}(\omega_n)$ is a thermodynamic function and the function defined by (13.10) is a *dynamical* correlation function related to the former by (1.30).

Carrying out all necessary integrations we get the following explicit expressions for the polarization operator Π^{00}:

$$\Pi^{(R)}(\Omega, q) = -\frac{p_F m}{\pi^2}\left[1 - \frac{\Omega}{2v_F q}\ln\frac{|\Omega + v_F q|}{|\Omega - v_F q|} + \frac{i\pi|\Omega|}{2v_F q}\theta(v_F q - |\Omega|)\right] \qquad D = 3$$

(13.27)

$$\Pi^{(R)}(\Omega, q) = -\frac{m}{\pi^2}\left[1 - \frac{|\Omega|\theta(|\Omega| - v_F q)}{\sqrt{\Omega^2 - v_F^2 q^2}} + \frac{i|\Omega|\theta(v_F q - |\Omega|)}{\sqrt{v_F^2 q^2 - \Omega^2}}\right] \qquad D = 2$$

(13.28)

$$\Pi^{(R)}(\Omega, q) = \frac{2}{\pi}\frac{v_F q^2}{(\Omega + i\delta)^2 - v_F^2 q^2} \qquad D = 1$$

(13.29)

where v_F is the Fermi velocity. In order to find the spectrum of excitations of the scalar potential one has to substitute the expressions for $\Pi^{(R)}$ into the Dyson equation and study the singularities of D^{00}:

$$[D^{00}]^{(R)} = \frac{4\pi e^2}{q^2 - 4\pi e^2\Pi(\Omega, q)}$$

(13.30)

The singularities occur when the denominator of the right-hand side of this expression vanishes. Substituting (13.27), (13.28) and (13.29) into (13.30) we find the following spectra:

$$\omega^2 = \omega_0^2 + v_F^2 q^2 + \mathcal{O}(q^4) \qquad \omega_0^2 = \frac{4\pi e^2 N}{m} \qquad D = 3$$

$$\omega^2 = \frac{2e^2 m v_F^2}{\pi} + \mathcal{O}(q^4) \qquad D = 2$$

$$\omega^2 = 8v_F e^2 + v_F^2 q^2 \qquad D = 1$$

(13.31)

These spectra always contain gaps; therefore at small frequencies one can neglect the Ω-dependence of the polarization loop Π^{00} and get the Debye formula for the screened Coulomb potential (13.12).

At this point I have to remind the reader that the spectra (13.31) are derived for the 'abstract' electrodynamics. For $D = 2$ (layered) or $D = 1$ (chain) metals these spectra must be modified by taking into account that fact that photons remain three-dimensional. The corresponding expressions are

$$\omega^2 = \frac{2e^2 m v_F^2}{\pi d}\frac{q_\perp^2}{q_\perp^2 + q_z^2} + \mathcal{O}(q^4) \qquad D = 2$$

$$\omega^2 = \frac{8v_F e^2}{d^2}\frac{q_z^2}{q_\perp^2 + q_z^2} + v_F^2 q_z^2 \qquad D = 1$$

(13.32)

where d is the interlayer (interchain) distance. Despite the anisotropy these interactions are effectively short range.

From (13.11) we know that for the vector potential we get quite a different picture. The corresponding self energy includes in this case two terms (see Fig. 13.3(b)), which cancel

Figure 13.4. Loop diagrams for the free energy.

each other at $\Omega, q = 0$:

$$\Pi^{ab}(\Omega_n, q) = \delta^{ab} \frac{N}{mc^2} + \frac{2T}{m^2c^2} \sum_m \int \frac{d^D q}{(2\pi)^D} G_0(\Omega_n + \omega_m, q + p)$$
$$\times G_0(\omega_m, p)(p^a + q^a/2)(p^b + q^b/2) \tag{13.33}$$

At $T = 0$ and $|\Omega| \ll v_F q$ we get for $D > 1$:

$$[\Pi^{(R)}]^{ab}(\Omega, q) = -(\delta^{ab} - q_a q_b/q^2)\left[i\frac{\pi\rho(\epsilon_F)v_F\Omega}{4c^2q} + e^{-2}\chi_D q^2\right] \tag{13.34}$$

Recall that in our notation $\rho(\epsilon_F)$ is the density of states on the Fermi surface. The last term contains the diamagnetic susceptibility χ_D (the Landau diamagnetism). Since the value of χ_D is determined by an integral over all filled states, this term depends on detailed band structure and is nonuniversal.

Substituting (13.34) into the Dyson equation for $\langle\langle A^a A^b\rangle\rangle$ and neglecting the term with $(\Omega/cq)^2$, we get the following expression at $|\Omega| \ll vq$:

$$[D^{(R)}(\Omega, q)]^{ab} = \frac{4\pi(\delta^{ab} - q^a q^b/q^2)}{iB\Omega/q + q^2(1 + 4\pi\chi_D)} \tag{13.35}$$

$$B = \frac{\pi^2 e^2 \rho(\epsilon_F)v_F}{c^2}$$

The obtained expression is related to the semiclassical expression (13.11):

$$\langle H_a(-\omega, -q)H_b(\omega, q)\rangle = q^2 D^{ab}(\omega, q)\frac{1}{\mu}$$

Equation (13.36) does not have poles on the real axis of Ω which means that the excitations are overdumped. The spectrum of the diffusive modes is purely imaginary: $\Omega \sim -iq^3$. This result holds for pure metals where the real part of the conductivity diverges at $q \to 0$: $\sigma(\omega, q) \sim q^{-1}$. According to (13.11), for impure metals (13.36) should be modified for $q < \lambda^{-1}$ (λ is the mean free path):

$$[D^{(R)}(\Omega, q)]^{ab} = \frac{4\pi(\delta^{ab} - q^a q^b/q^2)}{4i\pi\sigma\Omega/c^2 + q^2(1 + 4\pi\chi_D)} \tag{13.36}$$

As I have mentioned above, the fact that the magnetic field is screened only dynamically has far reaching consequences. To demonstrate this let us calculate the contribution of the field excitations to the specific heat.

The free energy is given by loop diagrams, several of which are depicted in Fig. 13.4. To get rid of the numerical coefficients one should differentiate with respect to e^2. Since

Figure 13.5. The most singular diagrams for the free energy containing scalar propagators.

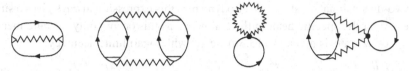

Figure 13.6. The most singular diagrams for the free energy containing vector propagators.

wavy and zigzag lines are singular at small frequencies and momenta, one should collect sequences of diagrams containing a minimal number of integrations over momenta of these lines. Applying these criteria we get the two sequences shown in Figs. 13.5 and 13.6. The first sequence corresponds to the contribution from the excitations of the scalar field (plasmons). Since these excitations have spectral gaps, nothing interesting can be expected. The contribution from the second sequence is more important at low temperatures because it comes from low energy excitations, the diffusive modes.

We have

$$\frac{\partial F}{\partial e^2} = e^{-2} T \sum_n \int \frac{d^D q}{(2\pi)^D} \Pi^{ab}(\omega_n, q) D^{ab}(\omega_n, q) \tag{13.37}$$

Integrating over e^2 and taking into account that

$$F(0) = -\frac{1}{2} \text{Tr} \ln D_0$$

we get

$$F(e^2, T) = -\frac{1}{2} \text{Tr} \ln D = -\frac{TV}{2} \sum_n \int \frac{d^D q}{(2\pi)^D} \ln \det D(\omega_n, q) \tag{13.38}$$

where D is given by (13.36) and V is the volume of the system. Since we know $D^{ab}(\omega)$ only on the real axis, it is convenient to use the Kramers–Kronig relation

$$\ln D(\omega_n) = \frac{1}{\pi} \int dy \frac{\Im m \ln D^{(R)}(y)}{-i\omega_n + y} = \frac{1}{\pi} \int dy \frac{\tan^{-1}[\Im m D^{(R)}(y)/\Re e D^{(R)}(y)]}{i\omega_n - y} \tag{13.39}$$

Using this relation we can calculate the sum over bosonic frequencies first, and thus get the following very convenient expression for the contribution from the bosonic modes:

$$F/V = - \int dy \coth(y/2T) \int \frac{d^D q}{(2\pi)^D} \tan^{-1}[\Im m D^{(R)}(y, q)/\Re e D^{(R)}(y, q)] \tag{13.40}$$

This expression is valid for any theory where we sum a sequence of bubble-type diagrams similar to the ones represented in Fig. 13.6.

Differentiating (13.40) with respect to T and substituting the expression for D given by (13.36), we get the following expression for the entropy:

$$S/V = \int_{-\infty}^{\infty} dy \frac{y}{2T^2 \sinh^2(y/2T)} \int \frac{d^D q}{(2\pi)^D} \tan^{-1}[4\pi\mu y\sigma(y,q)/c^2 q^2] \quad (13.41)$$

For the two-dimensional 'abstract' electrodynamics this expression always gives a singular coefficient γ for the specific heat; in three dimensions this occurs only for pure metals. In the latter case when σ is given by (13.36) we get with logarithmic accuracy:

$$S/N \approx \frac{\mu B}{2n\pi^2} \int_{-\infty}^{\infty} dy \frac{y^2}{2T^2 \sinh^2(y/2T)} \int_{\sim(\mu By)^{1/3}}^{p_F} \frac{dq}{q} \approx \frac{2\pi^2\alpha\mu}{3} \frac{T}{p_F c} \ln(\omega_0/T) \quad (13.42)$$

where $N = p_F^3 V/3\pi^2$ is the total number of electrons and

$$\omega_0 \sim \frac{1}{\alpha\mu} \frac{c}{v_F} \epsilon_F$$

and $\alpha = e^2/\hbar c$ is the fine structure constant. For real electrodynamics $\alpha \approx 1/137$, very small. Comparing the obtained expression with the entropy per electron for the free Fermi gas

$$S/N = 3\pi^2 mT/p_F^2$$

we find that the photon contribution will dominate at temperatures where

$$\ln(\omega_0/T) > \frac{9mc}{2\alpha\mu p_F} \sim 10^5\mu^{-1}$$

This temperature range may become realistic only in very pure ferromagnetic metals where $\mu \sim 10^4$. The necessary degree of purity is achieved when the inverse mean free path λ^{-1} is smaller than the characteristic wave vector $q \sim (\mu BT)^3$, which leads to the estimate

$$(p_F\lambda)^3 \gg \alpha^{-1}\frac{p_F c}{\mu T} \quad (13.43)$$

For $T \sim 1$ K, $\mu \sim 10^4$, and $c/v_F \sim 300$, this requires the mean free path to be of the order of 10 lattice constants.

For the two-dimensional 'abstract' model we have for the pure case

$$S \sim T^{2/3} \quad (13.44)$$

and for the impure case

$$S/V \approx \frac{4\pi\mu}{9c^2}\sigma T \ln(p_F^2 c^2/\mu\sigma T) \quad (13.45)$$

For a real layered metal we have to modify (13.41):

$$S/V = \int_{-\infty}^{\infty} dy \frac{y}{2T^2 \sinh^2(y/2T)} \int \frac{d^2 q_\perp dq_z}{(2\pi)^3} \tan^{-1}\left\{ \frac{4\pi y\sigma(y,q_\perp)}{c^2\left[q_\perp^2(1 + 4\pi\chi_D) + q_z^2\right]} \right\}$$

$$(13.46)$$

This expression yields a logarithmic singularity only for pure metals.

Figure 13.7. The interaction vertices for the effective theory of transverse photons.

Thus for both the abstract model and for the realistic electrodynamics of pure metals we get the singular coefficient in specific heat $\gamma = C_v/T \sim \ln T$. Such behaviour contradicts the standards of the Landau Fermi liquid theory described at the beginning of this chapter. As we have seen, this singularity originates from long-range unscreened interactions. In real metals, where the coupling constant $g = e^2 v_F \mu/\hbar c^2$ is numerically small, the singularities may occur only in pure ferromagnetic metals and at very low temperatures. It is especially remarkable that the Fermi liquid description loses its adequacy without a phase transition (the fact that this can happen through a phase transition is not at all surprising).

Let us show that the absence of screening is a consequence of the gauge symmetry and is robust under a renormalization. Higher-order corrections from the interaction and temperature effects affect only the constants B and χ_D in the expression for the correlation function (13.36). Therefore the excitation spectrum remains diffusive. As a result of our analysis we obtain the effective theory for the vector potential with the bare correlation function (13.36). The excitations interact and some of the vertices are represented in Fig. 13.7.

Due to gauge invariance the physical degree of freedom is not the vector potential, but the magnetic field. Therefore all vertices for **A**-fields must vanish at zero momenta, i.e. in each order in e^2 the corresponding diagrams cancel at $q = 0$, in the same manner as in (13.34). This statement has been checked explicitly for the A^3 and the A^4 terms in the effective action by Fukuyama *et al.* (1969) and Gan and Wong (1993) respectively. These issues are also addressed in the extended publication by Kim *et al.* (1994). Therefore when we formulate the effective theory of interacting magnetic fields in terms of the magnetic fields themselves, we get a theory with nonsingular vertices and the nonsingular propagator:

$$\langle\langle H_a(-\omega_n, -q)H_b(\omega_n, q)\rangle\rangle = \frac{4\pi(\delta_{ab} - q_a q_b/q^2)}{B|\omega_n|/q^3 + \mu^{-1}} \qquad (13.47)$$

where in the first order in e^2, $\mu_D = 1 - 4\pi e^2 \chi_D$. In this theory the perturbation expansion does not contain any singularities and all corrections to B and μ are finite.

Single-electron Green's function

A problem which we have not yet touched on is the problem of calculation of the electronic Green's function. The first remark to be made is that the standard expression for the single-particle Green's function is not gauge invariant and therefore does not have any sense. One

Figure 13.8. Lev Ioffe and Boris Altshuler.

can define a gauge invariant Green's function:

$$G(1,2;C_{12}) = \langle\langle \psi_\sigma(1) e^{i(e/c)\int_1^2 A_\mu dx_\mu} \psi_\sigma^+(2)\rangle\rangle \qquad (13.48)$$

This expression is gauge invariant, but depends on the path C_{12} connecting two points 1 and 2. This fact obscures the physical meaning of this object. It seems, however, that in angle-resolved X-ray photoemission experiments one measures this function with C_{12} being a straight line connecting the points 1 and 2. Indeed, these experiments measure absorption of X-rays with a given frequency ω and wave vector q. The absorption rate is proportional to the imaginary part of the current–current polarization bubble (see Fig. 13.3). The difference from our previous calculations is that now one of the electronic Green's functions has a very high frequency of the order of several electronvolts (from the absorbed X-ray quanta) and the frequency of the other is small. The high energy electron moves on a straight line and can be substituted in the path integral by the phase contribution

$$e^{i(e/c)\int_1^2 A_\mu dx_\mu}$$

where the integral goes along the classical trajectory of the electron, i.e. along the straight line connecting the two points.

The gauge invariant Green's function (13.48) with C_{12} represented by the straight line was calculated for the strictly two-dimensional model

$$\hat{H} = \int d^2x \left[\frac{1}{2m} \psi_\sigma^+ \left(i\nabla + \frac{e}{c}\mathbf{A} \right)^2 \psi_\sigma + \frac{1}{8\pi\mu}(\nabla \times \mathbf{A})^2 \right] \qquad (13.49)$$

by Altshuler and Ioffe (1992). I shall describe their result below. Now let us discuss a simpler problem directly related to the problem of the Green's function. Namely, let us calculate an average of the so-called Wilson exponent:

$$W(C) = \langle e^{i(e/c)\oint_C A_\mu dx_\mu} \rangle = \langle e^{i(e\mu/c)\int d^2x H_z} \rangle \qquad (13.50)$$

where C is some closed contour in the spatial (e.g. the (x, y)) plane. H_z is perpendicular to this plane. The quantity $W(C)$ is gauge invariant; its physical meaning will be explained later. The calculation is easy, since the effective action for the gauge fields is quadratic and the only nontrivial correlation function is given by (13.47). In such a case we can use the results of Chapter 4 and apply (4.6) with $A^{-1} = \langle\langle H_z H_z \rangle\rangle$. Let us consider a layered metal first and assume that the contour C lies on a layer. Let the z-direction be the direction perpendicular to the layers and d be the interlayer distance. Then the result is

$$
W(C) = \exp\left[-\frac{e^2 \mu^2}{2c^2} \int_C d^2 x_1 \int_C d^2 x_2 \langle\langle H(\tau; z, x_1) H(\tau; z, x_2) \rangle\rangle \right]
$$

$$
\approx \exp\left[-\frac{e^2 \mu^2}{2c^2} ST \sum_n \int_{-\pi/d}^{\pi/d} \frac{dq_z}{2\pi} \langle\langle H(-\omega_n, -q_z, -q_{\perp,0}) H(\omega_n, q_z, q_{\perp,0}) \rangle\rangle \right] \quad (13.51)
$$

where S is the area of the loop and $q_{\perp,0} \sim S^{-1/2}$. The appproximation works in the limit when the contour C is large. Then from (13.47) we get

$$
T \sum_n \langle\langle H(-\omega_n, -q_z, -q_{\perp,0}) H(\omega_n, q_z, q_{\perp,0}) \rangle\rangle \approx \frac{4\pi T}{d\mu} + O(S^{-1/2}) \quad (13.52)
$$

In the limit $S \to \infty$ the only nonvanishing contribution comes from $\omega_n = 0$. The result for the Wilson loop is[5]

$$
W(C) = \exp(-S/S_0)
$$

$$
S_0^{-1} = \frac{2\pi \mu T e^2}{d\hbar^2 c^2} \quad (13.53)
$$

Thus we have shown that the logarithm of the Wilson loop in the theory of layered metal scales like its area. This is called the area law, in contrast to the perimeter law where the Wilson loop behaves as the exponent of the contour length.

Exercise

Show that the perimeter law holds for the free electromagnetic field, and also that the area law holds for the strictly two-dimensional model, i.e. for the model where the gauge field exists in $(2 + 1)$ dimensions. Equation (4.6) can be used to calculate the Wilson loop.

The result (13.53) is of principal importance because the area law behaviour is associated with confinement (see Chapter 5 in Polyakov's book). The arguments are as follows: imagine that at some time a particle–antiparticle pair is produced and separated up to a distance R. It is assumed that the pair interacts by exchange of quanta of the gauge field A_μ. This interaction materializes as the potential energy of the pair; we denote this as $V(R)$. The pair propagates during a time τ, at a fixed separation R, and then annihilates. The corresponding probability amplitude is $\exp[-V(R)\tau]$ (do not forget that we are in Euclidean space-time).

---

[5] I restore Planck's constant in the expression for σ in order to make it clearer that its dimensionality is correct.

From the other side, this amplitude is equal to the average of the Wilson loop on the rectangle $R \times \tau$: $\exp[-V(R)\tau] = \exp(-\sigma R\tau)$. Therefore we find that the effective potential between the particle and the antiparticle is linear: $V(R) = \sigma R$. This is a confinement because the pair cannot escape from such a potential. The present situation differs, however, from the classical one in several respects. First, in the present case the area law holds for contours in the spatial plane. At the same time the above cited arguments work for contours with one of the sides along the temporal direction. Second, the string tension σ for QCD depends weakly on temperature.

Below I reproduce the result obtained by Altshuler and Ioffe for the Green's function of the model (13.49). The retarded Green's function in the frequency momentum representation is given by:

$$G^{(R)}(\omega, \mathbf{p}) = \tau_\phi^{-1/2} \int d\xi \frac{g(\xi)}{\left[\tau_\phi^{-1}\xi - \omega + \epsilon(p)\right]^{3/2}}$$

$$g(\xi) = -\frac{i^{1/3}}{4} \frac{\mathrm{Ai}(-i^{2/3}\xi)}{\mathrm{Ai}'(-i^{2/3}\xi)} \tag{13.54}$$

where

$$\tau_\phi^{-1} = 2\left[\pi^2\mu^2(e^2/d)^2\frac{T^2}{mc^2}\frac{v_F^2}{c^2}\right]^{1/3} \tag{13.55}$$

and $\mathrm{Ai}(x)$ is the Airy function. The integral can be expressed as a sum over zeros of the Airy function derivative $[\mathrm{Ai}'(\eta_n)] = 0$:

$$G^{(R)}(\omega, \mathbf{p}) = -\frac{\pi}{2\sqrt{\tau_\phi}} \sum_n \frac{(-i)^{1/3}\eta_n^{-1}}{\left[-\omega + \epsilon(p) + (-i)^{2/3}\tau_\phi^{-1}\eta_n\right]^{3/2}} \tag{13.56}$$

As a function of the complex variable $z = \tau_\phi[\omega - \epsilon(p)]$ the Green's function (13.56) has a series of cuts instead of a pole, as would be expected in the Fermi liquid theory.

As was mentioned above, this picture of the unusual behaviour of electrons interacting with an unscreened U(1) gauge field remains very attractive to theorists, and during recent years this has given rise to an outburst of creative activity. There are very serious arguments in favour of the point of view that gauge fields of nonelectromagnetic origin are generated effectively in strongly correlated metals; in particular, these fields can be generated by magnetic interactions (the idea of spin liquids and resonance valence bond states; see, for example, papers by Nagaosa and Lee (1990) and Ioffe and Wiegmann (1990)). Kalmeyer and Zhang (1992) and Halperin et al. (1993) suggested that a theory similar to that of two-dimensional 'abstract' quantum electrodynamics (13.49) with one fermion species could describe half-filled Landau levels in the quantum fractional Hall effect. The most complete analysis of this problem is given in the paper by Read (1998).

It was suggested that a system of two-dimensional electrons in an external magnetic field may generate its own magnetic field acting in the opposite direction, reducing the

Figure 13.9. Patrick Lee.

average value of the total field to zero. In this way the system lifts the enormous degeneracy of the Landau levels and benefits from the resulting decrease in the Coulomb energy. In the suggested approach real electrons are represented as bound states of Fermi quasi-particles and flux tubes of fluctuating internal magnetic field (see later in Chapter 15). The resulting effective action for low-lying excitations is very similar to the 'abstract' quantum electrodynamics (13.49) discussed in this chapter. This approach has found strong experimental support (see Willett *et al.*, 1993; Kang *et al.*, 1993). Thus the 'abstract' quantum electrodynamics can explain real experiments, being abstract only in quotation marks.

Some mechanisms producing gauge interactions of a nonelectromagnetic origin are discussed in Part III (see chapters about the Jordan–Wigner transformation). In all these scenarios the gauge fields propagate not with the speed of light, but with a speed characteristic of magnetic interactions, which is several orders of magnitude smaller. Since the dimensionless fine structure constant contains c in the denominator, such changes in the nature of the gauge fields dramatically increase the relevant energy scale.

References

Altshuler, B. L. and Ioffe, L. B. (1992). *Phys. Rev. Lett.*, **69**, 2979.
Fukuyama, H., Ebisawa, H. and Wada, Y. (1969). *Prog. Theor. Phys.*, **42**, 494.
Gan, J. and Wong, E. (1993). *Phys. Rev. Lett.*, **71**, 4226.
Halperin, B. I., Lee, P. A. and Read, N. (1993). *Phys. Rev. B*, **47**, 7312.
Holstein, T., Norton, R. E. and Pincus, P. (1973). *Phys. Rev. B*, **8**, 2649.
Ioffe, L. D. and Wiegmann, P. B. (1990). *Phys. Rev. Lett.*, **65**, 653.
Kalmeyer, V. and Zhang, S. C. (1992). *Phys. Rev. B*, **46**, 9889.

Kang, W., Stormer, H. L., Pfeiffer, L. N., Baldwin, K. W. and West, K. W. (1993). *Phys. Rev. Lett.*, **71**, 3850.

Kim, Y. B., Furusaki, A., Wen, X.-G. and Lee, P. A. (1994). *Phys. Rev. B*, **50**, 17917.

Nagaosa, Y. and Lee, P. A. (1990). *Phys. Rev. Lett.*, **64**, 2450.

Read, N. (1998). *Phys. Rev. B*, **58**, 16262.

Reizer, M. Yu. (1989). *Phys. Rev. B*, **40**, 11571; *Phys. Rev. B*, **44**, 5476 (1991).

Willet, R. L., Ruel, R. R., West, K. W. and Pfeiffer, L. N. (1993). *Phys. Rev. Lett.*, **71**, 3846.

14

Relativistic fermions: aspects of quantum electrodynamics

So far we have discussed only nonrelativistic fermions. Due to Lorentz invariance, relativistic fermions have certain special properties which are worth considering. Relativistic spectra appear not only in high energy physics, but also in condensed matter physics, where the Lorentz group can be present as an approximate symmetry for low energy excitations, which often occurs, for example, in $(1 + 1)$-dimensional metals and antiferromagnets. Therefore it makes sense not to restrict ourselves to $D = 3$, but to consider other dimensionalities as well.

Let us consider the Lagrangian for relativistic fermions introduced by Dirac. This Lagrangian describes fermions with mass M in $(D + 1)$-dimensional space-time interacting with the external gauge field A_μ:

$$L = \int d^D x [\bar{\psi} \gamma^\mu (i\partial_\mu + eA_\mu)\psi + M\bar{\psi}\psi] \tag{14.1}$$

The fermionic fields $\bar{\psi}$, $\psi\psi$ are s-dimensional spinors:

$$\psi = \begin{pmatrix} \psi_1 \\ \psi_2 \\ \cdot \\ \cdot \\ \cdot \\ \psi_s \end{pmatrix} \qquad \bar{\psi} = \psi^+ \gamma^0$$

The matrices γ^μ ($\mu = 0, \ldots, D$) are $s \times s$-dimensional matrices satisfying the Clifford algebra:

$$\{\gamma^\mu, \gamma^\nu\}_+ = 2g^{\mu\nu} \tag{14.2}$$

where $g^{00} = 1$, $g^{aa} = -1$ ($a = 1, \ldots, D$) for the Minkovsky space-time. In Euclidean space-time ($t = i\tau$) the metric tensor is just a unit matrix $g^{\mu\nu} = \delta^{\mu\nu}$. In even space-time dimensions there is an additional matrix

$$\gamma_S = i^{-(D+1)/2} \gamma_0 \gamma_2 \cdots \gamma_D \tag{14.3}$$

which anticommutes with all γ_μ and

$$\gamma_S^2 = 1$$

The Euclidean version of the Dirac action is given by (see, for example, Chapter A5 of the book by Zinn-Justin)

$$S = \int d\tau d^D x [\bar{\psi}\gamma^\mu(\partial_\mu + ieA_\mu)\psi + M_1\bar{\psi}\psi + iM_2\bar{\psi}\gamma_S\psi] \tag{14.4}$$

Here I have added the additional γ_S mass term which exists only if $D + 1$ is even.

Despite their vector-like appearance the fermion fields ψ are not vectors, but spinors. The reason for this is that the name 'vector' is reserved for entities transforming as space-time coordinates under Lorentz transformations. Spinors transform differently, according to their own representation of the Lorentz group, which is called 'spinor' representation. The γ^μ matrices transform as vectors.

An important difference between relativistic and nonrelativistic fermions comes from the fact that the role of the conjugate field in the path integral is played not by ψ^+, but by $\bar{\psi}$. The reason for this is that it is

$$\bar{\psi}\psi$$

and not

$$\psi^+\psi$$

which is a Lorentz invariant object. Since the measure of integration must be Lorentz invariant, it cannot be

$$D\psi^+ D\psi$$

as it is for nonrelativistic fermions, but

$$D\bar{\psi} D\psi$$

(In $(1 + 1)$ dimensions it is $d\bar{\psi}d\psi = d\psi_R^+ d\psi_L d\psi_L^+ d\psi_R$.) Correspondingly, the Green's function for relativistic electrons in Euclidean space-time is defined as

$$G_{ab}(1, 2) \equiv \langle\langle\psi_a(1)\bar{\psi}_b(2)\rangle\rangle = [(\gamma^\mu\partial_\mu + M_1 + iM_2\gamma_S)^{-1}]_{ab} \tag{14.5}$$

There are several points about the algebra of γ matrices which could be important to us. First, there are certain ambiguities in their choice. The dimensionality of γ matrices is uniquely defined only if $D + 1$ is even:

$$s = 2^{(D+1)/2}$$

Therefore for $D = 1$, where $s = 2$, one can represent the γ matrices by Pauli matrices. In Euclidean space-time we can make the following choice:

$$\gamma^0 = \sigma^x \qquad \gamma^1 = \sigma^y \qquad \gamma_S = \sigma^z$$

and for $D = 3$, where $s = 4$, the standard definition is

$$\gamma^0 = \sigma^y \otimes I \qquad \gamma^1 = \sigma^x \otimes \sigma^x \qquad \gamma^2 = \sigma^x \otimes \sigma^y \qquad \gamma^3 = \sigma^x \otimes \sigma^z \qquad \gamma_S = \sigma^z \otimes I$$

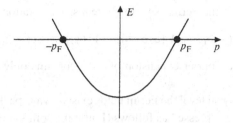

Figure 14.1. The qualitative form of electronic dispersion in a one-dimensional metal.

At the same time for $D = 2$ one can use two alternative representations: the two-dimensional and the four-dimensional ones. In the two-dimensional description the fermionic fields are two-dimensional spinors and the γ matrices can be chosen in two ways:

$$\gamma^\mu = \sigma^\mu$$

or

$$\gamma^x = \sigma^x \qquad \gamma^y = \sigma^y \qquad \gamma^z = -\sigma^z$$

which correspond to a left-handed and right-handed basis in three-dimensional space-time, respectively.

In what follows we shall consider relativistic fermions in $(1 + 1)$ and $(2 + 1)$ dimensions only. Let us discuss the one-dimensional case first. The Dirac Lagrangian appears naturally as a low energy effective Lagrangian for conduction electrons in one-dimensional metals. To show this let us consider the Euclidean action of nonrelativistic fermions interacting with an Abelian gauge field given by (13.14). The fermionic part of the action is (I put $c = 1$)

$$S = \int d\tau dx [\psi_\sigma^*(\partial_\tau + ie\phi + \epsilon_F)\psi_\sigma - \psi_\sigma^* \hat{\epsilon}(-i\partial_x - eA)\psi_\sigma] \qquad (14.6)$$

The dispersion law $E(p) = \epsilon(p) - \epsilon_F$ has the qualitative form depicted in Fig. 14.1.

Let us assume now that external fields ϕ, A have small amplitudes and are slow in comparison with $\exp(\pm 2i p_F x)$. Then all relevant processes take place near the Fermi points. There are two consequences of this. The first is that the only relevant Fourier components of the external fields are those with small wave vectors and wave vectors of the order of $\pm 2 p_F$. The former leave electrons near their Fermi points and the latter transfer them from one Fermi point to another. In real systems such fields are generated, for example, by lattice displacements (phonons). Because of the gauge symmetry we can move all fast terms into the scalar potential and write

$$\phi(\tau, x) = A_0(\tau, x) + \exp(2i p_F x)\Delta(\tau, x) + \exp(-2i p_F x)\Delta^*(\tau, x)$$
$$A(\tau, x) = A_1(\tau, x)) \qquad (14.7)$$

where the fields A_0, A_1, Δ, Δ^* are slow in comparison with $\exp[\pm 2i p_F x]$. The second fact is that since the relevant fermions are near the Fermi points we can linearize their spectrum:

$$E(p + p_F) \approx vp \qquad E(p - p_F) \approx -vp \qquad |p| \ll p_F \qquad (14.8)$$

where $v = \partial E/\partial p|_{p=p_F}$ is the Fermi velocity. In real space notation this looks as follows:

$$\exp(\pm i p_F x)\hat{E}(-i\partial_x - eA)[\exp(\mp i p_F x)f(x)] \approx \pm v(-i\partial_x - eA)f(x) \quad (14.9)$$

Here we assume that the Fourier expansion of $f(x)$ contains only harmonics with small wave vectors $|p| \ll 2p_F$.

In the vicinity of the Fermi level the Fermi fields exist as wave packets with average wave vectors $\pm p_F$. This can be expressed as follows (I omit the spin index):

$$\psi(\tau, x) \approx \psi_R(\tau, x)\exp(-i p_F x) + \psi_L(\tau, x)\exp(i p_F x) \quad (14.10)$$

where $\psi_R(x)$, $\psi_L(x)$ are fields slow in comparison with $\exp(\pm i p_F x)$. These fields represent right- and left-moving fermions near the right and left Fermi points, respectively. Substituting this expression together with (14.7) into the original action (14.6), using (14.9) and taking into account only nonoscillatory terms, we arrive at the Dirac action (14.4) with $M_1 = \Re e\Delta$, $M_2 = \Im m\Delta$.

The discussed equivalence works in two ways. First, it provides a very convenient regularization for relativistic theories. By representing them as limiting cases of nonrelativistic fermions, we eliminate ultraviolet divergences. Second, this equivalence discloses a hidden Lorentz invariance of the low energy sector in one-dimensional metals.

The Dirac action in two-dimenional Euclidean space-time is worth writing in the explicit form:

$$S = \int d\tau dx (\bar{\psi}_{rL}, \bar{\psi}_{rR}) \begin{pmatrix} \Delta & 2\partial + i\bar{A} \\ 2\bar{\partial} + iA & \Delta^* \end{pmatrix} \begin{pmatrix} \psi_L \\ \psi_R \end{pmatrix} \quad (14.11)$$

where

$$z = \tau + ix/v \qquad \partial = \frac{1}{2}(\partial_\tau - iv\partial_x)$$

$$\bar{z} = \tau - ix/v \qquad \bar{\partial} = \frac{1}{2}(\partial_\tau + iv\partial_x)$$

and $A = A_0 + iA_1$, $\bar{A} = A_0 - iA_1$.

In order to find the Green's function in zero field and constant Δ, we make a Fourier transformation of the fermionic variables. Then from (14.5) we find

$$G(\omega_n, p) = \frac{1}{\omega_n^2 + p^2 + |\Delta|^2} \begin{pmatrix} \Delta^* & -(i\omega_n + p) \\ -(i\omega_n - p) & \Delta \end{pmatrix} \quad (14.12)$$

For future purposes we need to have this expression in real space. At $T = 0$ we have

$$\frac{1}{(2\pi)^2} \int \frac{d^2 p e^{ipx}}{p^2 + |\Delta|^2} = \frac{1}{2\pi} K_0(|\Delta||x|) \quad (14.13)$$

where $K_0(x)$ is the zeroth Bessel function of the imaginary argument. Therefore we can write

$$G(1, 2) = \frac{1}{\pi} \begin{pmatrix} \Delta^* & 2\partial \\ 2\bar{\partial} & \Delta \end{pmatrix} K_0(|\Delta||z_{12}|) \quad (14.14)$$

It is especially instructive to calculate the zero mass limit of this expression. At small arguments we have

$$K_0(|\Delta||z_{12}|) \approx -\ln(|\Delta||z_{12}|) \tag{14.15}$$

and therefore

$$G_{\Delta=0}(1,2) = \frac{1}{2\pi}\begin{pmatrix} 0 & 1/z_{12} \\ 1/\bar{z}_{12} & 0 \end{pmatrix} \tag{14.16}$$

$(1+1)$-Dimensional quantum electrodynamics (Schwinger model)

As we have seen in the previous chapter, the presence of conducting electrons has a dramatic effect on the electromagnetic field. This field excites electron–hole pairs through the Fermi surface and becomes screened. As we have established above, in $(1+1)$ dimensions the relativistic electrodynamics of massless fermions is, in fact, an electrodynamics of metal. The corresponding screening of the electromagnetic field appears here as a broken gauge symmetry. This effect is called an *anomaly*. We begin the discussion from a formulation of a paradox. Let us consider massless fermions in Minkovsky space-time in an external potential of the following form:

$$A_0 + A_1 = \partial_x\Phi(x) \qquad A_0 - A_1 = \partial_x\Theta(x)$$

where Φ, Θ are some functions of x. The above potential appears to have no effect on the fermions at all because it can be removed by the canonical transformation

$$\hat{\psi}_R(x) \to e^{-i\Phi(x)}\hat{\psi}_R(x)$$
$$\hat{\psi}_L(x) \to e^{-i\Theta(x)}\hat{\psi}_L(x) \tag{14.17}$$

At the same time, from the calculations carried out in the previous chapter we know that the response function of the Fermi gas, given by the polarization loop (13.29), is finite. For spinless fermions we need to multiply (13.29) by $1/2$. Taking this into account, we get the following contribution to the free energy:

$$F = \frac{1}{2}\sum_q q^2(\Phi(-q) - \Theta(-q))\Pi(0,q)(\Phi(q) - \Theta(q)) = \int \frac{dx}{2\pi}[\partial_x(\Phi - \Theta)]^2 \tag{14.18}$$

The explanation of the paradox is that the transformation (14.17) does not leave the partition function invariant because it affects the combination

$$\bar{\psi}\psi \to e^{i[\Phi(x)-\Theta(x)]}\bar{\psi}_R\psi_L + e^{i[-\Phi(x)+\Theta(x)]}\bar{\psi}_L\psi_R$$

and therefore affects the measure in the path integral:

$$D\bar{\psi}D\psi$$

Technically the latter fact means that single fermion wave functions

$$\begin{pmatrix} f_{R,E}(x) \\ f_{L,E}(x) \end{pmatrix} \qquad \begin{pmatrix} f_{R,E'}(x) \\ f_{L,E'}(x) \end{pmatrix}$$

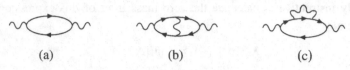

(a)   (b)   (c)

Figure 14.2.

with energies E, E', being orthogonal before the transformation, are not orthogonal after it. Thus the diagram expansion, which takes all these subtle properties of the measure into account, is a more reliable tool than operator transformations of Hamiltonians.

In general, *anomalies* appear in theories which allow transformations affecting only the measure of integration and not the action itself. In the particular case of $(1 + 1)$-dimensional massless quantum electrodynamics the anomaly leads to the illusion that the fermions decouple from the vector potential. They do not, as we have just found out. As we shall see in a moment, the anomaly allows us to integrate over the fermions and obtain an explicit expression for the effective action of the gauge field.

In the Hamiltonian the gauge field is coupled to the current operators:

$$S_{\text{int}} = i \int d\tau dx \, J^\mu(x) A_\mu \qquad J^\mu(x) = \bar\psi(x) \gamma^\mu \psi(x) \qquad (14.19)$$

and the free energy is expressed in terms of their correlation functions:

$$\ln Z[A] - \ln Z[0] = \sum_{n=0} \frac{(-i)^n}{n!} \int \prod_i d^2 x_i \langle\langle J^{\mu_1}(x_1) \cdots J^{\mu_n}(x_n)\rangle\rangle A_{\mu_1}(x_1) \cdots A_{\mu_n}(x_n)$$

$$(14.20)$$

Some of these correlation functions are represented in Fig. 14.2. It turns out that all these diagrams except the first one exactly cancel each other! Therefore the corresponding fermionic determinant can be calculated exactly:

$$\int D\bar\psi D\psi \exp\left[-\int d^2 x \bar\psi \gamma^\mu (\partial_\mu + iA_\mu)\psi\right] = \exp\{-S_{\text{eff}}[A]\}$$

$$(14.21)$$

$$S_{\text{eff}}[A] = -\frac{1}{2\pi} \int d^2 x d^2 y \, F_{\mu\nu}(x) \Delta^{-1}(x, y) F^{\mu\nu}(y)$$

The crucial observation leading to this result was made by S. Coleman (1975), who showed that the correlation functions of the fields $J^\mu(x)$ are the *same* as the correlation functions of the bosonic fields

$$J^\mu(x) \equiv \frac{1}{\sqrt\pi} \epsilon_{\mu\nu} \partial_\nu \Phi \qquad (14.22)$$

governed by the action

$$S = \frac{1}{2} \int d^2 x (\nabla \Phi)^2 \qquad (14.23)$$

This statement is based on the fact that the left and right sides of (14.22) have the same commutation relations. The proof includes a highly nontrivial moment. Let us consider a commutator of two right-moving current operators

$$J_R(q) = \sum_p \psi_R^+(p+q)\psi_R(p)$$

The straightforward calculation yields

$$[J_R(q), J_R(p)] = 0 \tag{14.24}$$

This result is absurd because it suggests that the pair correlation function of currents vanishes. However, we can calculate this correlation function using Feynman diagrams. It is more convenient to do this in real space. The current–current correlation function is given by the diagram in Fig. 14.2(a). According to (14.16), the single-electron Green's functions are given by

$$\langle\langle\psi_R(1)\psi_R^+(2)\rangle\rangle = G_R(z_{12}) = \frac{1}{2\pi z_{12}} \tag{14.25}$$

$$\langle\langle\psi_L(1)\psi_L^+(2)\rangle\rangle = G_L(\bar{z}_{12}) = \frac{1}{2\pi \bar{z}_{12}} \tag{14.26}$$

Thus Green's functions of right-moving particles depend only on $z = \tau + ix$ and the ones for left-moving particles contain only $\bar{z} = \tau - ix$. For this reason the corresponding components of currents can also be treated as analytical or antianalytical functions. Substituting these expressions into the diagram in Fig. 14.2(a), we get

$$\langle\langle J_R(z_1)J_R(z_2)\rangle\rangle = \frac{1}{4\pi^2 z_{12}^2} \tag{14.27}$$

If the correlation function of two Bose operators A, B is known, one can calculate their commutator at equal times according to the following rule:

$$\langle\langle[A(x), B(y)]\rangle\rangle = \lim_{\tau\to+0}[\langle\langle A(\tau, x)B(0, y)\rangle\rangle - \langle\langle A(-\tau, x)B(0, y)\rangle\rangle] \tag{14.28}$$

Let us show that this follows from the definition of the time ordered correlation function. Let us imagine that the operators $\hat{A}(z)$ and $\hat{B}(z_1)$ stand inside some correlation function, perhaps surrounded by other operators which we denote as $X(\{z_i\})$:

$$G(z, z_1, \{z_i\}) = \langle\langle A(z)B(z_1)X(\{z_i\})\rangle\rangle \tag{14.29}$$

Recall, that in our notation $z = \tau + ix$. Let us consider the situation when $\Re ez = \Re ez_1 \pm \delta$ where δ is a positive infinitesimal (see Fig. 14.3).

Without loss of generality we can put $z = ix \pm 0$, $z_1 = iy$, where x, y are real. Then due to the time ordering present in the correlation function the following is valid:

$$G(\tau - \tau' = +0; x, y) - G(\tau - \tau' = -0; x, y) = \langle\langle[A(x), B(y)]X(\{z_i\})\rangle\rangle \tag{14.30}$$

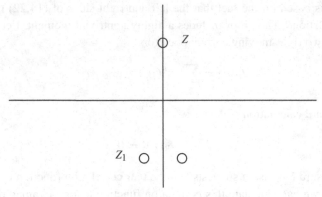

Figure 14.3.

which leads to (14.28). Substituting (14.27) into (14.28), we get

$$\langle\langle[J_R(x), J_R(y)]\rangle\rangle = \frac{1}{4\pi^2}\left\{\frac{1}{[0+i(x-y)]^2} - \frac{1}{[-0+i(x-y)]^2}\right\}$$

$$= \frac{1}{4\pi^2}\partial_x\left[\frac{1}{(-i0+x-y)} - \frac{1}{(i0+x-y)}\right] = \frac{i}{2\pi}\partial_x\delta(x-y) \quad (14.31)$$

A similar procedure for the left currents yields

$$\langle\langle[J_L(x), J_L(y)]\rangle\rangle = -\frac{i}{2\pi}\partial_x\delta(x-y) \quad (14.32)$$

One can say that there is a difference in definitions. The original commutator does not include any averaging and the one which I have just calculated does. Here we have an example of a commutator which is nonzero only for a system of an infinite number of particles and which vanishes for any finite system. Such commutators are called anomalous. There is a nice and clear discussion of the anomalous commutator of currents in Fradkin's book (see Section 4.3.1). An *important lesson* which follows from the present discussion is the following. In order not to miss anomalous commutators, it is always better not to calculate them straightforwardly, but to use a diagram expansion and identity (14.28).

Now it is straightforward to check that the commutation relations (14.31) and (14.32) are reproduced if we identify the fermionic currents with the derivative of the Bose field according to (14.22). This identification gives us the first example of *bosonization*, a topic we shall discuss extensively in Part IV.

Using the equivalence (14.22) one can write the generating functional of the A_μ-fields as follows:

$$\ln Z[A] = \int D\Phi \exp\left\{-S[\Phi] - \frac{i}{\sqrt{\pi}}\int d^2x\, A_\mu\epsilon_{\mu\nu}\partial_\nu\Phi\right\} \quad (14.33)$$

Figure 14.4. Lines of electrostatic field around a wire.

In order to calculate the above path integral it is convenient to integrate the second term in the exponent by parts:

$$\int d^2x\, A_\mu \epsilon_{\mu\nu} \partial_\nu \Phi = -\int d^2x\, F_{01}\,\Phi$$

Then integrating over Φ we get (14.21). Next we have to choose a bare action for the electromagnetic field. The standard action is

$$\frac{1}{8\pi e^2}\int d^2x\, E^2(x) \tag{14.34}$$

where $E = F_{01}$. This is really a one-dimensional action, so it is perfect as an abstract model of $(1+1)$-dimensional quantum electrodynamics, but entirely unrealistic as a model of one-dimensional electrons. Nevertheless, let us discuss this action first. Combining the action (14.34) with the action (14.21) we get the effective action of the electrodynamic field for $(1+1)$-dimensional quantum electrodynamics, the so-called Schwinger model:

$$S = \frac{1}{2\pi}\int d^2x\left[\frac{1}{4e^2}E^2(x) - E(x)\Delta^{-1}(x,y)E(y)\right] \tag{14.35}$$

The correlation function of the electromagnetic field is now given by:

$$\langle\langle E(-q)E(q)\rangle\rangle = \frac{\pi q^2}{q^2/4e^2 + 1}$$

The electromagnetic field acquires a mass gap due to screening by electric charges of massless electrons!

Let us discuss now a more realistic model. Suppose that our electrons belong to a thin metallic wire and the electromagnetic field is really $(3+1)$-dimensional. Such systems appear in computer technology as part of electronic devices. One wire cannot screen an electromagnetic field in three dimensions completely: the field lines avoid screening charges (see Fig. 14.4). Nevertheless the Coulomb interaction is modified. Let us consider its correlation functions right on the wire. Since the magnetic interactions are very weak, we shall discuss only the electrostatic (Coulomb) interaction. The Fourier transformation of the Coulomb potential in one dimension is

$$U_0(p) = \int dx\, \frac{e^2}{|x|}e^{ipx} \approx -2e^2\ln(pa) \tag{14.36}$$

where a is the thickness of the wire. As we have just established, the exact polarization loop for electrons coincides with the one for noninteracting electrons:

$$\Pi(\omega_n, p) = -\frac{1}{\pi} \frac{v_F p^2}{\omega_n^2 + p^2 v_F^2} \tag{14.37}$$

Therefore the screened potential is given by

$$U(\omega_n, p) = \frac{U(p)}{1 - U(p)\Pi(\omega_n, p)} = \frac{-2e^2 \ln(pa)(\omega_n^2 + v_F^2 p^2)}{\omega_n^2 + v_F^2 p^2[1 - (2e^2/\pi v_F)\ln(pa)]} \tag{14.38}$$

The spectrum of excitations of the electrostatic field on the wire (plasma waves) is determined by poles of this expression:

$$\omega^2 = v_F^2 p^2 \left[1 - \frac{2e^2}{\pi v_F} \ln(pa) \right] \tag{14.39}$$

Thus plasma waves have an almost linear spectrum (like phonons), but with the velocity logarithmically growing at small momenta. Long-distance Coulomb interactions are the cause of strong changes in the electronic Green's functions, a subject I discuss later in this book.

Reference

Coleman, S. (1975). *Phys. Rev. D*, **11**, 2088.

15

Aharonov–Bohm effect and transmutation of statistics

> My son, to transmute lead into gold take one measure of the mercury of philosophes
> and mix it with the Green Lion. Keep it in an atanar until Sirius will rise in the proper
> constellation. Then the work is done...
>
> Vasilius Vasilides

Quantum field theory in $(2 + 1)$ dimensions provides a very special mechanism for converting bosons into fermions and vice versa. This mechanism is based on the Aharonov–Bohm effect. To recall what this effect is let us consider a nonrelativistic electron on a two-dimensional plane in the field of a thin infinitely long solenoid directed perpendicular to the plane. The solenoid contains a magnetic flux Φ; the corresponding magnetic field is concentrated inside the solenoid and vanishes identically outside it. We assume that the electrons cannot penetrate the inside of the solenoid. From a naive point of view it seems that in this configuration the electrons are not affected by the magnetic field. Remarkably, this is not the case: the electrons respond to the magnetic field existing in an area to which they have no access! To prove this point let us consider the Schrödinger equation for an electron in the field of a circular solenoid. Since we have cylindrical symmetry, it is convenient to work in polar coordinates with the origin in the centre of the solenoid.

The Hamiltonian is given by

$$\hat{H} = \frac{1}{2m} \left[\left(\hat{p}_r + \frac{e}{c} A_r \right)^2 + \left(\hat{p}_\phi + \frac{e}{c} A_\phi \right)^2 \right] + U(r) \tag{15.1}$$

where $U(r)$ is a potential whose explicit form does not play any role. For example, one can imagine that $U(r)$ is zero outside the solenoid and infinite inside it. I choose the following gauge:[1]

$$A_r = 0 \qquad A_\phi = \frac{\Phi}{2\pi r}$$

Then the Schrödinger equation has the following form:

$$-\frac{\hbar^2}{2m} \left[\frac{1}{r} \partial_r (r \partial_r) + \frac{1}{r^2} (\partial_\phi + ie\Phi/2\pi c\hbar)^2 \right] \psi + U(r)\psi = E\psi \tag{15.2}$$

---

[1] Check that the magnetic field is zero everywhere except at $r = 0$!

Figure 15.1.

In terms of the cylindrical harmonics we have the following solutions:

$$\psi_m(r, \phi) = R_m(r) \exp\left[i\left(m + \frac{e\Phi}{2\pi c\hbar}\right)\phi\right] \tag{15.3}$$

where R_m does not depend on Φ explicitly. Since the obtained solutions must be periodic in ϕ, we have the new quantization rule:

$$m + \frac{e\Phi}{2\pi c\hbar} = \text{integer} \tag{15.4}$$

Thus it turns out that the magnetic field changes the quantization rules! This is the Aharonov–Bohm effect: a particle going around a flux Φ acquires a phase shift $(e\Phi/c\hbar)$.

Now let us consider how to apply the Aharanov–Bohm mechanism to the transmutation of statistics. Imagine that we have particles with flux tubes attached to them. Let every flux tube carry a half of the flux quanta, i.e. $(e\Phi/c\hbar) = \pi$. Consider a permutation of two particles.

Under such a permutation particles not only interchange their positions, but also go around flux tubes (see Fig. 15.2). Thus the many-body wave function picks an additional phase factor $\exp[i(e\Phi/c\hbar)] = -1$ which is equivalent to a change of statistics. Below I describe the realization of this procedure within the path integral approach. Let us consider a $(2 + 1)$-dimensional theory of particles (bosons or fermions, it does not matter) with a conserved current J_μ. In order to achieve transmutation of statistics one can modify the Lagrangian, adding to it the interaction with a fictitious gauge field a_μ whose Lagrangian

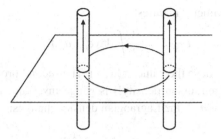

Figure 15.2.

consists solely of the so-called Chern–Simons term:

$$\delta L = i \int d^2x J_\mu a_\mu + L_{\text{Ch-S}} \tag{15.5}$$

$$L_{\text{Ch-S}} = -i\theta \int d^2x a_0 f_{xy} \tag{15.6}$$

where $f_{xy} = \partial_x a_y - \partial_y a_x$.[2] Since the entire action is linear in the field a_0, we can integrate it out:[3]

$$\int Da_0 \exp\left\{ i \int d\tau d^2x a_0(-J_0 + \theta f_{xy}) \right\} = \delta(-J_0 + \theta f_{xy}) \tag{15.7}$$

This delta function tells us that every particle (i.e. a unit 'charge' $\int d^2x J_0 = 1$) carries a flux $1/\theta$. As we know from the above discussion, transmutation of statistics occurs when

$$\int d^2x f_{xy} = \pi$$

which corresponds to $\theta = 1/\pi$ (I shall not discuss what happens at other values of θ; the curious reader is advised to read the book by Frank Wilczek listed in the select bibliography). Thus when we attach the Chern–Simons fictitious field with $\theta = 1/\pi$ to bosons, their many-body function becomes antisymmetric with respect to interchange of coordinates. When we attach it to fermions the wave function becomes symmetric. This does not mean, however, that the Chern–Simons procedure transforms free fermions into free bosons and vice versa. To understand this, let us consider free spinless fermions on a lattice. Since the Chern–Simons term cannot change the anticommutation relations on the same site, double occupation remains forbidden even after the transformation. The latter can be thought of as an infinite repulsion on the same site. Therefore the free fermions coupled to the gauge field with the Chern–Simons action are equivalent not to free but to hard-core bosons, i.e. to bosons with an infinite repulsion on the same site.

Before going any further I would like to mention several important properties of the Chern–Simons term. The first one is its Lorentz symmetry: the expression (15.6) can be

---

[2] This term remains purely imaginary in Minkovsky and in Euclidean space. Therefore here we have a theory which does not have any meaningful classical equivalent.

[3] Therefore the field a_μ is really 'fictitious'; it does not introduce any independent dynamics.

rewritten to make this symmetry manifest:

$$L_{\text{Ch-S}} = i\frac{\theta}{2} \int d^2x \epsilon_{\mu\nu\lambda} a_\mu f_{\nu\lambda} \tag{15.8}$$

(here the Greek subscripts have the values τ, x, y). The second property is that the Chern–Simons action (not the Lagrangian density!) is gauge invariant. Indeed, if we make the transformation $a_\mu \to a_\mu + \partial_\mu \Lambda$ the Lagrangian density changes:

$$\delta L_{\text{Ch-S}} = \frac{\theta}{2} \epsilon_{\mu\nu\lambda} \partial_\mu \Lambda f_{\nu\lambda}$$

but the action, i.e. the integral of this density over the entire space-time, remains invariant.[4] The third property is that the Chern–Simons term is not invariant under parity transformations. This follows from the presence in it of the absolutely antisymmetric tensor $\epsilon_{\mu\nu\lambda}$ which changes its sign when one changes the chirality of the coordinate basis. The combination of gauge invariance with parity violation indicates that if the Chern–Simons term arises dynamically from gauge invariant interactions of fermions, it may happen only through a breakdown of parity. Below I shall demonstrate that the Chern–Simons term is present in the effective action of the U(1) gauge field for the (2 + 1)-dimensional *chiral* massive quantum electrodynamics (i.e. for quantum electrodynamics where only fermions with certain chirality are present).

It is also worth mentioning that the Chern–Simons term can be introduced for non-Abelian gauge theories, i.e. for gauge theories where the vector potentials are matrices. The field strength for non-Abelian gauge theories is defined as $F_{\mu\nu} = \partial_\mu a_\nu - \partial_\nu a_\mu + [a_\mu, a_\nu]$; it is invariant with respect to the gauge transformations

$$a_\mu \to a_\mu + iG^{-1}\partial_\mu G \tag{15.9}$$

The globally gauge invariant Chern–Simons action is given by:

$$S_{\text{Ch-S}} = i\theta \int d^3x \epsilon_{\mu\nu\lambda} \text{Tr}\left[a_\mu \left(\partial_\nu a_\lambda + \frac{1}{3}[a_\nu, a_\lambda] \right) \right] \tag{15.10}$$

Under the transformation (15.9) this changes to:

$$\delta S_{\text{Ch-S}} = -\frac{\theta}{6} \int d^3x \epsilon_{\mu\nu\lambda} \text{Tr}(G^{-1}\partial_\mu G G^{-1}\partial_\nu G G^{-1}\partial_\lambda G) \tag{15.11}$$

Notice the difference from the Abelian case where the global change of the action vanishes! Here it does not vanish, but the global gauge invariance can nevertheless be preserved for certain values of θ. The reason is that the invariant quantity is not the action, but the partition function which contains $\exp(-\delta S_{\text{Ch-S}})$. The change of action given by the integral (15.11) is purely imaginary and if it is proportional to $2\pi i$, the partition function does not change. The integral (15.11) is the integral of the total derivative, the Jacobian of the transformation from space coordinates to the coordinates of the group, and as such depends only on the global topological properties of G (see the detailed discussion of this in Chapter 32). It is a

---

[4] Here I assume that the system is closed; in the presence of boundaries one has to include additional boundary terms.

well known fact that (15.11) is proportional to an integer number – the number of times the field G maps to the three-dimensional space: $\delta S_{\text{Ch-S}} = \mathrm{i}8\theta\pi^2 \times$ (integer). So, if $\theta = n/4\pi$, the theory is well defined.

As I have said, the Chern–Simons term may appear only if parity is broken. For instance, parity is broken in a magnetic field which provides a simple mechanism for generation of the Chern–Simons term. Let us consider free electrons with a quadratic spectrum on a plane in an external magnetic field $\mathbf{H}$ directed perpendicular to the plane. Besides this uniform magnetic field, there is a fluctuating electromagnetic field (it is always around), whose effective action we want to calculate. From elementary quantum mechanics, we know that in an external magnetic field electrons have a discrete spectrum consisting of equidistant levels (the Landau levels) and each level has the degeneracy

$$q = \frac{\Phi}{\Phi_0}$$

where Φ is the *total* magnetic flux through the plane and $\Phi_0 = 2\pi c\hbar/e$ is the flux quantum. Below we shall work in the system of units where $e = c = \hbar = 1$. Then for N filled Landau levels, the total number of electrons is equal to

$$n = N\frac{\Phi}{2\pi} = \frac{N}{2\pi}\left(HS + \int \mathrm{d}^2x\, F_{xy}\right) \tag{15.12}$$

where F_{xy} is the fluctuating electromagnetic field. However, the number of particles, being the electric charge, is given by

$$n = \mathrm{i}\frac{\partial F}{\partial A_0} \tag{15.13}$$

Comparing these two expressions we deduce that the effective action for the electromagnetic field in the long-wavelength limit, obtained as a result of integration over fermions, contains the Chern–Simons term with

$$\theta = -\frac{N}{2\pi} \tag{15.14}$$

Another example of a theory with a broken parity which generates a Chern–Simons term is $(2+1)$-dimensional quantum electrodynamics with *massive chiral* fermions (chiral quantum electrodynamics).[5] The word 'chiral' means that only fermions with definite parity are present. As was mentioned above, in $(2+1)$ dimensions one can choose two- or four-dimensional representations for γ matrices. Two-dimensional representations are not parity invariant; in fact they transform into each other under the parity transformation $t \to t$, $x \to x$, $y \to -y$. In chiral quantum electrodynamics, the fermions are represented by two-dimensional spinors and the Lagrangian has the following form:

$$L = \int \mathrm{d}^2x \left[\frac{F_{\mu\nu}F^{\mu\nu}}{8\pi} + \bar{\psi}\sigma(\nabla + \mathrm{i}e\,\mathbf{A})\psi + M\bar{\psi}\psi\right] \tag{15.15}$$

---

[5] $(2+1)$-Dimensional quantum electrodynamics has many condensed matter applications; see, for instance, Khveshchenko (2001), Franz and Tesanović (2001) and references therein.

To derive an effective action for the gauge field we need to integrate over the fermions. This is not possible to carry out exactly, as in $(1 + 1)$ dimensions, but only approximately, at small e^2. It is more convenient to perform the calculations in real space where the fermionic Green's functions have the following form:

$$\langle\langle\psi(1)\bar{\psi}(2)\rangle\rangle = \frac{1}{4\pi}(-\sigma\nabla_1 + M)\frac{e^{-M|r_{12}|}}{|r_{12}|} \tag{15.16}$$

In this notation it becomes very easy to calculate the polarization operators

$$\Pi_{\mu\nu} \equiv \langle\langle J_\mu(-q)J_\nu(q)\rangle\rangle = \pm\frac{1}{2\pi}i\epsilon_{\mu\nu\lambda}q^\lambda f_1(|q|/2M) + (\delta_{\mu\nu} - q_\mu q_\nu/q^2)|q|f_2(|q|/2M) \tag{15.17}$$

where

$$f_1(x) = \frac{\tan^{-1}x}{x}$$

The sign of the first term depends on the parity. The function $f_2(x)$ possesses the following properties:

$$f_2(x) \sim \begin{cases} x & x \ll 1 \\ 1 & x \to \infty \end{cases}$$

The integration over fermions cannot be performed exactly even at $M = 0$. Nevertheless, it can be shown that subsequent corrections do not introduce qualitative changes into the behaviour of the polarization loops. In particular, the coefficient in front of the Chern–Simons term does not renormalize. The resulting effective action at small q has the following form:

$$S_{\text{eff}}[A] = \frac{1}{2}\sum_q F_{\mu\nu}(-q)\left(\frac{1}{4\pi} + \frac{e^2}{|q|}f_2(|q|/2M)\right)F^{\mu\nu}(q) \pm \frac{e^2}{2\pi}\epsilon_{\mu\nu\lambda}F^{\mu\nu}(-q)A^\lambda(q) \tag{15.18}$$

It is instructive to calculate the pair correlation function of the electromagnetic fields with this effective action. Let us rewrite it in terms of the vector potentials:

$$S_{\text{eff}} = \frac{1}{8\pi}\sum_q A_\mu(-q)[4ie^2\epsilon_{\mu\nu\lambda}q_\lambda + (\delta_{\mu\nu} - q_\mu q_\nu/q^2)q^2\epsilon(q)]A_\nu(q) \tag{15.19}$$

where $\epsilon(q) = 1 + e^2|q|^{-1}f_2(q)$. The problem of finding the correlation functions is a little tricky due to the gauge invariance which makes it impossible to invert the matrix in the square brackets. It turns out, however, that one can find a matrix $D_{\mu\nu}(q)$ such that

$$[4i\epsilon_{\mu\nu\lambda}q^\lambda e^2 + (\delta_{\mu\nu} - q_\mu q_\nu/q^2)\epsilon q^2]D_{\nu\rho}(q) = (\delta_{\mu\rho} - q_\mu q_\rho/q^2) \tag{15.20}$$

I shall look for this matrix in the following form:

$$D_{\mu\nu}(q) = (\delta_{\mu\nu} - q_\mu q_\nu/q^2)D_1 + \epsilon_{\mu\nu\lambda}q^\lambda D_2 \tag{15.21}$$

and use the following identities:

$$(\delta_{\mu\nu} - q_\mu q_\nu/q^2)(\delta_{\nu\rho} - q_\nu q_\rho/q^2) = (\delta_{\mu\rho} - q_\mu q_\rho/q^2)$$
$$(\delta_{\mu\nu} - q_\mu q_\nu/q^2)\epsilon_{\nu\rho\lambda}q_\lambda = \epsilon_{\mu\rho\lambda}q_\lambda \tag{15.22}$$
$$\epsilon_{\mu\nu\lambda}q_\lambda\epsilon_{\nu\rho\eta}q_\eta = -q^2(\delta_{\mu\rho} - q_\mu q_\rho/q^2)$$

After simple calculations we get

$$D_1(q) = \frac{\epsilon q^2}{16e^4 + (\epsilon q^2)^2}$$

$$D_2(q) = -\frac{4ie^2}{16e^4 + (\epsilon q^2)^2} \tag{15.23}$$

So we see photons in this theory have a mass:

$$m^2 = 4e^2/\epsilon(0)$$

The general expression for the Chern–Simons term for $(2+1)$-dimensional noninteracting fermions in a uniform periodic potential with the Green's function $G(\omega, k)$ was derived by Volovik (1988) (see also Yakovenko, 1990):

$$2\pi\theta = \frac{\epsilon_{\mu\nu}}{8\pi^2}\text{Tr}\int d\omega d^2k\, G\frac{\partial}{\partial\omega}G^{-1}G\frac{\partial}{\partial k_\mu}G^{-1}G\frac{\partial}{\partial k_\nu}G^{-1} \tag{15.24}$$

This expression is a topological invariant of the matrix function $G(\omega, k)$ and takes integer values. This expression is very convenient for practical calculations.

The index theorem

In order to elucidate further the physical meaning of the Chern–Simons term, let us consider a simple problem of $(2+1)$-dimensional relativistic massive fermions in a static magnetic field. The Dirac equation has the following form:

$$[E\sigma^3 + \sigma^+(2\partial + i\bar{A}) + \sigma^-(2\bar{\partial} + iA) + M]\psi = 0 \tag{15.25}$$

where ψ is a two-component spinor and $A = A_x + iA_y$, $\bar{A} = A_x - iA_y$. Now I use the fact that any two-dimensional vector can be represented as

$$A_\mu = \partial_\mu\alpha + \epsilon_{\mu\nu}\partial_\nu\chi \tag{15.26}$$

such that (15.25) becomes

$$[E\sigma^3 + 2\sigma^+(\partial + i\partial\alpha - \partial\chi) + 2\sigma^-(\bar{\partial} + i\bar{\partial}\alpha + \bar{\partial}\chi) + M]\psi = 0 \tag{15.27}$$

The function α can be removed from the equation by the gauge transformation of the wave function:

$$\psi \to e^{-i\alpha}\psi \tag{15.28}$$

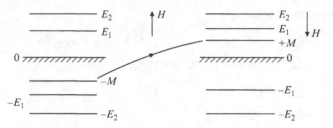

Figure 15.3. A schematic picture of the spectrum (15.30).

The function χ is determined by the magnetic field $H = \epsilon_{\mu\nu}\partial_\mu A_\nu$:

$$\chi = -[4\partial\bar{\partial}]^{-1} H \tag{15.29}$$

Rewriting the Dirac equation in components, we get

$$(M - E)\psi_L + 2(\bar{\partial} + \bar{\partial}\chi)\psi_R = 0$$
$$2(\partial - \partial\chi)\psi_L + (M + E)\psi_R = 0 \tag{15.30}$$

Suppose that on average $\chi > 0$ which is true if the total magnetic flux through the system is *positive*. Then provided $E \neq -M$, one can express ψ_R in terms of ψ_L from the second equation

$$\psi_R = -\frac{2}{M + E}(\partial - \partial\chi)\psi_L$$

and substitute the result in the first equation:

$$[M^2 - E^2 - 4(\bar{\partial} + \bar{\partial}\chi)(\partial - \partial\chi)]\psi_R = 0 \tag{15.31}$$

Since the obtained equation contains E^2, the eigenspectrum consists of an equal number of positive and negative energy levels. This is true for all levels but one. Namely, the above procedure does not work if $E = -M$, and the solution with $E = -M$ (a zero mode) does exist:

$$E = -M \qquad \psi_L = 0 \qquad \psi_R = f(z)e^{-\chi} \tag{15.32}$$

where $f(z)$ is an analytic function. The only condition on this function is that ψ_R is normalizable. One possible basis in the space of analytic functions is the basis of simple powers: $f_n(z) = z^n$. Then the condition of normalizability requires that the maximal power K is

$$K = \Phi/\Phi_0 \tag{15.33}$$

For the negative flux we also have a zero mode, but with the opposite energy and parity:

$$E = M \qquad \psi_R = 0 \qquad \psi_L = f(\bar{z})e^{\chi} \tag{15.34}$$

A schematic picture of the spectrum is shown in Fig. 15.3. When we change the sign of the flux, all states from the zero mode cross the chemical potential. Therefore the total charge of the system changes to an amount equal to the degeneracy of the zero mode.

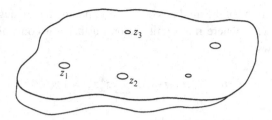

Figure 15.4. A plane with holes.

For a system with a trivial topology this degeneracy is equal to Φ/Φ_0 as we have already discussed. It is worth making a remark about nontrivial topologies. Suppose we are on a manifold with 'holes', where the wave function vanishes at certain points $(z_1, z_2, \ldots, z_N)$ (see Fig. 15.4); in a real system this can occur due to a strong potential scattering. Then the functions $f(z)$, $\bar{f}(\bar{z})$ vanish at $(z_1, z_2, \ldots z_N)$. Therefore N solutions disappear and the degeneracy reduces to

$$K = \Phi/\Phi_0 - N \qquad (15.35)$$

This means that for $\Phi/\Phi_0 < N$ there are no states at the lowest Landau level at all. Thus impurities have a very adverse effect on the Chern–Simons term.

Equation (15.35) is called the *index theorem*.

Relativistic-like spectra are not as frequent in two dimensions as in one. There are quite a few systems, however, where such spectra can exist. Among them are two-dimensional graphite sheets (Semenoff, 1984), semiconductor heterojunctions made between semiconductors with inverted band symmetry (Volkov and Pankratov, 1985) and two-dimensional d-wave superconductors. Many problems related to Chern–Simons terms and $(2+1)$-dimensional fermions are discussed in the books by Fradkin and by Wilczek, references to which are given in the select bibliography.

Quantum Hall ferromagnet

One interesting application of these ideas emerges in the so-called quantum Hall ferromagnet.

Let us first consider how the presence of finite magnetization affects the dynamics of electrons in a ferromagnet. I will also assume that the system is placed in an external magnetic field. The Lagrangian density is

$$\mathcal{L} = \psi_\sigma^+ \left[\partial_\tau - \frac{1}{2m^*} \left(i\nabla - \frac{e}{c}\mathbf{A} \right)^2 + \mu \right] \psi_s - \frac{g}{2}(\psi^+\sigma\psi)^2 \qquad (15.36)$$

A convenient treatment of the ferromagnetic state is achieved when one decouples the interaction by the Hubbard–Stratonovich transformation:

$$-\frac{g}{2}(\psi^+\sigma\psi)^2 \rightarrow \frac{\mathbf{M}^2}{2g} + \psi^+\sigma\mathbf{M}\psi \qquad (15.37)$$

Assuming that fluctuations of the modulus of the order parameter are negligible, I represent it as $\mathbf{M}(x, \tau) = M\mathbf{n}(x, \tau)$ where $\mathbf{n}$ is a unit vector. Further it is convenient to parametrize a unit vector as follows:

$$\mathbf{n} = z_\alpha^* \boldsymbol{\sigma}_{\alpha\beta} z_\beta \qquad \sum_{\alpha=1}^{2} z_\alpha^* z_\alpha = 1 \tag{15.38}$$

Now let us introduce new Grassmann fields $C_{1,2}$:

$$\psi_\alpha = g_{\alpha\beta} C_\beta \qquad g_{\alpha\beta} = \begin{pmatrix} z_\uparrow & -z_\downarrow^* \\ z_\downarrow & z_\uparrow^* \end{pmatrix} \tag{15.39}$$

where g is an SU(2) matrix chosen in such a way that

$$\psi^+ \boldsymbol{\sigma} \mathbf{M} \psi = M(C_2^+ C_2 - C_1^+ C_1) \tag{15.40}$$

We see that the transformation (15.39) puts electrons into a coordinate frame connected to the vector field $\mathbf{n}$: spins of C_1 particles are aligned parallel and spins of C_2 particles antiparallel to $\mathbf{n}$. This transformation diagonalizes the potential, but since $\mathbf{n}$ changes in space and time, derivatives of z, z^* emerge in the transformed Lagrangian. Indeed, in terms of these new fields the fermionic part of the Lagrangian (15.36) becomes

$$\mathcal{L}_{\text{ferm}} = C^+ \left[(\partial_\tau + g^{-1}\partial_\tau g) + \frac{1}{2m^*} \left(i\nabla + ig^{-1}\nabla g - \frac{e}{c}\mathbf{A} \right)^2 + \mu - M\sigma^3 \right] C \tag{15.41}$$

We see that the magnetization has two effects. As might be expected, it changes the chemical potential for the fermions with different spin, but besides that *inhomogeneous* changes of magnetization generate effective scalar and vector potentials.

An especially interesting situation arises in the so-called *quantum Hall ferromagnet* (the term coined by A. MacDonald). In this case the ferromagnetism develops in a two-dimensional electron gas placed in a strong magnetic field such that only the first Landau level is occupied. It is assumed that the cyclotron energy $\hbar He/2m^*c$ is much greater than the Zeeman splitting $g_L e H/mc$ which is valid for GaAs samples where the effective mass is much smaller than the bare electron mass: $m^* \approx 0.02m$. The Zeeman splitting can be further reduced; by applying pressure one can reduce the Landee factor g_L. In these conditions one can achieve a situation where the Zeeman splitting is the smallest energy scale in the problem, being not only much smaller than the distance between the Landau levels, but also much smaller than the exchange interaction. Then the spin degeneracy is lifted by creation of a ferromagnetic state with finite magnetization. Furthermore, in the situation where only one Landau level is filled the electron gas will be completely spin polarized, that is all electron spins are aligned along the magnetization (not along any fixed axis!). In other words all energy levels of C_2 fermions are empty. These fermions can be integrated out giving the following contribution to the effective action:

$$S_{\text{mag}} = \frac{1}{2} \Omega_\mu^+ \langle\langle C_1^+ C_2 C_2^+ C_1 \rangle\rangle \Omega_\mu^- = \frac{\rho}{2} \int d\tau d^2 x (\nabla \mathbf{n})^2 \tag{15.42}$$

where $\Omega_\mu^a = \frac{1}{2}i\text{Tr}\sigma^a g^+ \partial_\mu g$ (σ^a are the Pauli matrices) and ρ depends on the parameters of the bare Hamiltonian. This is not the only contribution; the fermions C_1 experience the net effective magnetic field

$$B_{\text{eff}} = B + i\epsilon_{\mu\nu}\partial_\mu(z_\sigma^* \partial_\nu z_\sigma) = B + \epsilon_{\mu\nu}\left(\mathbf{n}[\partial_\mu\mathbf{n} \times \partial_\nu\mathbf{n}]\right) \qquad (15.43)$$

Since according to (15.12) the net charge is proportional to the total flux, the deviation from the Landau level filling $\nu = 1$ must be accompanied by creation of a spin texture with a finite *topological charge*:

$$1 - \nu = \frac{1}{8\pi}\int d^2 x \epsilon_{\mu\nu}\left(\mathbf{n}[\partial_\mu\mathbf{n} \times \partial_\nu\mathbf{n}]\right) \qquad (15.44)$$

Therefore one can create a spin texture by changing the external magnetic field. The total action for the quantum Hall ferromagnet consists of the action for a *conventional* ferromagnet (see Chapter 16) with an additional contribution proportional to the topological charge. The Lagrange multiplier at this term should be chosen in such a way that condition (15.44) is satisfied. The reader can find more detailed information about quantum Hall ferromagnets in the original publications by Sondhi *et al.* (1993) and MacDonald *et al.* (1996).

References

Franz, M. and Tesanović, Z. (2001). *Phys. Rev. Lett.*, **87**, 257003.
Khveshchenko, D. V. (2001). *Phys. Rev. Lett.*, **87**, 246802.
MacDonald, A. H., Fertig, H. A. and Brey, L. (1996). *Phys. Rev. Lett.*, **76**, 2153.
Semenoff, G. (1984). *Phys. Rev. Lett.*, **53**, 2449.
Sondhi, S. L., Karlshede, A., Kivelson, S. A. and Rezayi, E. H. (1993). *Phys. Rev. B*, **47**, 16419.
Volkov, B. A. and Pankratov, O. A. (1985). *JETP Lett.*, **42**, 145.
Volovik, G. E. (1988). *Zh. Eksp. Teor. Fiz.*, **94**, 123 (*Sov. Phys. JETP*, **67**, 1804 (1988)).
Yakovenko, V. M. (1990). *Phys. Rev. Lett.*, **65**, 251.

III

Strongly fluctuating spin systems

Introduction

One of the fundamentally nonlinear problems we have not yet discussed is the problem of interacting quantum spins. The nonlinearity is embedded in the commutation relations of spin operators; in contrast with Bose or Fermi creation and annihilation operators, commutators of spin operators are not c-numbers, but are still operators. Therefore even apparently very simple spin models, such as the Heisenberg model, for which the Hamiltonian is quadratic in spins, may exhibit complicated dynamics. In fact the Heisenberg model describes the majority of phenomena occurring in magnets, such as various types of magnetic ordering, spin-glass transitions, etc. In the traditional approaches spins are treated as almost classical arrows weakly fluctuating around some fixed reference frame. This reference frame is defined by the existing global magnetic order. For example, in ferromagnets or antiferromagnets, where the global order specifies only one preferential direction (the direction of average magnetization or staggered magnetization), it is supposed that spins fluctuate weakly around this direction. In helimagnets there are two preferential directions describing a spiral; their spins fluctuate around the spiral configuration. When deviations of spins from average positions are small, the spin operators can be approximated by Bose creation and annihilation operators. This approach is called the spin-wave approximation. As I have already said, it is based on two assumptions: existence of a global reference frame and smallness of fluctuations. Difficulties arise when fluctuations become strong and destroy the global order.

In this part of the book I will primarily discuss disordered magnetic systems. However, I shall not touch on such complicated sources of disorder as quenched randomness. In my discussion of disordered magnetism I consider only magnets described by ideally periodic Hamiltonians. There are quite a few experimental examples of disordered magnets of this type. Some of them are shown in the table (I have slightly modified the table presented by Art Ramirez at the Kagome workshop at the NEC Research Institute in January 1992). In fact, many of the systems displayed there do order. The corresponding phase transitions, however, always occur at temperatures much smaller than the characteristic interaction between spins. Ramirez et al. (1991) suggested the characterization of magnets with 'delayed' magnetic ordering by a special number f, the frustration function, defined as the ratio of the Curie–Weiss temperature $-\Theta_{CW}$ to the temperature of actual ordering T_c:[1]

$$f = -\Theta_{CW}/T_c$$

---

[1] Since we are interested in antiferromagnets, Θ_{CW} is always negative.

The idea is that the Curie–Weiss temperature determined from the high temperature behaviour of the uniform magnetic susceptibility

$$\chi = \frac{A}{T - \Theta_{CW}}$$

measures the characteristic exchange integrals:

$$\Theta_{CW} = \frac{2}{3} Z S(S + 1) J_{nn}$$

(Z is the number of nearest neighbours and S is the value of spin). If $f \gg 1$ there is a region
$T_c \ll T \ll -\Theta_{CW}$ where the system exists in a strongly correlated phase without a spin
order. It is customary to call this region the 'spin fluid' state. Below I give an incomplete
list of spin fluids with $f > 10$.

The main subject of this book is application of quantum field theory to condensed matter
physics. As I have mentioned before, field theory is useful mostly when a continuous
description can be formulated. In other words it describes a part of the excitation spectrum
where energies are much smaller than the excitation bandwidth. Not all frustrated magnets
fall into this category; there are many where the spectral gap is of the order of the exchange
integral. I do not discuss such cases here.

Magnetic systems may remain disordered for various reasons. One source of disorder
is dimensionality: magnetic order associated with breaking of a continuous symmetry in
two spatial dimensions is destroyed by thermal fluctuations, and in one dimension also by
quantum fluctuations. Since real systems are never absolutely one- or two-dimensional,
but are made of chains or planes which always interact, three-dimensional effects can
eventually prevail. In this case the substance orders magnetically at some low temperature.
However, if the interactions are small, one can define a temperature region where a system
behaves like a low-dimensional magnet. For example, in a quasi-one-dimensional spin-1/2
antiferromagnet Sr_2CuO_3, the intrachain exchange integral is as large as $J \sim 1000$ K, but
the Néel temperature is just 5 K. Therefore in the temperature range $T > 10$ K this material
is effectively one-dimensional.

In this Part I concentrate primarily on two-dimensional systems. One-dimensional physics
will be considered separately in Part IV. Among the listed materials the most thoroughly
studied and understood is $LaCuO_4$, a two-dimensional analogue of Sr_2CuO_3. This compound is famous because under doping it becomes a superconductor with an unusually high
transition temperature $T_c \approx 30$ K. Without doping this material is an antiferromagnetic insulator with a very large inplane exchange integral $J \sim 1500$ K. The latter circumstance makes
experimental observations essentially easier. These reveal that the correlation length is very
large and strongly temperature dependent. For systems with a large correlation length one
can modify the spin-wave approach, introducing the local staggered magnetization which
precesses slowly in space. In the subsequent chapters we shall derive an effective action for
the staggered magnetization in one- or two-dimensional antiferromagnets with a large correlation length. This action is given by the familiar O(3) nonlinear sigma model. As we have
discussed in Chapter 9, the sigma models with non-Abelian symmetry in two dimensions

are asymptotically free theories, and studying them is very interesting and instructive. It is even more pleasant since the theoretical predictions are in very good agreement with the experimental data for $LaCuO_4$ and $Sr_2CuO_2Cl_2$.

Material	Lattice	T_c (K)	f	Ground state	Spin
Chain					
$KCuF_3$	—	39	~10	AF	1/2
$Ni(C_2H_8N_2)_2NO_2ClO_4$	—	—	>100	—	1
Sr_2CuO_3	—	5	100–200	AF	1/2
Layered					
$LaCuO_4$	square	250	~10	AF	1/2
VCl_2	triangular	35	13	AF	3/2
$NaTiO_2$	triangular	—	>10^3	—	3/2
$Ga_{0.8}La_{0.2}CuO_2$	triangular	0.7	16	SG	7/2
$SrCr_8Ga_4O_{19}$	Kagome	3	150	SG	3/2
$KCr_3(OH)_6(SO_4)_2$	Kagome	<2	>30	—	3/2
Three-dimensional					
$ZnCr_2O_4$	B-spinel	15	25	AF	3/2
K_2IrCl_6	FCC	3.1	10	AF	5/2
FeF_3	pyrochlore	15	16	AF	5/2
$CsNiFeF_6$	B-spinel	4.7	100	SG	1, 5/2
$MnIn_2Te_4$	dzblende	4	25	SG	5/2
$Gd_3Ga_5O_{12}$	garnet	<0.02	>100	—	7/2
Sr_2NbFeO_6	perovskite	28	30	SG	5/2
Ba_2NbVO_6	perovskite	15	30	SG	3/2

Another source of disorder is geometry. The materials with largest f among those shown in the table are geometrically frustrated: their lattices include triangles or pyramids as their elements. These lattices are especially unfavourable towards antiferromagnetic ordering. Classical spins on a triangular lattice can escape frustration, ordering themselves on 120° angles. For quantum spins the classical order becomes approximate; this approximation is particularly poor for quantum spins $S = 1/2$, since the total spin of a triad cannot be made less than 1/2, which has the same value as the elementary spin. Numerical calculations performed by Elstner et al. (1993) show that the spin $S = 1/2$ Heisenberg model with a nearest-neighbour exchange on a triangular lattice has a small but finite staggered magnetization at $T = 0$. The smallness of the staggered magnetization originates from strong quantum fluctuations typical for small values of spin. The latter is the third source of possible disorder. It is curious that, according to the table, even magnets with large values of spin do not order helimagnetically. It is quite possible that another scenario is realized: spins from the same triad create a state with the minimal possible spin $S = 1/2$ and the low energy sector is described as an effective spin $S = 1/2$ magnet.

The Kagome lattice.

Magnetic phase transitions are not the only ones which may occur in a system of spins. Such systems can undergo the so-called spin-Peierls ordering when the only broken symmetry is a translational one. At zero temperature we can have three different situations: (i) an order of some kind (magnetic, spin-Peierls etc.) exists at $T = 0$, (ii) the system remains disordered even at $T = 0$, (iii) there is a quantum critical point at $T = 0$. In one dimension, only the two latter possibilities remain. In all cases one can use the sigma model description provided the spin correlation length ξ is large compared to the lattice spacing a. In this case there is an intermediate scale $a \ll \Lambda^{-1} \ll \xi$, where one can introduce the local order parameter; for a uniaxial antiferromagnet this would be a unit vector $\mathbf{n}(x, \tau)$ proportional to the staggered magnetization averaged over the area Λ^{-2}. This area includes many lattice sites which allows the use of a continuous description. As I have mentioned above, the effective action for this slow variable is given by the O(3) nonlinear sigma model. There are cases when the slow variable is not staggered magnetization (as, for example, in helimagnets). Then the sigma model has a different symmetry.

In highly frustrated systems one can have other low-lying excitations apart from ordinary magnons. Such situations appear when there are several weakly interacting magnetic sublattices. The Abelian analogue of these excitations is optical phonons. The difference, however, is that non-Abelian modes are almost always strongly interacting. Therefore at low temperatures the sublattices can 'lock', which leads to breaking of the corresponding discrete symmetry and therefore occurs as a second-order phase transition. This phenomenon is discussed in Chapter 18. As a limiting case one can have a situation where the sublattices lock, but the global magnetic order does not emerge even at $T = 0$, i.e. the spin-spin correlation length is short. In this case the 'optical' branch lies below the spectrum of magnons. Apparently, such a situation is realized in the layered antiferromagnet $SrCr_8Ga_4O_{19}$ where magnetic Cr ions belong to the so-called Kagome lattice.

Inelastic neutron scattering in this material shows no singularity in the staggered magnetic susceptibility, but the specific heat measurement gives $C_v \propto T^2$ at low temperatures, which indicates the presence of excitations with a linear dispersion. The divergent susceptibility was discovered by Ramirez *et al.* (1990) who observed that the nonlinear susceptibility

$$\chi_3 = \frac{\partial^3 M}{\partial H^3}$$

is negative and diverges at $T_g \approx 10^{-2} J$. This was the first observation of the spin-glass-type behaviour in a system without quenched randomness. The quasi-elastic neutron scattering experiments performed by Broholm *et al.* (1990) give strong support to this idea, showing a marked increase in intensity at low temperatures which corresponds to the 'freezing' of spins. Spin-glass behaviour without disorder in highly degenerate systems has been proposed by De Sze (1977), Villain (1979) and Obradors *et al.* (1988). It has been suggested that macroscopic ground state degeneracy leads to destruction of thermodynamic equilibrium at low temperatures, as occurs in spin glasses. For Kagome antiferromagnets, mechanisms of spin glasses have been discussed by Chandra *et al.* (1993). The experiments in μSR (muon spin relaxation), however, demonstrate that below T_g the muon spin relaxation rate behaves in a way quite different from that of spin glasses (Uemura *et al.* (1993)). Thus, we have to admit that, despite intensive theoretical efforts, the problem of Kagome antiferromagnets still remains unsolved.

References

Broholm, C., Aeppli, G., Espinosa, G. P. and Cooper, A. S. (1990). *Phys. Rev. Lett.*, **65**, 3173; *J. Appl. Phys.*, **69**, 4968 (1990).

Chandra, P., Coleman, P. and Ritchey, I. (1993). *J. Phys. I (Paris)*, **3**, 591.

De Sze, L. (1997). *J. Phys. C: Solid State Phys.*, **10**, L353.

Elstner, N., Singh, R. R. and Young, A. P. (1993). *Phys. Rev. Lett.*, **71**, 1629.

Obradors, X., Labarta, A., Isalgue, A., Tejada, J., Rodriguez, J. and Pernet, M. (1988). *Solid State Commun.*, **65**, 189.

Ramirez, A. P., Espinosa, G. P. and Cooper, A. S. (1990). *Phys. Rev. Lett.*, **64**, 2070.

Ramirez, A. P., Jager-Waldau, R. and Siegrist, T. (1991). *Phys. Rev. B*, **43**, 10461.

Villain, J. (1979). *Z. Phys. B*, **33**, 31.

Uemura, Y. J., Keren, A., Le, L. P., Luke, G. M., Sternlieb, B. J. and Wu, W. D. (1993). *Proc. Int. Conf. on Muon Spin Rotations μSR-93, Miami, January 1993.*

16

Schwinger–Wigner quantization procedure: nonlinear sigma models

In this chapter I describe the most general path integral approach to spin systems which can be derived from the representation of spin operators suggested by Schwinger and Wigner. This path integral formulation leads naturally to a semiclassical approach to magnetism which encompasses the results of the spin-wave theory and survives the loss of ordered moment. Later in this chapter I shall derive the long-wavelengh effective actions for quantum antiferromagnets.

Schwinger and Wigner suggested representing spin S operators in terms of two Bose creation and annihilation operators b_α^+, b_α:

$$S^a(i) = \frac{1}{2} b_\alpha^+(i) \sigma_{\alpha\beta}^a b_\beta(i) \tag{16.1}$$

$$\sum_\alpha b_\alpha^+(i) b_\alpha(i) = 2S \tag{16.2}$$

(σ^a are the Pauli matrices).

Exercise

Check that expression (16.1) reproduces the commutation relations of the spin algebra and that the constraint (16.2) enforces the condition $\mathbf{S}^2(i) = S(S+1)$.

Now I want to dispense with the operators and introduce a path integral instead. The representations (16.1) and (16.2) suggest the following path integral for spin systems:

$$Z[\eta^a] = \int Db_\alpha^+(j) Db_\alpha(j) D\lambda(j) \exp\left(-\int d\tau \{i\lambda(j)[b_\alpha^+(j)b_\alpha(j) - 2S]\right.$$

$$\left. + b_\alpha^+(j)\partial_\tau b_\alpha(j) + H[S(b^+, b)] - \eta^a S^a\}\right) \tag{16.3}$$

where the constraint (16.2) is enforced by the functional delta-function. Since from now on b^+, b are not operators, but numbers, we can do whatever transformations we want. In particular, it is convenient to integrate over λ, which is achieved by resolving the constraint $b_\alpha^+(j)b_\alpha(j) = 2S$ explicitly in spherical coordinates:

$$b_1 = \sqrt{2S} \exp[i(\phi + \psi)/2] \cos(\theta/2)$$

$$b_2 = \sqrt{2S} \exp[i(\phi - \psi)/2] \sin(\theta/2) \tag{16.4}$$

$$\mathbf{S} = S(\cos\theta, \cos\psi \sin\theta, \sin\psi \sin\theta)$$

Notice that ϕ does not enter into the expression for spin, which reflects the gauge symmetry of the Schwinger–Wigner representation. Therefore the integration over ϕ in the path integral gives an infinite constant which can be safely omitted. The generating functional (16.3) in spherical coordinates acquires the following form:

$$Z[\eta] = \int D\theta D\psi \exp \left(-\int d\tau \{iS \cos \theta(j)\partial_\tau \psi(j) + H[S(\theta, \psi)] - \eta S\} \right) \quad (16.5)$$

An obvious disadvantage of this representation is that it is connected to the unique choice of coordinates which is not always convenient. For spin Hamiltonians with rotational symmetry it would be especially desirable to write the generating functional in a manifestly O(3)-symmetric form. It turns out, however, that the only possible O(3)-symmetric expression for the kinetic energy is nonlocal in spin. This expression is derived using the following trick. First, a new variable $0 < u < 1$ is introduced, such that we define a new continuous vector function $\mathbf{N}(\tau, u; j)$, a unit vector satisfying the following boundary conditions:

$$\mathbf{N}(\tau, u = 1; j) = \mathbf{S}(\tau, j)/S \qquad \mathbf{N}(\tau, u = 0; j) = (1, 0, 0)$$

Then the kinetic energy term in (16.5) can be written as an integral of a total derivative: the Jacobian of the transformation from spherical coordinates (θ, ψ) to plane coordinates (τ, u)

$$\begin{aligned} iS \int d\tau \cos \theta \partial_\tau \psi &= iS \int d\tau \int_0^1 du [\partial_u(\cos \theta \partial_\tau \psi) - \partial_\tau(\cos \theta \partial_u \psi)] \\ &= iS \int d\tau \int_0^1 du (\partial_\tau \psi \partial_u \cos \theta - \partial_\tau \cos \theta \partial_u \psi) \\ &= iS \int d\tau \int_0^1 du \, (\mathbf{N}[\partial_u \mathbf{N} \times \partial_\tau \mathbf{N}]) \end{aligned} \quad (16.6)$$

Since the integrals on the right-hand side depend only on values of (θ, ψ) on the boundary $u = 1$ (the contribution from the boundary $u = 0$ vanishes because $\mathbf{N}(u = 0)$ is just a constant vector), the u-variable is a pure dummy. Its introduction is justified by the fact that the second integral in (16.6) can be written as a local functional of $\mathbf{N}$. Since the kinetic energy (16.6) is purely imaginary, the problem of spin is fundamentally quantum mechanical and the corresponding path integral does not represent a partition function for any classical model.

Continuous field theory for a ferromagnet

When all spins are almost parallel one can easily pass from a lattice to a continuous description. In this case we just replace

$$-\frac{1}{2}JS^2 \sum_e \mathbf{N}(\mathbf{r}_j + a\mathbf{e})\mathbf{N}(\mathbf{r}_j) \approx \text{const} + \frac{\rho_s}{2} \int d^D x (\partial_\mu \mathbf{N})^2 \quad (16.7)$$

where $\rho_s = JS^2Z$ (Z is the number of nearest neighbours) and the effective action is ready:

$$S_{\text{ferro}} = \int d\tau d^D x \mathcal{L}$$

$$\mathcal{L} = iS \int_0^1 du \, (\mathbf{N}[\partial_u \mathbf{N} \times \partial_\tau \mathbf{N}]) + \frac{\rho_s}{2}(\partial_\mu \mathbf{N})^2$$

(16.8)

In spherical coordinates (16.4) this corresponds to

$$\mathcal{L} = iS \cos\theta \partial_\tau \psi + \frac{\rho_s}{2}[(\partial_\mu \theta)^2 + \sin^2\theta(\partial_\mu \psi)^2]$$

(16.9)

In the ordered state where there is a finite average magnetization $\langle \mathbf{N} \rangle$, one can choose its direction as the $\hat{x}$-axis and consider ψ and $\alpha = \pi/2 - \theta$ as small fluctuations. This choice of the quantization axis is dictated by the measure of the path integral (recall that it is proportional to $\sin\theta$). The Lagrangian density for small fluctuations around the ordered state is given by

$$\mathcal{L} = iS\alpha\partial_\tau\psi + \frac{\rho_s}{2}[(\partial_\mu\alpha)^2 + (\partial_\mu\psi)^2]$$

(16.10)

and yields the characteristic quadratic spectrum for the magnons:

$$\omega = S^{-1}\rho_s \mathbf{q}^2$$

(16.11)

Continuous field theory for an antiferromagnet

Now we shall discuss a more difficult problem. Namely, we shall derive the effective action for long-wavelength excitations in D-dimensional antiferromagnets in the vicinity of a ground state with collinear Néel order. This derivation leads to the nonlinear sigma model already familiar to us from Part I. Such a description was first introduced by Haldane (1983, 1985). Its advantage over the standard spin-wave approaches is that it does not require the existence of long-range order; it is enough that the order exists for distances larger than the lattice spacing. For typical magnets this condition is fulfilled in two dimensions at low temperatures $T \ll J$ and even in one dimension provided the spins are large.

The subsequent discussion closely follows the book by Sachdev.

Let us consider an antiferromagnet on a D-dimensional hypercubic lattice with only a nearest-neighbour exchange interaction J. If we are close to Néel order, the spins from neighbouring sites will almost be opposite to each other. Let $\mathbf{n}(\mathbf{r}, \tau)$ be a continuum vector field of unit length which describes the local orientation of the Néel magnetization. Then a local spin field can be written as follows:

$$\mathbf{N}(\tau, \mathbf{r}_j) = (-1)^{j_1 + \cdots + j_D}\mathbf{n}(\tau, r_j)\sqrt{1 - a^{2D}\mathbf{L}^2(\tau, r_j)} + a^D\mathbf{L}(\tau, r_j)$$

(16.12)

where a is the lattice spacing and

$$\mathbf{n}^2 = 1 \qquad (\mathbf{n}\mathbf{L}) = 0 \qquad \mathbf{L}^2 \ll a^{-2D}$$

In order to be able to use the form (16.6), we have to suppose that these relationships are continued onto the u-axis and hold for all values of u. The continued variables $\mathbf{n}(u)$, $\mathbf{L}(u)$

Figure 16.1. Subir Sachdev.

satisfy the same boundary conditions as $N(u)$. Naturally, the finite result is u-independent. The field $\mathbf{L}$ describes small deviations of the local antiferromagnetic (Néel) order. Substituting expression (16.12) into the expression for the energy and expanding the latter in gradients of $\mathbf{n}$ and powers of $\mathbf{L}$, we get

$$\frac{1}{2}JS^2 \sum_e \mathbf{N}(\mathbf{r}_j + a\mathbf{e})\mathbf{N}(\mathbf{r}_j) \approx \frac{1}{2}\int d^D x [\rho_s(\partial_\mu \mathbf{n})^2 + \chi_\perp S^2 \mathbf{L}^2] \qquad (16.13)$$

where the spin stiffness ρ_s and the transverse magnetic susceptibility $\chi_\perp$ in this particular model are given by

$$\rho_s = JS^2 a^{2-D} \qquad \chi_\perp = 2DJa^D \qquad (16.14)$$

The latter expression is model dependent; therefore I shall treat ρ_s and $\chi_\perp$ as formal parameters.

Substituting expression (16.12) into the kinetic energy (16.6) we get:

$$S_{\text{kin}} = S' + iS \int d^D x \int d\tau \int_0^1 du \{(\mathbf{n}[\partial_u \mathbf{n} \times \partial_\tau \mathbf{L}]) + (\mathbf{n}[\partial_u \mathbf{L} \times \partial_\tau \mathbf{n}]) + (\mathbf{L}[\partial_u \mathbf{n} \times \partial_\tau \mathbf{n}])\}$$

$$(16.15)$$

where

$$S' = iS \sum_j (-1)^{j_1+\cdots+j_D} \int d\tau \int_0^1 du \, (\mathbf{n}(j)[\partial_u \mathbf{n}(j) \times \partial_\tau \mathbf{n}(j)]) \qquad (16.16)$$

At first sight it seems that one can safely omit the term S' as it contains the rapidly oscillating prefactor $(-1)^{j_1+\cdots+j_D}$. After all, we have neglected the oscillating terms in (16.13). The term S' is quite special, however. The main difference comes from the fact that S' is purely imaginary. Since the partition function contains $\exp(-S')$, S' has to be calculated modulo $2i\pi$, i.e. with accuracy higher than required for real terms of the action. In one dimension the S'-term was calculated by Haldane (1983):

$$S' = iS \sum_j (-1)^j \int d\tau \int_0^1 du \, (\mathbf{n}(j)[\partial_u \mathbf{n}(j) \times \partial_\tau \mathbf{n}(j)])$$

$$\approx iS \sum_j \int d\tau \int_0^1 du \, ([(\mathbf{n}(2j) - \mathbf{n}(2j-1)][\partial_u \mathbf{n}(2j) \times \partial_\tau \mathbf{n}(2j)])$$

$$\approx i\frac{S}{2} \int dx \int d\tau \int_0^1 du \, (\partial_x \mathbf{n}[\partial_u \mathbf{n} \times \partial_\tau \mathbf{n}]) = -i\frac{S}{2} \int d\tau dx \, (\mathbf{n}[\partial_x n \times \partial_\tau \mathbf{n}]) \quad (16.17)$$

In this derivation I have used the fact that

$$\sum_j [f(2j) - f(2j-1)]g(2j) = \int \frac{dx}{2a} g(x)[a\partial_x f(x) + \mathcal{O}(a^2)] \approx \int dx g(x)\partial_x f(x)$$

Thus the lattice constant cancels in the final result and the action S' has a nonvanishing continuum limit in one dimension. The corresponding expression is called a *topological term*. Being an integral of the Jacobian, this term can be written as an integral over spherical angles:

$$S' = i\frac{S}{2} \int d\tau dx \, (\mathbf{n}[\partial_x \mathbf{n} \times \partial_\tau \mathbf{n}])$$

$$= i\frac{S}{2} \int_0^\pi d\theta \int_0^{2\pi k} d\psi \sin\theta = 2Si\pi k \qquad (16.18)$$

where k is an integer, a number of points on the (x, τ) plane transforming into the same value of

$$\mathbf{n} = (\cos\theta, \sin\theta \cos\psi, \sin\theta \sin\psi)$$

under the transformation $\theta(\tau, x)$, $\psi(\tau, x)$. For integer spins $\exp(-S') = 1$ and the entire term can be omitted from the path integral. For half-integer spins the situation is different: $\exp(-S') = (-1)^k$. It turns out that this difference is very important and leads to dramatically different behaviour for integer and half-integer spins (more about this later). In two dimensions the problem of S' has been addressed by several groups of authors (see the discussion in Chapter 5.9 of Fradkin's book). The conclusion is that this term makes no contribution if the ground state has an antiferromagnetic short-range order. In Part IV I discuss an example of a two-dimensional model where the topological term appears and plays a very important role.

The first term in S_{kin} can be simplified further if we take into account the fact that the vectors $\mathbf{L}$, $\partial_u \mathbf{n}$, $\partial_\tau \mathbf{n}$ are all perpendicular to $\mathbf{n}$ and therefore have a vanishing triple product. Therefore we have

$$S_{\text{kin}} = S' + iS \int d^D x d\tau \int_0^1 du \left\{ \partial_\tau \left(\mathbf{n}[\partial_u \mathbf{n} \times \mathbf{L}] \right) - \partial_u \left(\mathbf{n}[\partial_\tau \mathbf{n} \times \mathbf{L}] \right) \right\} \quad (16.19)$$

The total τ derivative yields zero as all fields are periodic in τ, while the total u derivative yields a surface contribution at $u = 1$. As a result we have:

$$S_{\text{kin}} = S' - iS \int d^D x d\tau \left(\mathbf{L}[\mathbf{n} \times \partial_\tau \mathbf{n}] \right) \quad (16.20)$$

Putting together (16.20) and (16.13) we get the following expression for the generating functional:

$$Z = \int D\mathbf{n} D\mathbf{L} \exp(-S_{\text{eff}}) \quad (16.21)$$

$$S_{\text{eff}} = S' + \frac{1}{2} \int d^D x d\tau \{ \rho_s (\partial_\mu \mathbf{n})^2 + \chi_\perp S^2 \mathbf{L}^2 + 2S\mathbf{L}(i[\mathbf{n} \times \partial_\tau \mathbf{n}] - \mathbf{h}) \} \quad (16.22)$$

where I have added an external uniform magnetic field $\mathbf{h}$. Since the action is quadratic in $\mathbf{L}$ the corresponding path integral can be easily calculated, yielding the effective action for the nonlinear sigma model:[1]

$$S_{\text{eff}} = S' + \frac{1}{2} \int d^D x d\tau \left\{ \rho_s \left[(\partial_\mu \mathbf{n})^2 + \frac{1}{c^2} (\partial_\tau \mathbf{n})^2 \right] + 2i\chi_\perp (\mathbf{h}[\mathbf{n} \times \partial_\tau \mathbf{n}]) \right\} \quad (16.23)$$

where $c = \sqrt{\rho_s \chi_\perp}$ is the spin-wave velocity.

In one dimension the change of coordinates $x_0 = c\tau$, $x_1 = x$ leaves us with the following effective action:

$$S_{\text{eff}} = \frac{1}{2g} \int d^2 x (\partial_\mu \mathbf{n})^2 + i \frac{S}{4} \int d^2 x \epsilon_{\mu\nu} (\mathbf{n}[\partial_\mu \mathbf{n} \times \partial_\nu \mathbf{n}]) \quad (16.24)$$

where $g = \sqrt{\chi_\perp / \rho_s}$ is a dimensionless coupling constant. As I have already said, the last term should be kept only if S is a half-integer. For the simple Heisenberg model with a nearest-neighbour exchange, $g = 2/S$. At $g \approx 1$ the present description becomes quantitatively poor. However, it can still remain qualitatively valid. In Part IV I shall describe alternative approaches to $S = 1/2$ and $S = 1$ cases. The sigma model (16.24) is exactly solvable. For an integer S (that is without the topological term) the solution was obtained by Zamolodchikov and Zamolodchikov (1979) and by Wiegmann (1985). The elementary excitations are massive triplets. The exact solution for half-integer S was obtained by Fateev and Zamolodchikov (1991). The spectrum is massless and excitations carry spin $1/2$.

---

[1] I have taken into account the fact that since $\mathbf{n}\partial_\tau \mathbf{n} = 0$, $[\mathbf{n} \times \partial_\tau \mathbf{n}]^2 = (\partial_\tau \mathbf{n})^2$.

The case of half-integer spins is the most peculiar. Since the topological term depends only on the class of the vector field $\mathbf{n}$, the partition function can be written as a sum of partition functions on different topological classes:

$$Z = \sum_{k=-\infty}^{\infty} (-1)^k Z_k$$

$$Z_k = \int D\mathbf{n}\, \delta\left\{ \frac{1}{8\pi} \int d^2x \epsilon_{\mu\nu}(\mathbf{n}[\partial_\mu \mathbf{n} \times \partial_\nu \mathbf{n}]) - k \right\} \exp\left[-\frac{1}{2g} \int d^2x (\partial_\mu \mathbf{n})^2 \right]$$

(16.25)

Each partial partition function Z_k does not change under small *continuous* variations of $\mathbf{n}$. Therefore the topological term does not affect the perturbation expansion of each Z_k in powers of g! It can be shown that the corresponding contributions to the Gell–Mann–Low function are proportional to $\exp(-2\pi/g)$. It is not at all clear whether it is legitimate to keep such nonanalytical terms alongside an expansion in powers of g. Therefore the effect of the topological term cannot be grasped within a perturbation theory. The standard renormalization group analysis in two loops yields the Gell–Mann–Low equation (recall the discussion in Chapter 8):

$$\frac{dg}{d\ln(\Lambda/k)} = \frac{g^2}{2\pi} + \frac{g^3}{(2\pi)^2} + \mathcal{O}(g^4)$$

$$g(|k| = \Lambda) = 2/S$$

(16.26)

the solution of which grows and becomes of the order of one at

$$|k| \sim \xi^{-1} \sim aS^{-1} \exp(\pi S)$$

(16.27)

At $g > 1$ the perturbation approach fails; it cannot tell us whether the coupling constant grows to infinity, which corresponds to the generation of a spectral gap, or whether the growth terminates at some finite value $g = g^* \sim 1$. In the latter case the low energy spectrum is gapless and the correlation functions decay as a power law. The difference between the two types of behaviour becomes manifest only on sufficiently large scales, $|x| \sim \xi$, where the effective coupling is already of the order of one. Haldane (1983) conjectured that Heisenberg chains with half-integer spins are critical. It is also believed that they are in the same universality class as the $S = 1/2$ chain for which the correlation functions are known exactly (the $S = 1/2$ model is discussed in great detail in Chapters 29 and 31). If this is indeed the case, the pair correlation function of vector fields decays at large distances, $x^2 + c^2\tau^2 \gg \xi^2$, as the correlation function of staggered magnetization in the $S = 1/2$ isotropic Heisenberg chain:

$$\langle \mathbf{n}(x, \tau)\mathbf{n}(0, 0) \rangle \sim \left(\frac{\xi^2}{x^2 + c^2\tau^2} \right)^{1/2}$$

(16.28)

Our confidence in Haldane's conjecture is supported by numerical studies on finite (but large) chains (Moreo, 1987; Ziman and Shultz, 1987).

From this equivalence we can figure out how the topological term affects the form of the Gell–Mann–Low function at $g \sim 1$. The Gell–Mann–Low equation for the $S = 1/2$

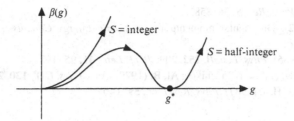

Figure 16.2. The Gell–Mann–Low function for a one-dimensional quantum antiferromagnet.

Heisenberg chain is given by

$$\frac{d\tilde{g}}{d\ln(\Lambda/k)} = -\frac{1}{2\pi}\tilde{g}^2$$

$$\tilde{g}(|k| = \Lambda) = \tilde{g}_0 \sim 1$$

(16.29)

with the solution

$$\tilde{g} = \frac{\tilde{g}_0}{1 + \frac{1}{2\pi}\tilde{g}_0 \ln(\Lambda/k)} \qquad g_0 \sim 1$$

(16.30)

which vanishes at small $|k|$. It is natural to suppose that $\tilde{g}$ is proportional to $g^* - g$. Then from (16.30) it follows that in the vicinity of the critical point we have the following beta-function:

$$\beta(g) = -A(g^* - g)^2$$

(16.31)

where A is some numerical constant. Putting this expression together with the perturbative beta-function given by (16.26) we can draw a graph for the beta-function in the entire interval of couplings (see Fig. 16.2). From this picture it is obvious that the strong coupling limit of the nonlinear sigma model does not depend on the topology: at $g > g^*$ the coupling constant always grows. This is, of course, what one can expect on general grounds. It is worth mentioning that there are isotropic spin $S = 1/2$ models which belong to the disordered phase $g > g^*$. One of the examples is the spin-1/2 Heisenberg antiferromagnet with interaction between nearest and next-nearest neighbours:

$$H = \sum_n \mathbf{S}_{n-1}(\mathbf{S}_n + \gamma \mathbf{S}_{n+1})$$

(16.32)

Its spectrum acquires a gap at $\gamma > 0.24$. At the so-called Majumdar–Ghosh point $\gamma = 1$ the ground state is known exactly (see Affleck *et al.*, 1988).

References

Affleck, I. K., Kennedy, T., Lieb, E. H. and Tasaki, H. (1988). *Commun. Math. Phys.*, **115**, 477.

Fateev, V. A. and Zamolodchikov, A. B. (1991). *Phys. Lett. B*, **271**, 91.

Haldane, F. D. M. (1983). *Phys. Lett. A*, **93**, 464; *Phys. Rev. Lett.*, **50**, 1153 (1983); *J. Appl. Phys.*, **57**, 3359 (1985).

Moreo, A. (1987). *Phys. Rev. B*, **36**, 8582.

Schwinger, J. (1952). On angular momentum. *US Atomic Energy Commission, NYO-3071*, unpublished report.

Wiegmann, P. B. (1985). *Phys. Lett. B*, **152**, 209; *JETP Lett.*, **41**, 95 (1985).

Zamolodchikov, A. B. and Zamolodchikov, Al. B. (1979). *Ann. Phys. (NY)* **120**, 253.

Ziman, T. and Shultz, H. J. (1987). *Phys. Rev. Lett.*, **59**, 140.

O(3) nonlinear sigma model in (2 + 1) dimensions: the phase diagram

This chapter is devoted to the analysis of the O(3) nonlinear sigma model in (2 + 1) dimensions. Why in (2 + 1) dimensions? As we have already seen in Chapter 16, the (1 + 1)-dimensional model has a disordered ground state. This ground state is formed by quantum fluctuations; at low temperatures the correlation length is temperature independent:

$$\xi = Aag \exp(2\pi/g) \tag{17.1}$$

where A is a nonuniversal numerical constant and a is the lattice distance (see the discussion around (9.3), (9.24) and (16.27)). Thus in (1 + 1) dimensions staggered magnetization does not exist. By contrast, in (3 + 1) dimensions, if staggered magnetization exists at $T = 0$, it persists to finite temperatures. The intermediate case of (2 + 1) dimensions is the most peculiar one because a continuous symmetry in this case can be broken at $T = 0$ and is restored by thermal fluctuations at finite T. Whether the system has a disordered or ordered ground state is determined by the parameters of the Hamiltonian; if the correlation length is much longer than the lattice spacing, one can still use the continuum description in terms of the O(3) nonlinear sigma model developed in the previous chapter:[1]

$$S_{\text{eff}} = \frac{\rho_s}{2} \int_0^{1/T} d\tau \int d^2x \left[(\partial_\mu \mathbf{n})^2 + \frac{1}{c^2} (\partial_\tau \mathbf{n})^2 \right] \tag{17.2}$$

In order to establish qualitative features of the phase diagram, I shall consider the $O(N)$-invariant generalization of this model and expand in $1/N$. Rescaling $\rho_s \to \rho_s/N$ and repeating the arguments of Chapter 9, we arrive at the following equation for the saddle point:

$$\rho_s^{-1} T \sum_n \int \frac{d^2k}{(2\pi)^2} \frac{1}{c^{-2}\omega_n^2 + k^2 + m^2} = 1 \tag{17.3}$$

It is easy to see that this integral diverges at large wave vectors, and therefore it is necessary to introduce an ultraviolet cut-off. In our discussion of regularization procedures in Chapter 6 we mentioned only lattice regularizations. There are other methods as well. For example, one can subtract from every diagram the same diagram with large mass. This method is

---

[1] Remember that there is no topological term in (2 + 1) dimensions.

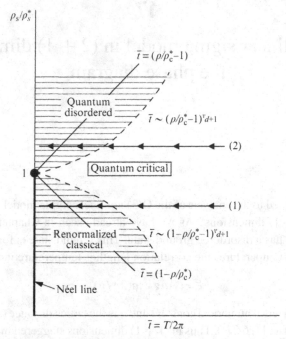

Figure 17.1. Crossover diagram for the O(3) nonlinear sigma model at $D = 2$ (from Chakravarty *et al.*, 1988, 1989). The lines marked (1) and (2) are two possible experimental paths for which the spin-spin correlation length will behave very differently. As the temperature is lowered along path (2) the correlation length becomes temperature independent. Along path (1) it becomes exponentially divergent.

called the Pauli–Willars regularization and it is the most convenient method in the given case. Following this method we rewrite the integral as follows:

$$\rho_s^{-1} T \sum_n \int_0^\infty \frac{d^2 k}{(2\pi)^2} \left(\frac{1}{c^{-2}\omega_n^2 + k^2 + m^2} - \frac{1}{c^{-2}\omega_n^2 + k^2 + \Lambda^2} \right) = 1 \qquad (17.4)$$

Performing the summation over frequencies and integrating over k we get

$$\frac{T}{2\pi\rho_s} \ln \left[\frac{\sinh(\Lambda/2T)}{\sinh(mc/2T)} \right] = 1 \qquad (17.5)$$

or, finally,

$$mc = 2T \operatorname{arcsinh} \left\{ \frac{1}{2} \exp[T^{-1}(\Lambda/2 - 2\pi\rho_s)] \right\} \qquad (17.6)$$

Here I have used the fact that $\Lambda \gg T$.

At small temperatures, (17.6) has three very different solutions. The corresponding phase, or rather crossover, diagram is shown in Fig. 17.1.

Figure 17.2. Bert Halperin.

Ordered phase

First, we have the *renormalized classical* region where the stiffness is large:[2]

$$\rho_s > \rho_s^* \equiv \frac{\Lambda}{4\pi}$$

Then the exponent in square brackets is small at low temperatures and we have

$$mc \approx T \exp\left[\frac{1}{T}(\Lambda/2 - 2\pi\rho_s)\right] = T \exp\left[-\frac{2\pi(\rho_s - \rho_s^*)}{T}\right] \tag{17.7}$$

A careful analysis, as performed by Chakravarty *et al.* (1988, 1989) and then refined by Hasenfratz and Niedermayer (1991), shows that at $N = 3$ this expression should be modified:

$$mc = Aca^{-1} \exp\left[-\frac{2\pi(\rho_s - \rho_s^*)}{T}\right]\left\{1 - \frac{1}{2}\left[\frac{T}{2\pi(\rho_s - \rho_s^*)}\right] + \mathcal{O}(T^2)\right\} \tag{17.8}$$

where A is a dimensionless parameter. Thus at $\rho_s > \rho_s^*$ the correlation length diverges as $T \to 0$ which indicates antiferromagnetic order at $T = 0$. At $T \ll (\rho_s - \rho_s^*)$ the spectral gap remains much smaller than the temperature. The divergences in the diagram expansion come from the region $c^{-1}\omega_n, k \sim m$; since $\omega_n = 2\pi T n$ is always much larger than m except for the case $n = 0$, the fluctuations on large scales are purely static (classical). Therefore to get the effective action for the long-wavelength fluctuations we just consider the time

---

[2] Notice that the critical value of ρ_s is nonuniversal. Therefore the continuous approach is unable to predict whether a given lattice model has a disordered ground state or not.

independent configurations $\mathbf{n}(\tau, x) = \mathbf{n}(x)$ in (17.2):

$$S_{\text{eff}} = \frac{(\rho_s - \rho_s^*)}{2T} \int d^2x (\partial_\mu \mathbf{n})^2 \tag{17.9}$$

and set a new cut-off $\tilde{\Lambda} \sim 1/c[T - 1/2\pi(\rho_s - \rho_s^*)]$. The new cut-off and the renormalized stiffness are the only consequences of quantum fluctuations in this regime. Thus the critical dynamics is described by the two-dimensional O(3) nonlinear sigma model with $g^{-1} = (\rho_s - \rho_s^*)/T$ and the temperature dependent cut-off $\tilde{\Lambda}(T)$. Substituting these quantities into the expression for the correlation length of the two-dimensional O(3)-symmetric nonlinear sigma model (17.1) we reproduce (17.8). Numerical calculations for spin $S = 1/2$ antiferromagnets on a square lattice with a nearest-neighbour exchange J give the following estimates for the parameters of (17.8) (see the discussion in Greven $et\ al.$ (1994)):

$$2\pi(\rho_s - \rho_s^*) = 1.15J \qquad A^{-1} = 0.493a$$

Substituting these values into (17.8) we get the following expression for the reduced correlation length $\xi/a \equiv (mc)^{-1}$:

$$\xi/a = 0.493e^{1.15J/T}[1 - 0.43(T/J) + \mathcal{O}(T^2/J^2)] \tag{17.10}$$

The experiments on the two-dimensional spin $S = 1/2$ antiferromagnet $Sr_2CuO_2Cl_2$ are in excellent agreement with the theoretical expression (17.10) (see Fig. 17.3).

Quantum disordered phase

Now let $\rho_s < \rho_s^*$. At low temperatures the exponent in (17.6) becomes large and the gap becomes almost temperature independent:

$$mc \approx \frac{4\pi}{N}(-\rho_s + \rho_s^*) + \mathcal{O}[\exp(-T^{-1})] \tag{17.11}$$

We are in the strong coupling limit described in Chapter 11. The low-lying excitations are spin $S = 1$ quanta with a spectral gap mc. The $1/N$ corrections change the expression (17.11) (see Sachdev's book referenced in the Select bibliography):

$$mc \sim (\rho_s^* - \rho_s)^\nu$$
$$\nu = 1 - \frac{32}{3\pi^2 N} \tag{17.12}$$

Quantum critical region

Now let us consider the region $|\rho_s - \rho_s^*| \sim T$. The correlation length in this region is of the order of $1/T$. At $N = \infty$ when (16.6) is exact we have $mc \sim T$; this result remains robust at finite N. As usual for critical theories, all physical quantities obey scaling laws; for example, the staggered susceptibility χ_s, the uniform magnetic susceptibility χ_u and the

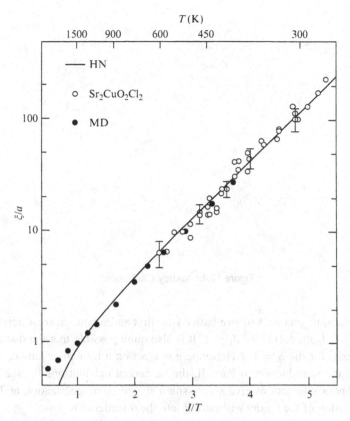

Figure 17.3. Semilog plot of the reduced magnetic correlation length ξ/a versus J/T in $Sr_2CuO_2Cl_2$ (from Greven *et al.* 1994). The solid line is the theoretical prediction of (17.10).

specific heat C_v are given by

$$\chi_s(\omega, q) = T^{-2+\eta} F_s(\omega/T, cq/T, T/mc)$$

$$\chi_u(\omega, q) = T F_u(\omega/T, cq/T, T/mc) \tag{17.13}$$

$$C_v = T^2 F_c(T/mc)$$

where all scaling functions are nonzero at zero arguments. The $1/N$-expansion performed by Chubukov *et al.* (1994) yields the following scaling dimension η:

$$\eta = \frac{8}{3\pi^2 N} - \frac{512}{27\pi^4 N^2} \tag{17.14}$$

The scaling function for the staggered susceptibility is given by

$$F_s(x, y; \infty) = \frac{A}{x^2 - y^2} \tag{17.15}$$

where A is some numerical constant.

It follows from recent numerical calculations by Shultz and Ziman (1992) that the critical and the quantum disordered phases of the O(3) nonlinear sigma model are realized in the

Figure 17.4. Andrey Chubukov.

$S = 1/2$ antiferromagnet on a square lattice with first and second nearest-neighbour interaction (the $J_1 - J_2$ model) at $J_2/J_1 \sim 1$. It is also quite possible that the quantum critical
regime can exist for the $S = 1/2$ Heisenberg magnet on a triangular lattice. As has been
mentioned in the introduction to Part III, the numerical calculations for the model with
nearest-neighbour interactions give a very small staggered magnetization at $T = 0$ and a
slight modification of the model will push it into the disordered regime.

Topological excitations: skyrmions

Let us consider a classical two-dimensional O(3) sigma model and look for minimal
energy configurations with a definite topological charge. Such configurations correspond
to a ground state in the quantum Hall ferromagnet (see Chapter 15) with the Landau level
filling $\nu \neq 1$. The following parametrization turns out to be convenient:

$$w = \cot(\theta/2)\mathrm{e}^{i\psi} = \frac{z_\uparrow}{z_\downarrow} \tag{17.16}$$

In terms of this new field we have

$$(\partial_\mu \mathbf{n})^2 = \frac{\partial w \, \bar{\partial} w^* + \bar{\partial} w \, \partial w^*}{(1 + |w|^2)^2} \tag{17.17}$$

The topological charge,

$$q = \frac{1}{\pi} \int \frac{\partial w \, \bar{\partial} w^* - \bar{\partial} w \, \partial w^*}{(1 + |w|^2)^2} \mathrm{d}^2 x \tag{17.18}$$

is an integer for every configuration with finite energy. Combining (17.17) and (17.18) we can rewrite the energy of the O(3) sigma model as a sum of two nonnegative terms:

$$E = 4\pi q \rho_s + 8\rho_s \int d^2x \frac{|\bar{\partial}w|^2}{(1 + |w|^2)^2} \tag{17.19}$$

For a given q this functional achieves its minimum when the second term vanishes, that is when w is a purely analytic or antianalytic function:

$$w = w(z) \quad \text{or} \quad w = w(\bar{z}) \tag{17.20}$$

A convenient way to parametrize an analytic function which reaches a constant at infinity is in terms of its zeros and poles:

$$w(z) = h \prod_{j=1}^{N} \frac{(z - a_j)}{(z - b_j)} \tag{17.21}$$

Such a solution is called a multi-*skyrmion* solution. The topological charge of this solution is equal to N; its energy is $E = 4\pi\rho N$. In a quite remarkable way it does not depend on the parameters $\{a, b\}$. In other words, in the classical limit skyrmions do not interact. This interaction is generated by fluctuations around the classical solution. In the limit of large skyrmion density

$$\frac{N}{S}\xi^2 \gg 1$$

(here S is the area occupied by the system and ξ is the correlation length of the O(3) sigma model) it is sufficient to take into account only Gaussian fluctuations. The resulting expression for the partition function is (Fateev *et al.* (1979), see also the book by Polyakov):

$$Z = \sum_{N} \frac{e^{-(4\pi\rho - \mu)N/T}}{N!} \int d^2a_j d^2b_j \frac{\prod_{i>j} |a_i - a_j|^2 |b_i - b_j|^2}{\prod_{i,j} |a_i - b_j|^2} \tag{17.22}$$

where μ is the chemical potential of skyrmions. This partition function describes a part of the phase diagram from the quantum Hall ferromagnet described in Chapter 15 (see Green *et al.* (1996) for more detail).

Superficially it may appear that skyrmions are bosons. The following argument shows that this is not the case. We have already established that solutions with nonzero topological charge appear when one changes occupancy of the Landau level in a quantum Hall ferromagnet. As follows from (15.44), changing the net charge by one (that is introducing one electron or hole into the system) changes the topological charge of the **n**-field also by one. This means that by creating an electron one creates a skyrmion, and therefore the skyrmion is a *fermion*. Since a single skyrmion is parametrized by two parameters, in that sense it can be viewed as a combination of two *merons*. The merons however are always confined (this follows from a detailed analysis of the partition function (17.22)).

References

Chakravarty, S., Halperin, B. I. and Nelson, D. R. (1988). *Phys. Rev. Lett.*, **60**, 1057; *Phys. Rev. B*, **39**, 2344 (1989).

Chubukov, A., Sachdev, S. and Ye, J. (1994). *Phys. Rev. B*, **49**, 11919.

Fateev, V. A., Frolov, I. V. and Schwarz, A. S. (1979). *Nucl. Phys. B*, **154**, 1.

Green, A. D., Kogan, I. I. and Tsvelik, A. M. (1996). *Phys. Rev. B*, **53**, 6981.

Greven, M., Birgeneau, R. J., Endoh, Y., Kastner, M. A., Keimer, B., Matsuda, M., Shirane, G. and Thurston, T. R. (1994). *Phys. Rev. Lett.*, **72**, 1096.

Hasenfratz, P. and Niedermayer, F. (1991). *Phys. Lett. B*, **268**, 231.

Shultz, H. and Ziman, T. (1992). *Europhys. Lett.*, **18**, 355.

18

Order from disorder

Some magnetic Hamiltonians have not only continuous symmetry, associated with spin rotations, but also discrete symmetry. This occurs if the system has several equivalent magnetic sublattices; then the discrete symmetry is a symmetry with respect to permutation of the sublattices. If there is a new symmetry, it can be spontaneously broken, thus opening the possibility of a new phase transition. This situation is especially interesting for us because, as will be shown later, the symmetry breaking is driven by the terms of the effective action absent on the classical level. These terms are generated by short-range quantum fluctuations. This mechanism was discovered by Villain (1977, 1980) for the Ising spins and later generalized for continuous spins by Shender (1982) and Henley (1989). The predicted excitation spectrum was observed experimentally by neutron scattering in the spin $S = 5/2$ garnet $Fe_2Ca_3(GeO_4)_3$ (Gukasov *et al.*, 1988). In order not to keep the discussion abstract, let us consider a particular example: the Heisenberg model with spins on two square sublattices shifted with respect to each other in such a way that sites of one sublattice are in the centres of plaquettes of the other (see Fig. 18.1).

This model is not at all unrealistic and describes a two-layered lattice where the two sublattices belong to different layers separated in space in the direction perpendicular to the plane and shifted with respect to each other along the diagonal of the square. The Hamiltonian is given by

$$\hat{H} = \frac{1}{2}J_1 \sum_{x,e_\mu} [S(x)S(x + e_\mu) + S(x + a)S(x + a + e_\mu)] + \frac{1}{2}J_2 \sum_{x,e_\mu^*} S(x)S(x + e_\mu^*)$$

(18.1)

where $e_1 = (1, 0)$, $e_2 = (0, 1)$, $e_1^* = (1/\sqrt{2}, 1/\sqrt{2})$, $e_2^* = (1/\sqrt{2}, -1/\sqrt{2})$. The Hamiltonian is Z_2-invariant, which corresponds to invariance with respect to permutations of the sublattices.

The Hamiltonian (18.1) describes a frustrated system: there is no spin configuration which simultaneously minimizes the energies of all magnetic bonds. Therefore the ground state is achieved via a competition between different configurations. Let us consider the case when the interaction between the sublattices is weak: $\eta = |J_2/J_1| \ll 1$. In this case one can be certain that each sublattice at $T = 0$ develops a Néel order characterized by the staggered magnetization n_i $(i = 1, 2)$. In the limit $\eta = 0$ the sublattices are independent. It is easy to figure out, however, that the classical ground state energy does not depend on

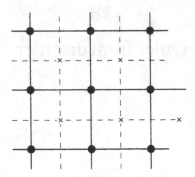

Figure 18.1.

mutual orientation of $\mathbf{n}_i$, even at finite η! Indeed, each spin is surrounded by a quadrangle of spins from a different sublattice whose classical contributions to the energy are cancelled. Of course, the excitation spectrum is affected; gradients of $\mathbf{n}_i$ interact even on the classical level. A dependence of the ground state energy on the orientation arises indirectly, due to the contribution of quantum fluctuations. These quantum fluctuations do depend on $(\mathbf{n}_1\mathbf{n}_2)$ and contribute to the effective action a term $\sim -(\mathbf{n}_1\mathbf{n}_2)^2$. Thus on the quantum level the classical degeneracy is lifted; moreover, the appearance of the new interaction can lead to locking of the sublattices via a spontaneous breaking of the Z_2-symmetry. For the present model this was demonstrated by Chandra et al. (1990), who also showed that thermal fluctuations produce a similar effect.

The principle behind the calculation of the effective action is quite general. Suppose we have a magnet consisting of several sublattices. Suppose further that we have antiferromagnetic order on each sublattice. Then we have to fix the orientations of staggered magnetization on each sublattice and calculate the spin-wave spectrum using a $1/S$-expansion. This spectrum contains zero modes, reflecting the fact that on the classical level these sublattices do not interact and can be rotated with respect to each other. The spectrum of the remaining modes depends on the mutual orientations of the sublattices $\sigma_{ij} = (\mathbf{n}_i\mathbf{n}_j)$ which we consider as slow variables. The effective action for the slow variables arises from quantum fluctuations of the remaining modes:

$$F(\sigma_{ij}) = T \int \frac{d^D q}{(2\pi)^D} \ln\{\sinh[\omega(\sigma_{ij}, q)/2T]\} \qquad (18.2)$$

Below I shall modify this approach to avoid standard spin-wave calculations. We shall see that this modification has a certain disadvantage, but I accept it since it is shorter. Let us separate fast and slow variables in the problem in the same fashion as was done for the ordinary antiferromagnet. That is for each sublattice we write (16.12):

$$\mathbf{N}_i(j) = (-1)^{x+y+i}\mathbf{n}_i(r_j)\sqrt{1 - a^4\mathbf{L}_i^2(r_j)} + a^2\mathbf{L}_i(r_j) \qquad (18.3)$$

where $i = 0, 1$, a is the lattice spacing and

$$\mathbf{n}_i^2 = 1 \qquad (\mathbf{n}_i\mathbf{L}_i) = 0 \qquad \mathbf{L}_i^2 \ll a^{-4}$$

Substituting $\mathbf{S} = S\mathbf{N}$ with $\mathbf{N}$ given by (18.3) into the Hamiltonian (18.1) we get the expression for the potential energy:

$$E = \frac{JS^2}{2} \int d^2x \left[\sum_i (\partial_\mu \mathbf{n}_i)^2 + \eta(\partial_x \mathbf{n}_0 \partial_y \mathbf{n}_1 + \partial_y \mathbf{n}_0 \partial_x \mathbf{n}_1) \right]$$

$$+ \frac{JS^2 a^2}{2} \int d^2x \left[\sum_i \mathbf{L}_i^2 + 2\eta(\mathbf{L}_1 \mathbf{L}_0) \right] \tag{18.4}$$

The kinetic energy can be obtained from the Berry phase (16.6):

$$S_{\text{kin}} = -iS \sum_i \int d^2x d\tau \left(\mathbf{L}_i [\mathbf{n}_i \times \partial_\tau \mathbf{n}_i] \right) \tag{18.5}$$

This is not so interesting for us, however, because the kinetic energy contains time derivatives and vanishes for static configurations which give strongest divergences in two dimensions.

At first sight it seems that ferromagnetic fluctuations are completely decoupled. This is wrong, however, because of the constraints $(\mathbf{n}_i \mathbf{L}_i) = 0$. Let us be careful and recall our measure of integration. The slow variables are $\mathbf{n}_0, \mathbf{n}_1$. Let us introduce the locally orthonormal system of coordinates $\mathbf{e}_i$ ($i = 1, 2, 3$) such that

$$\mathbf{n}_0 = \mathbf{e}_1 \qquad \mathbf{n}_1 = \mathbf{e}_1 \cos\theta + \mathbf{e}_2 \sin\theta \tag{18.6}$$

Then the measure of integration is given by

$$D\mathbf{n}_0 D\mathbf{n}_1 = D\mathbf{e}_1 D\mathbf{e}_2 D\cos\theta \, \delta(\mathbf{e}_1^2 - 1) \delta(\mathbf{e}_2^2 - 1) \delta(\mathbf{e}_1\mathbf{e}_2) \tag{18.7}$$

Now we can write $\mathbf{L}_i$ in the orthonormal basis and satisfy all constraints:

$$\mathbf{L}_1 = a\mathbf{e}_2 + b\mathbf{e}_3 \qquad \mathbf{L}_2 = c(\sin\theta \, \mathbf{e}_1 - \cos\theta \, \mathbf{e}_2) + d\mathbf{e}_3 \tag{18.8}$$

Substituting (18.8) into (18.4) we get

$$\sum_i \mathbf{L}_i^2 + 2\eta(\mathbf{L}_1 \mathbf{L}_0) = a^2 + b^2 + c^2 + d^2 - 2\eta \cos\theta \, ac \tag{18.9}$$

Now the integration over a, b, c, d gives us a nontrivial determinant:

$$\int DaDc \exp\left[-\int d^2x d\tau (a^2 + c^2 - 2ac\eta \cos\theta) \right]$$

$$= \exp\left[-\frac{1}{2} \text{Tr} \ln(I + \sigma^x \eta \cos\theta) \right] \approx \exp\left[\frac{\eta^2}{2(\Delta\tau)a^2} \int d^2x d\tau (\cos\theta)^2 \right] \tag{18.10}$$

where $\Delta\tau$ is some characteristic minimal time interval. This quantity appears because in order to calculate the determinant, one needs to discretize the time integral. The present calculation cannot give the value of $1/\Delta\tau$. The only thing one can be sure about is that $1/\Delta\tau \sim JS$, the magnon bandwidth. The accurate calculation performed by Chandra et al. (1990) gives

$$1/\Delta\tau = 0.26JS$$

Combining together (18.4) and (18.10) we get the following expression for the effective action of the static modes:

$$S_{st} = \frac{1}{2T} \int d^2x \left\{ \rho_s \left[\sum_i (\partial_\mu \mathbf{n}_i)^2 + \eta(\partial_x \mathbf{n}_0 \partial_y \mathbf{n}_1 + \partial_y \mathbf{n}_0 \partial_x \mathbf{n}_1) \right] - 2g(\mathbf{n}_0 \mathbf{n}_1)^2 \right\} \quad (18.11)$$

with $\rho_s = JS^2$, $g = 0.26J Sa^{-2}\eta^2$. We see that even for small η the last term is dominant for small wave vectors $|q|a < \sqrt{ga^2/\rho_s} \sim \eta/\sqrt{S}$.

The simplest way to treat the action (18.11) is as a $1/N$-expansion. The partition function for the generalized $O(N)$-symmetric model is given by

$$Z = \int D\lambda_0 D\lambda_1 D\sigma \, Dn_0 Dn_1 \exp\left[-\int d^2x L(\lambda_i, \sigma, \mathbf{n}_i) \right] \quad (18.12)$$

$$L = \sum_i N\lambda_i \left(\mathbf{n}_i^2 - 1 \right) + 2\sigma N(\mathbf{n}_0 \mathbf{n}_1) + TN\sigma^2/g$$

$$+ \frac{N\rho_s}{T} \left[\sum_i (\partial_\mu \mathbf{n}_i)^2 + \eta(\partial_x \mathbf{n}_0 \partial_y \mathbf{n}_1 + \partial_y \mathbf{n}_0 \partial_x \mathbf{n}_1) \right] \quad (18.13)$$

In the leading order in $1/N$ we substitute fields λ_i, σ with their saddle point values determined from the self-consistency equations:

$$1 = N \int \frac{d^2q}{(2\pi)^2} G_{ii}(q) \, T\langle\sigma\rangle/g = N \int \frac{d^2q}{(2\pi)^2} G_{01}(q) \quad (18.14)$$

where the saddle point Green's function is given by

$$G_{ij}(q) = \frac{1}{NT} \begin{pmatrix} \rho_s \mathbf{q}^2 + T\langle\lambda_0\rangle & T\sigma + \eta\rho_s(q_x q_y) \\ T\sigma + \eta\rho_s(q_x q_y) & \rho_s \mathbf{q}^2 + T\langle\lambda_1\rangle \end{pmatrix} \quad (18.15)$$

After the necessary integrations we get from (18.14) the following equations for the parameters of the saddle point $\langle\lambda_0\rangle = \langle\lambda_1\rangle \equiv \lambda$ and σ:

$$\lambda^2 - \sigma^2 = (\rho_s \Lambda/T)^2 \exp[-8\pi\rho_s/T] \quad (18.16)$$

$$\sigma = \frac{g}{8\pi\rho_s} \ln\left(\frac{\lambda + \sigma}{\lambda - \sigma}\right) \quad (18.17)$$

The solution of these equations exists at

$$g/4\pi\rho_s \gg \frac{\rho_s \Lambda}{T} \exp(-4\pi\rho_s/T) \quad (18.18)$$

and is given by

$$|\sigma| \approx (g/4\pi\rho_s) \ln\left(gT/4\pi\rho_s^2 \Lambda\right) + g/T \quad (18.19)$$

$$\lambda - |\sigma| \approx \frac{1}{2|\sigma|}(\rho_s \Lambda/T)^2 \exp(-8\pi\rho_s/T) \quad (18.20)$$

Thus $\langle \sigma \rangle$ appears below some critical temperature T_c. Below T_c the system chooses between two mutual orientations of the sublattices: $\langle (\mathbf{n}_0 \mathbf{n}_1) \rangle = \pm 1$. Since the transition breaks the Z_2-symmetry one can suggest that it belongs to the universality class of the two-dimensional Ising model.

Well below the transition the system has two different correlation lengths: one is equal to $R = \pi(\lambda + |\sigma|)^{-1} \sim |\sigma|^{-1}$ and the other is $\xi = \pi(\lambda - |\sigma|)^{-1} \gg R$. The first one governs the 'optical' mode, i.e. the mode where sublattice magnetizations deviate from each other. This mode is associated with the order parameter σ. Since the correlation length $R \sim T/g$ becomes very short at low temperatures, these fluctuations become quantum at $T/g \approx T^{-1}$ and the current approach can no longer be used. The second correlation length is a length for those modes where sublattice magnetizations fluctuate coherently. This length grows exponentially with $1/T$; at small temperatures the system is described by the standard O(3) nonlinear sigma model with renormalized ρ_s and Λ.

Now recall that the phase transition we are studying occurs at finite temperatures when the average staggered magnetization is zero. Therefore its existence is not related to the fact that the corresponding Heisenberg model has an ordered ground state. In their paper Chandra *et al.* (1990) came up with the suggestion that one can have a magnetic phase transition without a magnetic moment. We have discussed so far only one possibility, namely when spins of different sublattices couple together creating a scalar order parameter $\sigma = \langle (\mathbf{n}_0 \mathbf{n}_1) \rangle$. The large-$N$ approximation we have used is very favourable to such an order parameter because it includes summation over indices. In general, there can be tensorial order parameters

$$Q_{ab} = \langle S_a(i) S_b(j) \rangle - \frac{1}{3} \delta_{ab} \langle S_a(i) \rangle \langle S_b(j) \rangle$$

which can, in principle, survive a loss of local magnetic moment. Below I consider such a model.

Model of spin nematic

Let us suppose that we have a two-dimensional magnet with three sublattices (the Z_3-symmetry). Each sublattice favours an antiferromagnetic order and on the classical level the sublattices do not interact. The effective action for each sublattice is given by (17.9). Quantum fluctuations generate the interaction between the sublattices; symmetry arguments suggest the following form:

$$S_{\text{int}} = \int d\tau d^2 x \left\{ \lambda (\mathbf{n}_1 [\mathbf{n}_2 \times \mathbf{n}_3])^2 + \alpha \sum_{i,j} (\mathbf{n}_i \mathbf{n}_j)^2 \right\} \tag{18.21}$$

Suppose that $|\alpha| \ll \lambda > 0$ and $\lambda \gg T$. Then quantum fluctuations favour coplanar spin configurations making the spin system rigid. Thus all staggered magnetizations $\mathbf{n}_i$ locally belong to the same plane. This plane is defined by two mutually perpendicular unit vectors $\mathbf{e}_1, \mathbf{e}_2$. Thus we can choose the following parametrization:

$$\mathbf{n}_i = \mathbf{e}_1 \cos \theta_i + \mathbf{e}_2 \sin \theta_i \tag{18.22}$$

with the following gauge fixing condition:

$$\sum_{i=1}^{3} \theta_i = 0 \tag{18.23}$$

Substituting this expression into (16.9) and (18.21) we get

$$\sum_i (\partial_\mu \mathbf{n}_i)^2 = (\partial_\mu \mathbf{e}_1)^2 \sum_i \cos^2 \theta_i + (\partial_\mu \mathbf{e}_2)^2 \sum_i \sin^2 \theta_i + \sum_i (\partial_\mu \theta_i)^2 \tag{18.24}$$

$$\alpha \sum_{i,j} (\mathbf{n}_i \mathbf{n}_j)^2 = \alpha \cos^2 \theta_{ij} \tag{18.25}$$

Suppose now that α is so small that the correlation length of the $\mathbf{e}_1$-, $\mathbf{e}_2$-fields is much shorter than the correlation length of the θ-fields. In this situation we can safely replace the $\sin^2$, $\cos^2$ terms by their averages. The resulting approximate *classical* effective action is

$$S = S(e_1, e_2) + S(\theta_i) \tag{18.26}$$

$$S(e_1, e_2) = \frac{1}{2g} \int d^2x [A(\partial_\mu \mathbf{e}_1)^2 + B(\partial_\mu \mathbf{e}_2)^2] \tag{18.27}$$

$$S(\theta_i) = \frac{1}{2} \int d^2x \left[\frac{1}{g} \sum_i (\partial_\mu \theta_i)^2 - T^{-1}\alpha \sum_{i,j} \cos 2\theta_{ij} \right] \tag{18.28}$$

where

$$A = \left\langle \sum_i \cos^2 \theta_i \right\rangle \qquad B = \left\langle \sum_i \sin^2 \theta_i \right\rangle \tag{18.29}$$

In the limit when the correlation length of θ-fields is infinite we have $\langle \sin 2\theta_i \rangle = \langle \cos \theta_i \rangle = 0$ and $A = B = 3/2$.

   The model $S(e_1, e_2)$ belongs to the class of nonlinear sigma models discussed in Chapter 9.[1] Its correlation length is finite and exponentially large in $1/g$. So our main concern will be about the model (18.28). Substituting into (18.28)

$$\theta_1 = \sqrt{g/2}\left(\frac{1}{\sqrt{3}}\phi_1 + \phi_2\right)$$

$$\theta_2 = \sqrt{g/2}\left(\frac{1}{\sqrt{3}}\phi_1 - \phi_2\right) \tag{18.30}$$

$$\theta_3 = -\sqrt{2g/3}\phi_1$$

we get the action in the canonical form:

$$S(\phi_i) = \frac{1}{2} \int d^2x \sum_{i=1}^{2} (\partial_\mu \phi_i)^2 + \frac{\alpha}{2T}\{\cos(\sqrt{8g}\phi_2)$$

$$+ \cos[\sqrt{g/2}(\sqrt{3}\phi_1 + \phi_2)] + \cos[\sqrt{g/2}(\sqrt{3}\phi_1 - \phi_2)]\} \tag{18.31}$$

---

[1] In fact, it is a particular case of the so-called principal chiral field model.

This effective action is a generalization of the sine-Gordon action discussed in Chapters 23 and 34. Applying to this action the criteria of irrelevance discussed in the same chapters, we can show that the cos terms renormalize to zero if $g > \pi$. At this region the fields ϕ_i are free massless fields. The correlation length is infinite. As is shown in Chapter 23, the relevant fields in the sine-Gordon model give rise to a finite correlation length. Thus at $g < \pi$ we are in a state where all sublattices are locked. Since $g \propto T/J$, this region corresponds to low temperatures. It appears that at sufficiently high temperatures the fluctuations have infinite correlation length and we are in a critical state. This is not true, however, because in our treatment we have not taken into account an important relevant perturbation arising from disclinations (vortices) of the ϕ_i-fields. These perturbations are described in Chapter 24 where we discuss the Kosterlitz–Thouless transition. In the present model vortices are relevant at $g > \pi$. Therefore there is only one point where the model is critical: $g = \pi$. At this temperature the second-order phase transition associated with breaking of the Z_3-symmetry occurs. Thus we have a classical example of a finite temperature phase transition. It can be shown that at the critical point the two-point correlation functions of $Q_{ij} = (\mathbf{n}_i \mathbf{n}_j)$ decay as a power law with the following exponent:

$$\langle Q_{ij}(x) Q_{ij}(y) \rangle \sim \frac{1}{|x - y|} \qquad (18.32)$$

References

Chandra, P., Coleman, P. and Larkin, A. I. (1990). *Phys. Rev. Lett.*, **66**, 88.

Gukasov, A. G., Bruckel, T., Dorner, B., Plakhty, V. P., Prandl, W., Shender, E. F. and Smirnov, O. P. (1988). *Europhys. Lett.*, **7**, 83; Bruecel, Th., Dorner, B., Gukasov, A. G., Plakhty, V., Prandl, W., Shender, E. F. and Smirnov, O. P., *Z. Phys. B*, **72**, 477 (1988).

Henley, C. L. (1989). *Phys. Rev. Lett.*, **62**, 2056.

Shender, E. F. (1982). *Sov. Phys. JETP*, **56**, 178.

Villain, J. (1977). *J. Phys. (Paris)*, **38**, 26; Villain, J., Bidaux, R., Carton, J. P. and Conte, R., *J. Phys. (Paris)*, **41**, 1263 (1980).

19
Jordan–Wigner transformation for spin $S = 1/2$ models in $D = 1, 2, 3$

In this chapter and in the following one I discuss nonsemiclassical approaches to spin systems. All these approaches are unique for spin $S = 1/2$ and are based on the fermionic representation of spin operators. This fact supports the original idea of Anderson (1987) that the $S = 1/2$ case may be special.

Let us consider a simple Heisenberg Hamiltonian

$$\hat{H} = J \sum_{\langle i,j \rangle} \left[\frac{1}{2}(S_i^+ S_j^- + S_i^- S_j^+) + \Delta S_i^z S_j^z \right] \tag{19.1}$$

where the indices i, j denote sites of a D-dimensional lattice, the summation includes nearest neighbours and S^+, S^-, S^z are spin $S = 1/2$ operators. In the following discussion I shall consider $D \le 3$ only.

The spin $S = 1/2$ operators have the following remarkable property: taken on the same site they anticommute

$$\{S_j^+, S_j^-\} = 1$$

which suggests an analogy with Fermi creation and annihilation operators

$$S_j^+ = \psi_j^+ \qquad S_j^- = \psi_j \qquad S_j^z = \psi_j^+ \psi_j - 1/2 \tag{19.2}$$

This analogy fails for different sites: spin operators on different sites do not anticommute, but *commute*. It turns out, however, that it is possible to modify the definitions and reproduce the correct commutation relations of spins in terms of fermions. In general, one should modify (19.2), introducing the phase factors:

$$S_j^+ = \psi_j^+ U(j) \qquad S_j^- = U^+(j)\psi_j \qquad S_j^z = \psi_j^+ \psi_j - 1/2 \tag{19.3}$$

where $U(j)$ is a nonlocal functional of ψ.

For a one-dimensional lattice ($D = 1$) the phase factors were found by Jordan and Wigner back in 1928:

$$S_j^+ = \psi_j^+ \exp\left(i\pi \sum_{k<j} \psi_k^+ \psi_k\right)$$

$$S_j^- = \exp\left(-i\pi \sum_{k<j} \psi_k^+ \psi_k\right) \psi_j \qquad (19.4)$$

$$S_j^z = \psi_j^+ \psi_j - 1/2$$

The Jordan–Wigner transformation establishes equivalence between the spin-1/2 Heisenberg chain and the theory of interacting spinless fermions:

$$\hat{H} = \frac{J}{2} \sum_j [\psi_j^+ \psi_{j+1} + \psi_{j+1}^+ \psi_j + 2\Delta(\psi_j^+ \psi_j - 1/2)(\psi_{j+1}^+ \psi_{j+1} - 1/2)] \qquad (19.5)$$

This equivalence will play an important role in the further discussion of the Heisenberg chain in the chapters devoted to the physics of one-dimensional magnets and metals.

The generalization of the Jordan–Wigner transformation for higher dimensions has been suggested only recently; for $D = 2$ it was done by Fradkin (1989) (some inconsistencies of the original work were corrected by Eliezer and Semenoff (1992)) and for $D > 2$ by Huerta and Zanelli (1993). For $D = 2$ the phase factor is given by

$$U_{2D}(j) = \exp\left[i \sum_{k\neq j} \arg(k, j) \psi_k^+ \psi_k\right] \qquad (19.6)$$

where $\arg(k, j)$ is the angle between $k - j$ and an arbitrary space direction on the lattice. Substituting (19.6) into (19.3) and then (19.3) into the original Heisenberg Hamiltonian (19.1), we get the Hamiltonian of the two-dimensional spin $S = 1/2$ Heisenberg model in the fermionic representation:

$$\hat{H} = \frac{J}{2} \sum_{i,\mu} \left\{ \left[\psi^+(i + e_\mu)e^{iA_\mu(i)}\psi(i) + \text{c.c.}\right] + 2\Delta[\psi^+(i)\psi(i) - 1/2] \right.$$

$$\left. \times [\psi^+(i + e_\mu)\psi(i + e_\mu) - 1/2] \right\} \qquad (19.7)$$

$$A_\mu(i) = \sum_k{}' [\arg(k, i) - \arg(k, i + e_\mu)] \psi_k^+ \psi_k \qquad (19.8)$$

where j marks lattice sites of a two-dimensional lattice, e_μ ($\mu = 1, 2$) are basis vectors of the Brave lattice and $\sum'$ means that the sum does not include terms where the sites in the argument function coincide.

To develop a path integral formulation of the Hamiltonian (19.7) one can use a lattice version of the transmutation of statistics device described in the previous chapters. To generalize the Chern–Simons construction for a lattice, let us introduce a vector potential

$A(i, i + e_\mu) \equiv A_\mu(i)$ defined on links and a scalar potential $A_0(i)$ defined on sites. The corresponding field strength ('magnetic field') is then defined as follows:

$$F_{\mu\nu}(i) = A_\nu(i + e_\mu) - A_\nu(i) - A_\mu(i + e_\nu) + A_\mu(i)$$

This construction is unambiguous only for a rectangular lattice where there is a unique choice of basis vectors e_μ. For a triangular lattice, for example, one has to choose between different possible sets of e_μ, which can lead to certain formal difficulties.

It is convenient to rewrite the above expression for $F_{\mu\nu}$ in Fourier components:

$$F_{\mu\nu}(q) = \epsilon_{\mu\nu} t_\mu^* A_\nu(q)$$

$$t_\mu(q) = e^{iqe_\mu} - 1 \tag{19.9}$$

Then the Lagrangian for the Hamiltonian (19.7) is given by:

$$L = \sum_i \left\{ \bar{\psi}(i)[\partial_\tau + A_0(i)]\psi(i) - \frac{1}{\pi} A_0(i) F_{xy}(i) \right\} - \frac{J}{2} \sum_{i,\mu} \{ [e^{iA_\mu(i)} \bar{\psi}(i + e_\mu)\psi(j) + \text{c.c.}]$$

$$+ 2\Delta[\bar{\psi}(i)\psi(i) - 1/2][\bar{\psi}(j)\psi(j) - 1/2] \} \tag{19.10}$$

where $\bar{\psi}$ is a Grassmann variable corresponding to the operator ψ^+ in the Hamiltonian approach. Indeed, since the Lagrangian is linear in A_0 one can integrate it out and get the constraint:

$$J_0(i) \equiv \bar{\psi}(i)\psi(i) = -\pi F_{xy}(i) \tag{19.11}$$

Substituting here the expression from (19.9) we get the equation for the Fourier components of A_μ:

$$\epsilon_{\mu\nu} t_\mu^*(q) A_\nu(q) = -\frac{1}{\pi} J_0(q) \tag{19.12}$$

This equation can be solved by the following trick. Any two-dimensional vector can be written as a sum of a gradient and a curl:

$$A_\mu(q) = t_\mu^*(q)\phi(q) + \epsilon_{\mu\nu} t_\nu(q)\chi(q) \tag{19.13}$$

Substituting this into (19.12) we get

$$|t_\mu(q)|^2 \chi(q) = -\frac{1}{\pi} J_0(q) \tag{19.14}$$

and eventually

$$A_\mu(q) = -\frac{1}{\pi} \epsilon_{\mu\nu} t_\nu(q) |t_\lambda(q)|^{-2} J_0(q) + \text{gradient terms} \tag{19.15}$$

Although the latter expression coincides with (19.8) only in the continuous limit, it has the same properties with respect to the transmutation of statistics.

In fact, what we have done with the fermions is a gauge transformation: the U rotate the phase of fermions on each site in a prescribed manner. Since it is a gauge transformation, it is natural that the field strength is zero almost everywhere. The nontriviality comes from this 'almost': the field strength is nonzero only at the point where a particle appears. Therefore the corresponding vector potential cannot be removed by a continuous transformation. The main difference between $D = 1, 2$ and $D = 3$ cases is that the three-dimensional Jordan–Wigner transformation requires an extended Hilbert space and non-Abelian gauge transformations. Here, on each site we introduce two noncommuting triads of operators $S_\alpha^a(i)$ (the Greek subscripts take values ± 1; the English superscripts take values $\pm z$). The commutators $S_{\alpha,\beta}(i) = (1/2)[S_\alpha^+(i), S_\beta^-(i)]$ for noncoinciding indices define new operators; in fact, the described operators form a part of the fundamental representation of the SU(4) group. The one-site basis includes four states: $|0\rangle$, no particles; $|\pm\rangle$, one particle with $\alpha = \pm 1$; and $|+-\rangle$, a state with two particles. The fermions also have isotopic indices and the Jordan–Wigner transformation is given by

$$S_\alpha^+(j) = \psi_\alpha^+(j) \exp\left[i \sum_k Q_a(k)\omega^a(k, j; \mathbf{n}_0) \right] \qquad (19.16)$$

$$S_{\alpha,\beta}^z(j) = \frac{1}{2}[1 - \rho(j)]\delta_{\alpha,\beta} - \frac{1}{2}Q_a(j)\tau_{\alpha,\beta}^a \qquad (19.17)$$

where

$$Q_a(j) = \psi^+(j)_\alpha \tau_{\alpha,\beta}^a \psi_\beta(j)$$

is the SU(2) isospin operator and τ^a are the matrices of spin $1/2$. The vector $\omega^a(k, j; \mathbf{n}_0)$ is defined as follows:

$$\omega^a(k, j; \mathbf{n}_0) = \arg(k, j; \mathbf{n}_0) e^a(k, j; \mathbf{n}_0) \qquad (19.18)$$

with $\mathbf{e}$ being a unit vector perpendicular to $j - k$ and $\mathbf{n}_0$. One can check that for $\alpha = \beta$ we have the familiar expression for σ^z:

$$S_{\alpha,\alpha}^z = 1/2 - \psi_\alpha^+ \psi_\alpha$$

(no summation over α).

In applications, where one almost always deals not with the SU(4), but with the SU(2) group, one can get rid of one of the fermions (ψ_2, for example) by adding to the Hamiltonian a term

$$\sum_j h S_{--}^z(j)$$

with $h \to \infty$. Then only the ψ_1 operators remain. The ordinary Heisenberg model (19.1) on a three-dimensional lattice in the fermionic representation has the same form (19.7) as in two dimensions, but with a different vector potential. Since the corresponding expression on the lattice is too bulky, I write the explicit expression for A_μ only in the continuum limit:

$$A_\mu(x) = \int d^3 y \, \epsilon_{\mu\nu a} \frac{(x - y)^\nu}{|x - y|^2} \psi_1^+(y) \tau^a \psi_1(y) \qquad (19.19)$$

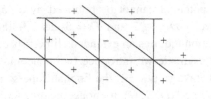

Figure 19.1.

The point non-Abelian charge $\psi_1^+(y)\tau^a\psi_1(y) = e^a\delta^{(3)}(y)$ creates the following vector potential:

$$A_\mu^a(x) = \epsilon_{\mu\nu a}\frac{x^\nu}{|x|^2}$$

It is worth playing a little with the $(2 + 1)$-dimensional model described by the fermionic Lagrangian (19.10) with $\Delta = 0$. Let us write down a simple mean field theory for this Lagrangian, treating the vector potential as time independent. Its configuration will be determined self-consistently as one which minimizes the ground state energy. The idea is that a wave function for fermions in a large magnetic field will be a good approximation for the ground state of a spin fluid. It may seem improbable that charged fermions benefit from large magnetic fields and therefore can generate them spontaneously, but the explicit calculations first performed by Hasegawa et $al.$ (1989) support this idea, demonstrating that the ground state energy of lattice fermions $E(\nu, \phi)$, where ν is the band filling factor and ϕ is the number of flux quanta per elementary plaquette,[1] has its minimum at $\phi = \nu$. For example, for a triangular lattice at half-filling ($\nu = 1/2$) the ground state energy in zero field is equal to $E(1/2, 0) = -0.988J$ and the ground state energy at $\phi = 1/2$ [2] is equal to $E(1/2, 1/2) = -1.201J$.

Since a triangular lattice is one of the most frustrated, I concentrate the discussion on it. Suppose that the fermionic density is uniform, i.e. A_0 is coordinate independent. Then our fermions move in a constant magnetic field in which the flux per elementary plaquette is equal to the band filling (see footnote 1). If the total moment is zero, as we expect it to be, the band is half filled. As I have mentioned above, the ground state energy does have its minimum at $\phi = 1/2$. Let us take a look at what kind of fermionic spectrum we have in this case. It is convenient to represent the triangular lattice as a square lattice with diagonals and choose the gauge where the only negative links are vertical links with even x-coordinate (see Fig. 19.1):

$$\exp[iA_x(i)] = 1$$

---

[1] Notice that the elementary plaquette is an area defined for any two basis vectors of the lattice e_i, e_j as a closed contour spanned by $four$ lattice vectors e_i, e_j, $-e_i$, $-e_j$. Therefore for a triangular lattice, for example, an elementary plaquette contains two triangles.

[2] In the case of a triangular lattice my definition of ϕ differs from that accepted by Hasegawa et $al.$ (1989): $\phi = 2\phi_{\text{Hasegawa}}$.

and

$$\exp[iA_y(i)] = (-1)^{i_x}$$

It turns out that the gauge field on diagonals is not included in the Chern–Simons term (recall the discussion around (19.9)). Since the main purpose of this discussion is to illustrate the ideas, I shall not discuss this subtlety in detail. Let us assume for a moment that the transmutation of statistics procedure can be used for the triangular lattice and study the consequences.

The mean field Hamiltonian for the lattice fermions in such a magnetic field has the following form:

$$
\begin{aligned}
\hat{H} = \sum_i \{ & \psi_A^+(i)[\psi_B(i) + \psi_B(i - x) + \psi_B(i - y) + \psi_B(i - x + y) \\
& - \psi_A(i + y) - \psi_A(i - y)] + \psi_B^+(i)[\psi_A(i) + \psi_A(i + x) \\
& + \psi_A(i + y) + \psi_A(i + x - y) + \psi_B(i + y) + \psi_B(i - y)] \}
\end{aligned}
\tag{19.20}
$$

The summation goes over elementary cells where each cell contains two sites A and B. The Hamiltonian is diagonalized by the Fourier transformation:

$$
\hat{H} = J \sum_k (\psi_A^+(k), \psi_B^+(k)) \begin{pmatrix} -2\cos k_y & 1 + e^{-ik_x} + e^{-i(k_x + k_y)} \\ 1 + e^{ik_x} + e^{i(k_x + k_y)} & 2\cos k_y \end{pmatrix} \begin{pmatrix} \psi_A(k) \\ \psi_B(k) \end{pmatrix}
$$

The fermionic spectrum is given by

$$E^2(k) = J^2[6 - 2\cos 2k_y - 2\cos(k_x + 2k_y) - 2\cos 2k_y] \tag{19.21}$$

and has a large gap $2\sqrt{3}J$ at $\mathbf{k} = 0$. Thus single-particle excitations have spectral gaps. It is reasonable to expect that the collective modes, i.e. excitations of the gauge field, are also gapful. We know already that the excitation spectrum of a U(1) gauge field may contain gaps if the effective action includes the Chern–Simons term. We already have such a term, but its coefficient is chosen in such a way that this term converts statistics of massive particles. Since excitations of the gauge field should be bosons, there should be some mechanism which changes the coefficient. Indeed, as was shown by Yang et al. (1993), the integration over fermions in this particular case doubles the coefficient. This integration also gives rise to the ordinary $F_{\mu\nu}^2$ terms. Thus the collective modes are described by the lattice version of the Chern–Simons quantum electrodynamics. Since the spectral gap is very large and therefore the corresponding correlation length is of the order of the lattice distance, the continuous limit is ill defined. It is reasonable to suggest, however, that the Chern–Simons quantum electrodynamics can describe a hypothetical spin fluid with a small spectral gap:

$$S = \int d\tau d^2 x \left[\frac{\epsilon E^2}{8\pi} + \frac{B^2}{8\mu\pi} + \frac{2}{\pi} A_0 B \right] \tag{19.22}$$

This effective action for spin fluids was first conjectured by the author and Reizer in 1991. Excitations of this model are massive bosons.

References

Anderson, P. W. (1987). *Science*, **235**, 1196.

Eliezer, D. and Semenoff, G. W. (1992). *Phys. Lett. B*, **286**, 118.

Fradkin, E. (1989). *Phys. Rev. Lett.*, **63**, 322.

Hasegawa, Y., Lederer, P., Rice, T. M. and Wiegmann, P. B. (1989). *Phys. Rev. Lett.*, **63**, 907.

Huerta, L. and Zanelli, J. (1993). *Phys. Rev. Lett.*, **71**, 3622.

Jordan, P. and Wigner, E. (1928). *Z. Phys.*, **47**, 631.

Tsvelik, A. M. and Reizer, M. (1991). *Proc. University of Miami Workshop on Electronic Structure and Mechanisms for High Temperature Superconductivity, January 1991.*

Yang, K., Warman, L. K. and Girvin, S. (1993). *Phys. Rev. Lett.*, **70**, 2641.

20
Majorana representation for spin $S = 1/2$ magnets: relationship to Z_2 lattice gauge theories

Imagine that we have a disordered spin $S = 1/2$ magnet with very short spin-spin correlation length. In this case, local magnetization is a fast variable and cannot be used to characterize the ground state. For this case Anderson (1973) suggested a dual description, where a spin-wave function is characterized not in terms of spin states on a site but in terms of states of spin pairs. This description has a better chance of being adequate for disordered systems with antiferromagnetic correlations than the standard one, because even in a disordered material spins prefer to remain in singlet with their neighbours (a singlet state is a state not of a single spin, but of a pair). Suppose we have a state which is a spin singlet. Then as a building block of the multi-spin wave function we use a pair singlet function of spins on sites i, j:

$$|(ij)\rangle = \frac{1}{\sqrt{2}}(|\uparrow_i\downarrow_j\rangle - |\downarrow_i\uparrow_j\rangle) \tag{20.1}$$

Let us connect pairs of lattice sites with bonds; each such connection determines a partition P. Then we can define a *valence bond* state for a given partition as the product of singlet states:

$$|VB_P\rangle = \prod_{\text{pairs}} |(i_k j_k)\rangle \tag{20.2}$$

and represent an arbitrary singlet as

$$|\Psi\rangle = \sum_P A(P) \prod_{\text{pairs}} |(i_k j_k)\rangle \tag{20.3}$$

where the sum includes all partitions of the lattice into sets of pairs. For general functions such decomposition works very badly because valence bond states are not orthogonal and their basis is overcomplete. The picture can work only in a case when the true eigenstates are close to some valence bond state.

The valence bond picture strongly resembles lattice gauge theories, as in these theories the relevant variables are defined on links. Since each link can either be full (singlet state) or empty (the state is not specified), this suggests that the gauge field description of spin fluids should include Z_2 gauge fields. Below we consider such a description based on a special fermionic representation of spin $S = 1/2$ operators.

Let us consider real fermions $\gamma_a(r)$ $(a = 1, 2, 3)$ on a lattice of N sites with the following anticommutation relations:

$$\{\gamma_a(r), \gamma_b(r')\} = \delta_{ab}\delta_{r,r'} \tag{20.4}$$

The fermions are real in the sense that

$$\gamma^+(r) = \gamma(r) \tag{20.5}$$

We have already met this algebra before when Dirac fermions were discussed. This algebra is just the algebra of γ matrices of the Dirac equation in $3N$-dimensional space. Mathematicians call it the Clifford algebra and all its properties are well known. For example, the number of states in the irreducible representation is $2^{[3N/2]}$, where $[x]$ is the integer part of x. Those who are still uncomfortable with the concept of real fermions should refer to Chapter 28 where I explain the link between Majorana and conventional fermions.

We can express the spin-1/2 operators in terms of these new fermions. One can check that the following representation

$$S^a(r) = -\frac{i}{2}\epsilon_{abc}\gamma_b(r)\gamma_c(r) \tag{20.6}$$

faithfully reproduces the commutation relations of the spin operators

$$[S^a(r), S^b(r')] = i\epsilon_{abc}S^c(r)\delta_{r,r'} \tag{20.7}$$

together with the constraint on the spin-1/2 representation:

$$\mathbf{S}^2 = 3/4 \tag{20.8}$$

The latter relation holds due to the property of the γ matrices (see (20.4) and (20.5)):

$$\gamma_a^2(r) = 1/2 \tag{20.9}$$

Despite the fact that this fermionic representation does not introduce nonphysical states with incorrect $\mathbf{S}^2$, the number of states is $2^{[N/2]}$ times too many. In practical calculations these extra states are removed by gauge fixing.

The described Majorana representation of spins has a venerable history: it was suggested by Martin in 1959, described by Mattis in his book on magnetism (1964), and rediscovered in high energy physics by Casalbuoni (1976) and Berezin and Marinov (1977). Recently it was applied to condensed matter problems (see Coleman et al., 1992, 1993).

Here we discuss the Majorana representation within the path integral approach. It is straightforward to see that the following Lagrangian reproduces the anticommutation relations (20.5) under quantization:

$$L = \frac{1}{2}\sum_r \gamma_{a,r}\partial_\tau\gamma_{a,r} + H[S] \tag{20.10}$$

Here $H[S]$ is a Hamiltonian which, as we assume, depends only on spins. The measure of integration is given by

$$D\mu_0[\gamma] = \prod_{a,r} D\gamma_{a,r}(\tau) \tag{20.11}$$

One cannot expect anything different for real fermions. Thus the generating functional for spins 1/2 has the following form:

$$Z[h_a] = \int D\gamma_a \exp\left\{-\int d\tau \sum_r \left[\frac{1}{2}\sum_r \gamma_{a,r}\partial_\tau\gamma_{a,r} + H[S] + h_r^a S_{a,r}\right]\right\} \quad (20.12)$$

Now let us consider how this approach works for a spin $S = 1/2$ Heisenberg antiferromagnet far from magnetic instability. The Hamiltonian has the following form:

$$\hat{H} = \sum_{r,r'} J_{r,r'} S^a(r)S^a(r') \quad (20.13)$$

Substituting the spin operators by their fermionic representations we get

$$\hat{H} = \frac{1}{2}\sum_{r,r'} J_{r,r'} \sum_{a\neq b}[\gamma_a(r)\gamma_a(r')][\gamma_b(r)\gamma_b(r')] \quad (20.14)$$

Thus we get a fermionic theory where fermions have only potential energy and no bare dispersion. Thus it is the opposite to standard fermionic theories where fermions propagate in a broad band. The idea is, however, that in the spin fluid phase fermions acquire a dispersion self-consistently. Let us introduce the following averages on links (remember the discussion at the beginning of the chapter!)

$$\Delta_0^a(r,r') = iJ_{r,r'}\langle\gamma_a(r)\gamma_a(r')\rangle \quad (20.15)$$

and rewrite the Hamiltonian as follows:

$$\hat{H} = \hat{H}_0 + \hat{H}_{int} - \sum_{r,r'}\sum_{a\neq b}\frac{\Delta_0^a(r,r')\Delta_0^b(r,r')}{J_{r,r'}} \quad (20.16)$$

$$\hat{H}_0 = -i\sum_{r,r'}\left[\sum_{b\neq a}\Delta_0^b(r,r')\right] : \gamma_a(r)\gamma_a(r') : \quad (20.17)$$

$$\hat{H}_{int} = \frac{1}{2}\sum_{r,r'} J_{r,r'} \sum_{a\neq b} : [\gamma_a(r)\gamma_a(r')] :: [\gamma_b(r)\gamma_b(r')] : \quad (20.18)$$

Now (20.17) gives the kinetic energy and the term $\hat{H}_{int}$ describes interactions in the Majorana band.

The idea is to treat $\hat{H}_{int}$ as a perturbation. Since the model does not have any small parameters, the success of such an approach is not obvious. We hope, however, that this approximation can encapsulate the basic physics of the spin fluid state. Therefore it is worth spending some time illustrating its basic principles.

Since we are going to use the perturbation theory, it is convenient to write down a path integral representation. This representation will also make more clear the link between spin models and lattice gauge theories. In path integral language the auxiliary field $\Delta(r, r')$ is introduced via the Hubbard–Stratonovich transformation (this requires that all $J_{r,r'}$ are positive):

$$\exp\left(-\int d\tau H\right) = \int D\Delta(\tau; r, r') \exp\left\{-\int d\tau \left[\sum_{\text{links}} \frac{\Delta^a(r, r')\Delta^b(r, r')}{2J_{r,r'}}\right.\right.$$

$$\left.\left. -i\sum_{r,r'}\sum_{a\neq b}\Delta^b(r, r')\gamma_a(r)\gamma_a(r')\right]\right\} \tag{20.19}$$

After this transformation the partition function of the magnet is expressed as the partition function of the lattice gauge theory:

$$Z = \int D\Delta(r, r')D\gamma_a(r)\exp\left[-\int d\tau L(\Delta, \gamma_a)\right]$$

$$\tag{20.20}$$

$$L = \sum_{(r,r')} \frac{\Delta^a(r, r')\Delta^b(r', r)}{2J_{r,r'}} + \sum_i \frac{1}{2}\gamma_a(r, \tau)\partial_\tau\gamma_a(r, \tau) + i\sum_{r,r'}\Delta^b(r', r)\gamma_a(r)\gamma_a(r')$$

This path integral is invariant with respect to the local *time independent* gauge transformations:

$$\gamma_a(r) = (-1)^{\sigma(r)}\gamma_a(r)$$

$$\Delta^b(r, r') = (-1)^{\sigma(r)+\sigma(r')}\Delta^b(r, r') \tag{20.21}$$

($\sigma(i) = 0, 1$) composing the gauge group of the theory. This group is isomorphic to the Z_2 group. This is the Z_2 gauge symmetry discussed in the beginning of this chapter.

In the mean field approximation corresponding to the Hamiltonian (20.16)–(20.18) we expand the path integral around a static configuration of $\Delta(r, r')$ realizing the minimal energy. Then the fermion Green's function in the mean field approximation is equal to

$$[G^{-1}(\omega_n)]_{r,r'} = i\omega_n\delta_{r,r'}\delta_{ab} + 2i\Delta_0^b(r, r')(1 - \delta_{ab}) \tag{20.22}$$

(it is important to remember that $\Delta(r, r') = -\Delta(r', r)$). In the simplest case when $\Delta_0^b(r, r') = \Delta_0(r - r')$,[1] one can make a Fourier transformation and write down

$$G_{ab}(\omega_n, k) = [i\omega_n - \epsilon(k)]^{-1}\delta_{ab}$$

$$\epsilon(k) = \sum_r \Delta_0(r)\sin(\mathbf{kr}) \tag{20.23}$$

which looks like the Green's function for conventional fermions. The important property of the Majorana Green's function is

$$G_{ab}(-\omega_n, -k) = -G_{ab}(\omega_n, k) \tag{20.24}$$

The diagram expansion of the corresponding partition function contains the elements shown in Fig. 20.1, where the fat solid line is the fermionic propagator $G(r, r')$ and the wavy line is the exchange integral. The diagram Fig. 20.1(b) is forbidden because it is already taken into account in the definition of the fat solid line.

---

[1] In fact, this is a rather strong assumption which requires a careful choice of exchange integrals. In the majority of cases the minimal energy solution breaks the translational invariance which corresponds to the spin-Peierls transition.

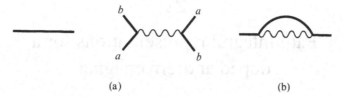

(a)   (b)

Figure 20.1.

The perturbation expansion takes into account small fluctuations of $\Delta(r, r')$. There are nonperturbative aspects of the problem related to sign fluctuations of this field which are not addressed by this approach. The problem is what to do with time dependent gauge transformations when the fermionic variable on some site r_0 abruptly changes its sign in time:

$$\gamma_a(r, \tau) = (-1)^{\delta_{r,r_0}\theta(\tau - \tau_0)}\gamma_a(r, \tau) \tag{20.25}$$

When we substitute the above expression into (20.20) the time derivative gives the extra contribution

$$\int d\tau \gamma_a(r_0, \tau)\delta(\tau - \tau_0)\gamma_a(r_0, \tau) \tag{20.26}$$

Here we have an ill defined expression because we multiply zero ($\gamma_a(r_0, \tau)^2 = 0$, do not forget γ are not operators, but Grassmann numbers!) to infinity. Therefore sign fluctuations need to be treated with care, as has been done by Coleman *et al.* (1995).

References

Anderson, P. W. (1973). *Mater. Res. Bull.*, **8**, 153.

Berezin, F. A. and Marinov, M. S. (1977). *Ann. Phys. NY*, **104**, 336.

Casalbuoni, R. (1976). *Nuovo Cimento A*, **33**, 389.

Coleman, P., Miranda, E. and Tsvelik, A. M. (1992). *Phys. Rev. Lett.*, **69**, 2142; *Phys. Rev. Lett.*, **70**, (1993) 2960.

Coleman, P., Ioffe, L. and Tsvelik, A. M. (1995). *Phys. Rev. B*, **52**, 6611.

Martin, J. L. (1959). *Proc. R. Soc. London Ser. A*, **251**, 536.

Mattis, D. C. (1964). *Theory of Magnetism*. Harper & Row.

21

Path integral representations for a doped antiferromagnet

The coexistence of conducting electrons and spins is one of the most difficult problems in condensed matter physics. In fact, this general heading covers two distinctly different classes of physical systems. In the first class conducting electrons with a wide band $\epsilon(k)$ are coupled antiferromagnetically to local spins, presented either as diluted magnetic impurities or as a periodic array. The standard model for these materials is the Anderson model:

$$\hat{H} = \sum_{k,j} \epsilon(k) c_{kj}^+ c_{kj} + \sum_{r,j} \epsilon_d d_{r,j}^+ d_{r,j} + \frac{U}{2} \sum_{r,j \neq j'} d_{r,j}^+ d_{r,j} d_{r,j'}^+ d_{r,j'}$$
$$+ \Omega^{-1/2} \sum_{k,j,r} V(k)(c_{k,j}^+ d_{r,j} e^{i\mathbf{k}\mathbf{r}} + \text{H.c.}) \tag{21.1}$$

where $c_{k,j}^+$, $c_{k,j}$ are creation and annihilation operators of a conduction electron with wave vector $\mathbf{k}$ and spin projection j and $d_{r,j}^+ (d_{r,j})$ creates (annihilates) a localized electron on a site r. Ω is the volume of the system. The quartic term describes a strong Coulomb repulsion on the localized orbitals. The Anderson model incorporates two competing processes, making this model so interesting and complicated. The first one is the formation of local magnetic moments caused by the Coulomb repulsion. Since the Coulomb repulsion is usually very strong (several electronvolts), a multiple occupation of localized orbitals is effectively forbidden. When $\epsilon_d < 0$, $\epsilon_d + U > 0$ the f-sites are predominantly singly occupied and therefore magnetic. The second process is a screening of the magnetic moments by conduction electrons (the Kondo effect). The screening is caused by hybridization which opens for localized electrons a possibility to escape into the conduction band (see Fig. 21.2).

The competition between the Coulomb forces and the hybridization leads to very beautiful effects. Namely, at high temperatures we observe conduction and localized electrons behaving quite independently. The latter appear as free local moments: their contribution to the magnetic susceptibility follows the Curie law $\chi \sim 1/T$. When the temperature decreases, the local moments are screened by conduction electrons. The system gradually transforms into something resembling an ordinary metal. This 'ordinary' metal, however, has a very high linear specific heat and magnetic susceptibility (up to 1000 times larger than copper). For a single impurity problem these exciting low energy properties are caused by

Figure 21.1. Philip Anderson.

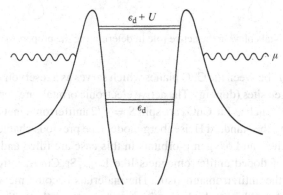

Figure 21.2. Schematic representation of the energy levels in the Anderson model.

the formation of a narrow resonance close to the Fermi level (the Abrikosov–Suhl resonance) with the characteristic width

$$T_k \approx \sqrt{DJ}\exp[-1/\rho(\epsilon_F)J] \qquad J = -V^2 U/\epsilon_d(\epsilon_d + U) \qquad (21.2)$$

For systems with periodic arrays of spins, the low temperature singularities are ascribed to the formation of a new band of heavy particles crossing the Fermi level.

The single impurity Anderson model is exactly solvable and we have an almost complete description of its properties (see, for example, the review articles by Tsvelik and Wiegmann (1983) and Schlottmann (1989)). Unfortunately, our understanding of the periodic Anderson model is far from being so clear. The majority of available results have been derived by the $1/N$-expansion,[1] with N being the number of spin components, which in reality is almost always equal to 2. In application to the periodic Anderson model this method is well discussed in the extended publication by Coleman (1987).

The second class of materials consists of doped antiferromagnets. This class is well represented by copper oxide compounds. A typical material of this sort consists of planes made of

---

[1] Below in Chapter 37 I resort to other nonperturbative methods.

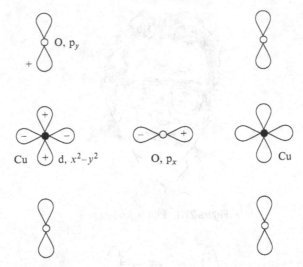

Figure 21.3. The orbitals playing the active role in determining the properties of CuO materials.

CuO and some 'stuff' between the CuO planes which serves as a reservoir of holes removing electrons from oxygen sites (doping). The active electronic orbitals are depicted in Fig. 21.3.

Undoped materials such as La_2CuO_4 are spin $S = 1/2$ antiferromagnets whose properties are well explained by the standard Heisenberg model (see previous chapters). The spins are located on copper sites, and oxygen p-orbitals in this case are filled and inert. As follows from observations of doped antiferromagnets like $La_{2-x}Sr_xCuO_4$, introduction of holes rapidly suppresses the antiferromagnetism. The materials become metals and eventually even superconductors with very high transition temperatures (high-T_c superconductors). Apparently, the antiferromagnetism is suppressed due to mobility of holes. Moving through the lattice the holes flip localized spins and create dynamical frustration.

The most obvious difference between these two classes of materials is that in the second case the Fermi energy of the holes, being proportional to their concentration, is always smaller than or of the order of the Cu–Cu exchange integral. Despite the obvious difference in the physics, the mathematical description shares certain common features on which we are going to concentrate.

In order to take doping into account, we need to modify the path integral representation of spins. The derivation I am going to present and the subsequent analysis will be rather brief. For a more detailed discussion I refer the reader to the book by Auerbach listed in the select bibliography.

Magnetic moments in metals are usually created by ions of transition or rare-earth elements. Let us consider a Hilbert space of a single atomic orbital of such an ion; it includes three states $|a\rangle = |0\rangle, |\uparrow\rangle, |\downarrow\rangle$ corresponding to the empty site and spin-up (spin-down) sites. The state with the double occupancy is eliminated by the large Coulomb repulsion. Transitions between the states are described by the operators $X_{ab}(i)$ introduced by Hubbard:

$$X_{ab}(i)|c\rangle = \delta_{bc}|a\rangle \tag{21.3}$$

On a single site each X_{ab} operator is represented as a 3×3 matrix with only one nonzero element equal to

$$(X_{ab})^{\alpha\beta} = \delta_{a\alpha}\delta_{b\beta} \tag{21.4}$$

The Hubbard operators are related to the creation and annihilation operators:

$$X_{\sigma 0} = d_\sigma^+(1 - d_{-\sigma}^+ d_{-\sigma}) \qquad X_{0\sigma} = (1 - d_{-\sigma}^+ d_{-\sigma})d_\sigma$$

For different sites there is a problem with the statistics. Since transitions between empty and occupied states include a change in the fermionic number, the operators $X_{\sigma 0}(i)$, $X_{0\sigma}(i)$ anticommute on different sites and the operators $X_{\sigma\sigma'}(i)$, $X_{00}(i)$ commute. Therefore it is natural to call the operators of the first kind Fermi and those of the second kind Bose. Thus we have an algebra whose elements have different statistics. Such algebras are called *graded* algebras. Taking these facts together with (21.4) we can derive the following commutation relations:

$$[X_{ab}(i), X_{cd}(j)]_\pm = \delta^{ij}[X_{ad}(i)\delta_{bc} \pm X_{cb}(i)\delta_{ad}] \tag{21.5}$$

where $(+)$ should be used only if both operators are fermionic. These commutation relations define the double graded algebra called Spl(1,2). The term 'double graded' refers to the fact that the algebra includes generators of different kinds – some of them are bosonic and some fermionic. One can easily generalize the notation for Spl(N_b, N_f) algebra allowing N_b 'empty' (i.e. bosonic) states $|a)$ and N_f fermionic single occupied states $|\sigma)$. The commutation relations (21.5) remain unchanged.

One can use the described representation to rewrite the Anderson Hamiltonian (21.1). Namely, in the case $U \to \infty$ one has to replace the $d_{r,j}$ operators by $X_{0j}(r)$ and $d_{r,j}^+$ operators by $X_{j0}(r)$. The resulting Hamiltonian is

$$\hat{H} = \sum_{k,j} \epsilon(k)c_{kj}^+ c_{kj} + \sum_{r,j} \epsilon_d X_{jj}(r) + \Omega^{-1/2} \sum_{k,j,r} V(k)(c_{k,j}^+ X_{0j}(r)e^{ikr} + \text{H.c.}) \tag{21.6}$$

The standard model for a doped antiferromagnet of the CuO type is the so-called $t - J$ model:

$$H = \sum_{\langle r,r'\rangle} \left[-t X_{\sigma 0}(r)X_{0\sigma}(r') + \frac{1}{2}J X_{\sigma\sigma'}(r)X_{\sigma'\sigma}(r') \right] + \sum_r \epsilon_d X_{\sigma\sigma}(r)$$
$$+ \frac{e^2}{2} \sum_{r \neq r'} \frac{X_{00}(r)X_{00}(r')}{|r - r'|} \tag{21.7}$$

The last term representing the Coulomb repulsion between holes from different sites is usually omitted in the literature. However, this term is crucial in preventing the system from phase separation when all holes gather together creating a single 'lump'. In the limit of zero doping $\epsilon_d \to -\infty$, this model is equivalent to the spin-1/2 Heisenberg model. In the literature one can find many discussions of the generalization of the $t - J$ model for higher symmetry groups where $\sigma = 1, \ldots, N$.

The purpose of this chapter is to formulate the path integral for double graded algebras and in particular for the models (21.1) and (21.7). This derivation is a generalization of the path integral for spins described in Chapter 16. As the first step we introduce an analogue of the Schwinger–Wigner representation for the algebra (21.5). For this purpose we define a fake vacuum vector $|\Omega\rangle$ and represent the states as follows:

$$|a\rangle = f_a^+|\Omega\rangle \qquad |\sigma\rangle = b_\sigma^+|\Omega\rangle \tag{21.8}$$

where f^+ and b^+ are ordinary creation operators. Then the X operators are

$$X_{ab} = f_a f_b^+ \qquad X_{\sigma\sigma'} = b_\sigma b_{\sigma'}^+ \qquad X_{a\sigma} = f_a b_\sigma^+ \tag{21.9}$$

$$\sum_{a=1}^{N_b} f_a^+ f_a + \sum_{\sigma=1}^{N_f} b_\sigma^+ b_\sigma = Q \tag{21.10}$$

The constraint is introduced to eliminate unphysical states such as

$$|\Omega\rangle, \; b_{\sigma_1}^+ b_{\sigma_2}^+|\Omega\rangle \qquad \text{etc.}$$

Since transitions between empty and occupied states include a change in the fermionic number, the operators f and b must have different statistics. At this point we have a choice: we can choose f to be Fermi and b Bose or vice versa. From the mathematical point of view it does not make any difference. I consider both possibilities.

Let us consider the representation where f_a are Fermi and b_σ are Bose operators first. For $N_b = 1$, $N_f = 2$ this derivation was done by Wiegmann (1988). The path integral for the constrained fermions and bosons has the following measure:

$$D\mu \exp\left(\int d\tau L\right) = Df_a^+ Df_a Db_\sigma^+ Db_\sigma \delta(f_a^+ f_a + b_\sigma^+ b_\sigma - Q)$$

$$\times \exp\left[-\int d\tau(f^+\partial_\tau f + b_\sigma^+\partial_\tau b_\sigma)\right] = Df_a^+ Df_a Db_\sigma^+ Db_\sigma D\lambda$$

$$\times \exp\left\{-\int d\tau[f_a^+(\partial_\tau + \lambda)f_a + b_\sigma^+(\partial_\tau + \lambda)b_\sigma - \lambda Q]\right\} \tag{21.11}$$

For the case $N_b = 1$, $N_f = 2$, $Q = 2S$ one can simplify this measure resolving the constraint (21.10) in the explicit form:

$$b_1 = S\left(1 - \frac{1}{2}f^+ f\right)\exp[i(\phi + \psi)/2]\cos(\theta/2)$$

$$b_2 = S\left(1 - \frac{1}{2}f^+ f\right)\exp[i(\phi - \psi)/2]\sin(\theta/2) \tag{21.12}$$

$$\mathbf{S} = \frac{1}{2}(1 - f^+ f)(\cos\theta, \cos\psi\sin\theta, \sin\psi\sin\theta)$$

Here I have taken advantage of the fact that inside the path integral f^+, f and b^+, b are not operators, but Grassmann and ordinary numbers, respectively. Therefore $(f^+ f)^2 = 0$

and $(1 - \frac{1}{2} f^+ f)^2 = (1 - f^+ f)$. Substituting (21.12) into (21.11) and repeating all the manipulations of Chapter 16 we get

$$
D\mu = Df^+ Df D\mathbf{n} D\phi \delta(\mathbf{n}^2 - 1)
$$

$$
\times \exp \left\{ \int d\tau \left[iS \int_0^1 du(\mathbf{n}[\partial_u \mathbf{n} \times \partial_\tau \mathbf{n}]) + f^+ (\partial_\tau - iA_\tau^z) f \right] \right\} \quad (21.13)
$$

where

$$
A_\tau^z = \partial_\tau \phi + \cos\theta \, \partial_\tau \psi \quad (21.14)
$$

The fermionic term in the measure does not contain an integral over u. The measure thus defined is gauge invariant: it is invariant with respect to the *time-dependent* transformations

$$
f(\tau) = e^{i\alpha(\tau)} f(\tau) \qquad \phi(\tau) = \phi(\tau) + \alpha(\tau) \quad (21.15)
$$

The gauge field A^z has the following remarkable property on the (u, τ) plane, namely its field strength is equal to the density of the topological charge of the **n**-field:

$$
F_{\tau u} = \partial_\tau A_u^z - \partial_u A_\tau^z = \sin\theta(\partial_u \theta \partial_\tau \psi - \partial_\tau \theta \partial_u \psi) = (\mathbf{n}[\partial_u \mathbf{n} \times \partial_\tau \mathbf{n}]) \quad (21.16)
$$

It turns out that for $S = 1/2$ one can eliminate the interaction of the spinless fermions with the gauge field A_τ^z by a canonical transformation. This was done by Wang and Rice (1994). As a result they obtained the following equivalent representation for the Hamiltonian of the $t - J$ model:

$$
\hat{H} = \sum_{r \neq r'} \frac{e^2}{2|r - r'|} f_r^+ f_r f_{r'}^+ f_{r'} - t \sum_{\langle r, r' \rangle} f_r f_{r'}^+ \left(\frac{1}{2} + 2\mathbf{S}_r \mathbf{S}_{r'} \right)
$$

$$
+ \frac{J}{2} \sum_{\langle r, r' \rangle} (1 - f_r^+ f_r) \left(\mathbf{S}_r \mathbf{S}_{r'} - \frac{1}{4} \right) (1 - f_{r'}^+ f_{r'}) \quad (21.17)
$$

where S_i^a are spin $S = 1/2$ operators and f_r^+, f_r are creation and annihilation operators of spinless fermions on the site r with the standard fermionic anticommutation relations (holons). The advantage of this representation is that it makes clear that mobile holes introduce frustration favouring ferromagnetic ordering. In the limit $J = 0$ the holes align the spins in such a way as to maximize their own bandwidth, which corresponds to the ferromagnetic order. At finite J this order disappears at small concentrations $x < x_c \sim J/t$.

Let us consider now another representation of Spl(1, N), namely the representation where the spin operators are expressed in terms of fermions. This is the so-called 'slave boson' representation introduced by Coleman (1985).

This representation is more convenient for the Anderson model. I accept the following parametrization:

$$
X_{00} = 1 - f_j^+ f_j \qquad X_{jj'} = f_j^+ f_{j'} \qquad X_{0j} = b^+ f_\sigma \quad (21.18)
$$

$$
b = \sqrt{\rho} e^{i\theta} \qquad b^+ = \sqrt{\rho} e^{-i\theta} \quad (21.19)
$$

$$
\rho^2 + f_j^+ f_j = N/2 \quad (21.20)
$$

Figure 21.4. Piers Coleman.

The measure of integration is given by

$$D\mu = Df_j^+ Df_j D\theta D\rho \delta(\rho + f_j^+ f_j - Q) \qquad (21.21)$$

The partition function with this measure diverges because the action contains f and b in bilinear combinations (see (21.20)) and is therefore invariant with respect to the gauge transformations:

$$\theta(\tau) = \theta(\tau) + \alpha(\tau) \qquad f_j(\tau) = f_j(\tau) e^{i\alpha(\tau)} \qquad (21.22)$$

In order to obtain a correct measure we should fix the gauge. One popular choice of gauge fixing is the *radial* gauge $\theta = 0$. Substituting the expression (21.20) into the measure (21.21) we get

$$D\mu = Df_j^+ Df_j D\rho \int D\lambda \exp\left\{-\int d\tau[f_j^+(\partial_\tau + i\lambda)f_j + i\lambda(\rho - Q)]\right\} \qquad (21.23)$$

where the auxiliary variable $\lambda(\tau, r)$ is introduced to secure the delta-function constraint.

Using this result we can write down the partition functions for the Anderson and the $t - J$ models. For the Anderson model we have

$$Z_A = \int Df_j^+ Df_j D\rho D\lambda \exp\left(-\int d\tau L_A\right)$$

$$L_A = \sum_k c_{kj}^+[\partial_\tau - \epsilon(k)]c_{kj} + \sum_r f_{r,j}^+(\partial_\tau + i\lambda_r - \epsilon_d)f_{r,j} + i\lambda_r(\rho_r - Q) \qquad (21.24)$$

$$- \Omega^{-1/2} \sum_{k,j,r} V(k)\sqrt{\rho_r}(c_{k,j}^+ f_{r,j} e^{ikr} + \text{c.c.})$$

The partition function for the $t - J$ model has the same measure, but a different Lagrangian:

$$L_{t-J} = \sum_r f_j^+(r)[\partial_\tau + i\lambda(r) - \mu]f_j(r) + i\lambda(r)[\rho(r) - Q]$$

$$+ \sum_r \sum_{\langle r,r'\rangle}\left[-tf_j^+(r)f_j(r')\sqrt{\rho(r)\rho(r')} - \frac{J}{2}f_j^+(r)f_j(r')f_k^+(r')f_k(r)\right] \quad (21.25)$$

The standard approach to these models uses the $1/N$-expansion. In its framework one expands the partition function around the saddle point

$$\rho(\tau, r) = \rho_0 + \delta\rho(\tau, r) \qquad i\lambda(\tau, r) = u_0 + iu(\tau, r) \quad (21.26)$$

where ρ_0 and u_0 are determined by the self-consistency conditions (cancellation of tadpole diagrams)

$$\langle f_{r,j}^+ f_{r,j}\rangle + \rho_0 + \langle\delta\rho_r\rangle = Q$$
$$\langle f_{r,j}^+ f_{r,j}\rangle = \text{number of d-electrons} \quad (21.27)$$

and $\delta\rho$ and u are considered as corrections small in $1/N$. The effective Lagrangians at the saddle points are

$$L_A^{\text{mf}} = \sum_k \{c_{kj}^+[\partial_\tau - \epsilon(k)]c_{kj} + f_{kj}^+(\partial_\tau + \tilde{\epsilon}_d)f_{k,j} + \tilde{V}(k)(c_{k,j}^+ f_{k,j} + \text{c.c.})\} \quad (21.28)$$

for the Anderson model, where $\tilde{\epsilon}_d = -\epsilon_d + u_0$, $\tilde{V}(k) = \sqrt{\rho_0}V(k)$, and (Larkin and Ioffe, 1989)

$$L_{t-J}^{\text{mf}} = \sum_k\left\{f_{kj}^+(\partial_\tau - \tilde{t}(k) - \mu)f_{kj} - \frac{1}{2}\sum_{q,p}J(p-k)f_{k+q/2,j}^+ f_{k-q/2,j}f_{p-q/2,j'}^+ f_{p+q/2,j'}\right\}$$

$$+ \text{ Coulomb interaction} \quad (21.29)$$

for the $t - J$ model. In the latter case

$$\tilde{t}(k) = t\sqrt{\rho_0}\sum_{\langle r\rangle}\cos(\mathbf{kr})$$

The effective mean field Lagrangian for the Anderson model is quadratic. The excitation spectrum is given by

$$E_\pm = \frac{\epsilon(k) + \tilde{\epsilon}_d}{2} \pm \sqrt{\left(\frac{\epsilon(k) - \tilde{\epsilon}_d}{2}\right)^2 + \tilde{V}^2(k)} \quad (21.30)$$

and is represented in Fig. 21.5. The spectrum contains almost flat portions responsible for the mass enhancement in heavy fermionic systems.

Fluctuations around the saddle point cause interaction between the quasi-particles. Thus the picture emerging from the large N approach is rather peaceful: the spin degrees of freedom are incorporated into the band and the remaining interactions are small. In other words, we have a classical Fermi liquid, though perhaps with very small Fermi energy $\tilde{\epsilon}_F \sim \tilde{V}^2/\epsilon_F$. Despite the fact that this falls short of being a complete description of the world of heavy

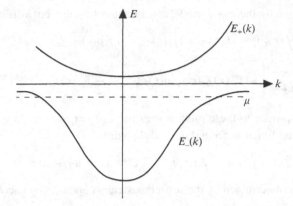

Figure 21.5. A schematic picture of the spectrum of the Anderson lattice.

Figure 21.6. Gabriel Kotlyar.

fermion materials, it does capture certain important features. This approximation has been successfully used to explain de Haas–van Alphen experiments in heavy fermion materials (Wasserman *et al.*, 1989; Springford, 1991). This is due to the important property of the radial gauge, that the $1/N$-expansion remains qualitatively correct for the physical case $N = 2$. Namely, as we see from (21.24), due to the positiveness of $\sqrt{\rho}$, conduction electrons always hybridize with the f-quanta, however strong the fluctuations of $\sqrt{\rho}$ may be. This makes the spectrum (21.30) robust for any N, unless it is destroyed by other processes such as superconductivity or magnetic correlations. In particular, as will be shown in Chapter 37, this approach does not work for a one-dimensional Kondo lattice where the antiferromagnetic correlations play a dominant role.

The situation with the $t - J$ model is much more complicated because it remains the interacting theory even for $N \to \infty$. The existing analysis (see, for example, Grilli and Kotliar (1990)) shows that for small $J/\tilde{t}$ when the ground state is nonmagnetic, the saddle point has a Fermi liquid character. The latter is the most disappointing feature of the $1/N$ approach to the $t - J$ model which invariably demonstrates its inability to explain the manifest violations of the Fermi liquid behaviour in copper oxides. Despite enormous theoretical effort in recent years, our understanding of doped antiferromagnets remains very vague...

References

Coleman, P. (1985). Proc. Fourth Int. Conf. on Valence Fluctuations, *J. Magn. Mater.*, **47–48**, 323.

Coleman, P. (1987). *Phys. Rev. B*, **35**, 5072.

Grilli, M. and Kotliar, G. (1990). *Phys. Rev. Lett.*, **64**, 1170.

Larkin, A. I. and Ioffe, L. B. (1989). *Phys. Rev. B*, **39**, 8988.

Schlottmann, P. (1989). *Phys. Rep.*, **181**, 2.

Springford, M. (1991). *Physica B*, **171**, 151.

Tsvelik, A. M. and Wiegmann, P. B. (1983). *Adv. Phys.*, **32**, 453.

Wang, Y. R. and Rice, M. J. (1994). *Phys. Rev. B*, **49**, 4360.

Wasserman, A., Springford, M. and Hewson, A. C. (1989). *J. Phys. Condens. Matter*, **1**, 2669.

Wiegmann, P. B. (1988). *Phys. Rev. Lett.*, **60**, 821.

IV

Physics in the world of one spatial dimension

Introduction

This part of the book is fully devoted to the physics of $(1 + 1)$-dimensional quantum or two-dimensional classical models. As usual, I shall distinguish between classical and quantum systems only if it is necessary. Those models where the difference between quantum and classical is not relevant will be called 'two-dimensional'; those where for some reason or other I want to emphasize the quantum aspect will be called 'one-dimensional'.

The reason I spend so much time discussing two-dimensional physics is that here one finds a kind of paradise for strong interactions and nonperturbative effects. Those effects which are tricky or even impossible to achieve in higher dimensions appear in almost every two-dimensional model. To our greatest satisfaction there are theoretical tools at our disposal which allow us to solve many of the corresponding problems and describe these phenomena. The availability of these strong mathematical tools is a unique feature of two-dimensional physics. Even if it turns out to be impossible to generalize these tools for higher dimensions, I hope that the reader will be rewarded for the time spent by the pleasure obtained from contemplating their beauty.

The world we are about to enter has certain distinct features which are worth mentioning in the introduction. The first one concerns the second-order phase transitions. In $(1 + 1)$-dimensional systems such transitions may occur only at zero temperature. Sometimes such a transition is associated with the formation of a finite order parameter (for instance, at $T = 0$ a one-dimensional ferromagnet has a finite magnetization which vanishes at any nonzero T); sometimes the order parameter is not formed even at zero temperature (for instance, in a one-dimensional spin-$1/2$ antiferromagnet with easy plane anisotropy). So since phase transitions may occur only at $T = 0$, there are no 'low temperature' phases where the presence of a nonzero order parameter gives rise to various remarkable changes of the excitation spectrum. Recall that spontaneous symmetry breaking leads to the following spectral changes: (i) creation of gapless Goldstone modes associated with transverse fluctuations of the order parameter and (ii) formation of spectral gaps for the longitudinal modes. This mechanism of gap formation is modified in $(1 + 1)$ dimensions; we will see that interactions do generate spectral gaps though no order parameter appears. Gapless collective modes also appear, but not as Goldstone bosons, rather as *critical fluctuations*.

Another distinct feature of the $(1 + 1)$-dimensional world is that *topological* excitations appear here as point-like objects (in other words, they become particles). Such excitations appear as *domain walls* between degenerate ground sates. In higher dimensions walls are extended objects and, as such, carry a macroscopic energy. In one dimension a 'wall' is just a point and the related energy is microscopic. As we shall see, the appearance of topological excitations leads to a remarkable phenomenon of quantum number *fractionalization*. Among the particular manifestations of this phenomenon is the existence of electrically neutral spin-$1/2$ excitations in spin $S = 1/2$ antiferromagnets (see Chapter 29) and deconfinement electrons in neutral spin-$1/2$ excitations (spinons) and spinless charge e particles (holons). The latter effect is called spin and charge separation (see Chapter 28). Many remarkable one-dimensional phenomena have been experimentally observed (see the corresponding references in later chapters). The rapid progress in material synthesis makes available such new systems which in former days the theorist might see only in a dream. Therefore one may be confident that the future will bring us more things to marvel at and more puzzles to solve.

Now let me say several words about the general approach employed by the theory of strong interactions. All successes of the theory are related to our ability to reformulate it in such a way that interactions are either eliminated altogether or somehow localized. For instance, there is a considerable class of nontrivial two-dimensional models which are equivalent to the theory of the free massless bosonic field (see (22.1) below). In these models free particles populating the world which only our mind can penetrate into generate strong correlations in the world of physical observables. Such a trick is successfully performed when physical observables depend nonlinearly on free fields; a corresponding example will be given in the following chapter. Later in this part I shall consider many models disguising free particles: among them there are the spin $S = 1/2$ Heisenberg chain and various models of interacting fermions. There are nontrivial theories which are described by truncated models of the free massless bosonic field, where some states of free bosons are projected out. A list of these models includes the Ising and the three-state Potts models at criticality, and many others. A relatively new subject is non-Abelian bosonization; it also helps to solve certain problems, such as the problem of the $S = 1$ Heisenberg chain (Chapter 36), the problem of the one-dimensional Kondo lattice (Chapter 37) and the multi-channel Kondo problem (not described in this book). There are solvable models where the interactions cannot be eliminated; instead they are 'localized' in the sense that it turns out to be possible to include them in the commutation relations of the particle creation and annihilation operators. Such models constitute a very important class of integrable theories. I shall discuss them in Chapter 34.

I would like to emphasize again that the world of one-dimensional physics exists not only on paper. There are many systems which can be described as one-dimensional, at least in a certain temperature range. Whenever it is appropriate, I give references to experimental papers.

22

Model of the free bosonic massless scalar field

This chapter, along with several others, is devoted to the apparently trivial theory of the free bosonic massless scalar field in a two-dimensional Euclidean space. The action of this model is defined as

$$S = \frac{1}{2} \int_A d^2x (\nabla \Phi)^2 \tag{22.1}$$

where A is some area of the infinite plane. So far we have considered only the case when A is a rectangle: $A = (0 < \tau < \beta, \, 0 < x < L)$. Now I shall not be so restrictive.

The model (22.1) appears in many applications. In particular, as we have seen in Part I, it arises as an effective theory for *transverse fluctuations* of the *two*-component vector field Φ in the low temperature phase $(\tau^* < 0)$ of the O(2) symmetric vector model. This fact maintains a formal contact between the present chapter and the previous discussion. But the significance of the model (22.1) is indeed much wider.

One might think that nothing could be more boring than the model of free bosons, which is just a bunch of harmonic oscillators. However, as quantum mechanics grows from the theory of the harmonic oscillator, the theory of free bosons acts as a seed for QFT in two dimensions. Therefore I advise the reader to be patient, and promise that this patience will be rewarded in due course. Those who are really bored with the formal developments can try to go ahead and read the chapters where I discuss the applications of the theory (22.1) to various two-dimensional models.

The first nontrivial fact about the model (22.1) is that some of its correlation functions are quite remarkable. This illustrates my previous statement that even trivial theories can have nontrivial correlation functions. In order to calculate these functions we need to recall the generating functional of Φ-fields. This functional was found as early as Chapter 3 (3.11) and in a more general form in Chapter 4 (see (4.6) and (4.9)). It was found that

$$Z[\eta(x)]/Z[0] = \exp \left[\frac{1}{2} \eta(\xi) G(\xi, \xi') \eta(\xi') \right] \tag{22.2}$$

where $G(\xi, \xi')$ is the Green's function of the Laplace operator on the area A.

There are two novel features in the approach taken in Part IV.

(i) We shall use the real space representation of correlation functions.
(ii) Instead of the expansion of $Z[\eta]$ around $\eta = 0$, we shall calculate the entire generating functional explicitly for certain *finite* $\eta(x)$.

I have mentioned above that the discussion can be held for a general area A. However, for the sake of simplicity let us first consider the case when A = (infinite plane). In order to make (22.2) well defined it is necessary to regularize the action (22.1) at both large and small distances. For this purpose we introduce the lattice spacing a and treat the plane as a very large disc with $R \to \infty$. Then G satisfies the Laplace equation on the infinite plane:

$$-\left(\partial_x^2 + \partial_\tau^2\right)G(x, \tau; x', \tau') = \delta(x - x')\delta(\tau - \tau') \tag{22.3}$$

with the above mentioned regularization. It will be more convenient for later purposes to use the complex coordinates $z = \tau + ix, \bar{z} = \tau - ix$ where

$$\Delta = 4\partial_z\partial_{\bar{z}} \qquad \partial_z = \frac{1}{2}(\partial_\tau - i\partial_x) \qquad \partial_{\bar{z}} = \frac{1}{2}(\partial_\tau + i\partial_x)$$

The Green's function has the following well known form:

$$G(z, \bar{z}) = \frac{1}{4\pi} \ln\left(\frac{R^2}{z\bar{z} + a^2}\right) \tag{22.4}$$

Now let us consider a particular choice of η, namely

$$\eta(\xi) = \eta_0(\xi) \equiv i\sum_{n=1}^{N} \beta_n\delta(\xi - \xi_n) \tag{22.5}$$

where β_n are some numbers. The generating functional (22.2) for this particular choice of η coincides with the correlation function of bosonic exponents:

$$Z[\eta_0]/Z[0] \equiv \langle \exp[i\beta_1 \Phi(\xi_1)] \cdots \exp[i\beta_N \Phi(\xi_N)] \rangle \tag{22.6}$$

Substituting (22.5) into (22.2) we get the *most important* formula of Part IV:

$$Z[\eta_0]/Z[0] = \exp\left[-\sum_{i>j} \beta_i\beta_j G(\xi_i; \xi_j)\right] \exp\left[-\frac{1}{2}\sum_i \beta_i^2 G(\xi_i; \xi_i)\right] \tag{22.7}$$

The terms with the Green's functions of coinciding arguments are singular in the continuous limit. However, since we have regularized the theory, they are finite. Substituting the expression for G from (22.4) into (22.7) we get the following expression for the correlation function of the exponents:

$$\prod_{i>j}\left(\frac{z_{ij}\bar{z}_{ij}}{a^2}\right)^{(\beta_i\beta_j/4\pi)} \left(\frac{R}{a}\right)^{-\left(\sum_n \beta_n\right)^2/4\pi} \tag{22.8}$$

This expression does not vanish at $R \to \infty$ only if

$$\sum_n \beta_n = 0 \tag{22.9}$$

Thus we have a general expression for correlation functions of bosonic exponents. In fact, (22.7) holds not only for a plane, but for any area A, provided G is chosen properly. Later we shall discuss this point in great detail. Since the bosonic exponents form a basis for local functionals of Φ, one can calculate correlation functions of any local functional $F(\Phi)$ by expanding it as the Fourier integral

$$F(\Phi) = \int d\beta \, \tilde{F}(\beta) e^{i\beta\Phi}$$

and using (22.8). In fact, we can achieve even more. Let us come back to (22.8) and rewrite its right-hand side as a product of the analytic and the antianalytic functions:

$$\langle \exp[i\beta_1 \Phi(\xi_1)] \cdots \exp[i\beta_N \Phi(\xi_N)] \rangle = G(z_1, \ldots, z_N) \, G(\bar{z}_1, \ldots, \bar{z}_N) \delta_{\sum \beta_n, 0} \quad (22.10)$$

where

$$G(\{z\}) = \prod_{i>j} \left(\frac{z_{ij}}{a} \right)^{(\beta_i \beta_j / 4\pi)}$$

Since the correlation functions are factorized into the product of the analytic and the antianalytic parts, the latter ones may be studied *independently*. Since the factorization of correlation functions is a general fact, it can be formally written as a factorization of the corresponding fields: *under the* $\langle \cdots \rangle$ *sign* one can rewrite $\Phi(z, \bar{z})$ as a sum of independent *analytic* and *antianalytic* fields:

$$\Phi(z, \bar{z}) = \varphi(z) + \bar{\varphi}(\bar{z}) \quad (22.11)$$

$$\exp[i\beta\Phi(z, \bar{z})] = \exp\left(\frac{i}{8}\beta^2 \right) \exp[i\beta\varphi(z)] \exp[i\beta\bar{\varphi}(\bar{z})] \quad (22.12)$$

I emphasize that this decomposition should be understood only as a property of correlation functions and *not* as a restriction on the variables in the path integral. The attentive reader understands that the expression (22.10) is obtained by integration over *all* fields Φ. The origin of the phase factor will be explained later (see the discussion around (25.20)).

For many purposes it is convenient to use the 'dual' field $\Theta(z, \bar{z})$ defined as

$$\Theta(z, \bar{z}) = \varphi(z) - \bar{\varphi}(\bar{z}) \quad (22.13)$$

The dual field satisfies the following equations:

$$\partial_\mu \Phi = -i\epsilon_{\mu\nu}\partial_\nu \Theta \quad (22.14)$$

or, in components,

$$\partial_z \Phi = \partial_z \Theta$$
$$\partial_{\bar{z}} \Phi = -\partial_{\bar{z}} \Theta \quad (22.15)$$

In order to study correlation functions of the analytic and the antianalytic fields, I define the fields

$$A(\beta, z) \equiv \exp\left\{\frac{i}{2}\beta[\Phi(z, \bar{z}) + \Theta(z, \bar{z})]\right\}$$

$$\bar{A}(\bar{\beta}, \bar{z}) \equiv \exp\left\{\frac{i}{2}\bar{\beta}[\Phi(z, \bar{z}) - \Theta(z, \bar{z})]\right\}$$

(22.16)

with, generally speaking, *different* β, $\bar{\beta}$. With the operators $A(\beta, z)$, $\bar{A}(\bar{\beta}, \bar{z})$ one can expand local functionals of mutually nonlocal fields Φ and Θ. Let us construct a complete basis of bosonic exponents for a space of local periodic functionals. Suppose that

$$F(\Phi, \Theta)$$

is a local functional periodic both in Φ and Θ with the periods T_1 and T_2, respectively. This functional can be expanded in terms of the bosonic exponents:

$$F(\Phi, \Theta) = \sum_{n,m} \tilde{F}_{n,m} \exp[(2i\pi n/T_1)\Phi$$

$$+ (2i\pi m/T_2)\Theta] \sum_{n,m} \tilde{F}_{n,m} A(\beta_{nm}, z)\bar{A}(\bar{\beta}_{nm}, \bar{z})$$

(22.17)

where

$$\beta_{nm} = 2\pi\left(\frac{n}{T_1} + \frac{m}{T_2}\right)$$

$$\bar{\beta}_{nm} = 2\pi\left(\frac{n}{T_1} - \frac{m}{T_2}\right)$$

(22.18)

Are T_1, T_2 arbitrary? No, they are related to each other. The reason for this lies in the fact that the correlation functions must be uniquely defined on the complex plane. We can see how this argument works using the pair correlation function as an example:

$$\langle A(\beta_{nm}, z_1)\bar{A}(\bar{\beta}_{nm}, \bar{z}_1)A(-\beta_{nm}, z_2)\bar{A}(-\bar{\beta}_{nm}, \bar{z}_2)\rangle = (z_{12})^{-\beta_{nm}^2/4\pi}(\bar{z}_{12})^{-\bar{\beta}_{nm}^2/4\pi}$$

$$= \frac{1}{|z_{12}|^{2d}}\left(\frac{z_{12}}{\bar{z}_{12}}\right)^S$$

(22.19)

where I introduce the quantities

$$d = \Delta + \bar{\Delta} = \frac{1}{8\pi}(\beta^2 + \bar{\beta}^2)$$

and

$$S = \Delta - \bar{\Delta} = \frac{1}{8\pi}(\beta^2 - \bar{\beta}^2)$$

which are called the 'scaling dimension' and the 'conformal spin', respectively.

The two branch cut singularities in (22.19) cancel each other and give a uniquely defined function only if

$$2S = \beta_{nm}^2/4\pi - \bar{\beta}_{nm}^2/4\pi = \text{integer}$$

(22.20)

i.e. physical fields with uniquely defined correlation functions must have integer or half-integer conformal spins. This equation suggests the relation

$$T_2 = \frac{4\pi}{T_1} \equiv \sqrt{4\pi K} \tag{22.21}$$

as the minimal solution. Here I introduce the new notation K for future convenience. The normalization is such that at $K = 1$ the periods for the fields Φ and Θ are equal. The relationship (22.21) specifies the exponents of the multi-point correlation functions. It is universally accepted that these exponents, divided by 2, are called 'conformal dimensions'. In the case of model (22.1) the conformal dimensions of the basic operators are given by:

$$\Delta_{nm} \equiv \beta_{nm}^2/8\pi = \frac{1}{8}\left(n\sqrt{K} + \frac{m}{\sqrt{K}}\right)^2$$

$$\bar{\Delta}_{nm} \equiv \bar{\beta}_{nm}^2/8\pi = \frac{1}{8}\left(n\sqrt{K} - \frac{m}{\sqrt{K}}\right)^2 \tag{22.22}$$

Bosonization

Correlation functions of operators with integer conformal spins are invariant under permutation of coordinates; therefore such operators are bosonic. In contrast, when one permutes operators with half-integer conformal spins, the correlation function changes its sign. Therefore a field with a half-integer conformal spin is fermionic. As we have seen, both types of fields can be represented by bosonic exponents. Therefore for a fermionic theory with a linear spectrum, the fermionic operators can be represented in terms of a massless bosonic field and its dual field. The corresponding procedure is called bosonization. This was suggested by Coleman (1975) and Mandelstam (1975). We have already discussed this procedure in Chapter 14 in the section devoted to $(1 + 1)$-dimensional quantum electrodynamics. One can also find a thorough and up-to-date discussion of bosonization in Chapter 4 of Fradkin's book.

Let us consider pair correlation functions of the bosonic exponents at the special point:

$$\beta = \sqrt{4\pi}$$

Then we have

$$D(z) \equiv \frac{1}{2\pi a}\langle \exp[i\sqrt{4\pi}\,\varphi(z_1)]\exp[-i\sqrt{4\pi}\,\varphi(z_2)]\rangle = \frac{1}{2\pi z_{12}} \tag{22.23}$$

Let us regularize this function:

$$D_a(z, \bar{z}) = \frac{1}{2\pi z} = \lim_{a\to 0}\frac{1}{2\pi}\partial_z \ln(a^2 + |z|^2)$$

which then satisfies the equation

$$2\partial_{\bar{z}} D(z, \bar{z}) = \delta^2(x, y) \tag{22.24}$$

for the Green's function of the massless fermionic field with right chirality! (Recall the discussion around (14.22).) It is obvious that the corresponding antianalytic function

Table 22.1. Bosonization

Massless bosons	Massless fermions
Action	Action
$\dfrac{1}{2}\displaystyle\int d^2x(\nabla\Phi)^2$	$2\displaystyle\int d^2x(R^+\partial_{\bar z}R + L^+\partial_z L)$
Operators	Operators
$\pm\dfrac{1}{\sqrt{2\pi a}}\exp[\pm i\sqrt{4\pi}\,\varphi(z)]$	$R,\ R^+$
$\dfrac{1}{\sqrt{2\pi a}}\exp[\mp i\sqrt{4\pi}\,\bar\varphi(\bar z)]$	$L,\ L^+$
$\dfrac{1}{\pi a}\cos[\sqrt{4\pi}\,\Phi(z,\bar z)]$	$i(R^+L - L^+R)$
$\dfrac{i}{\sqrt{\pi}}\partial\Phi$	$:R^+R:$
$-\dfrac{i}{\sqrt{\pi}}\bar\partial\Phi$	$:L^+L:$

coincides with the Green's function of relativistic massless fermions with left chirality. So we can establish a correspondence between operators of the two theories which have the same correlation functions, as shown in Table 22.1.

This equivalence between the current operators has been discussed in Chapter 14.

Given the bosonization table, one can solve many highly nontrivial models. The simplest example is given by the following action:

$$S = \int d^2x\left\{\frac{1}{2}(\nabla\Phi)^2 - \frac{M}{\pi a}\cos(\sqrt{4\pi}\,\Phi)\right\} \equiv \int d^2x(\bar\psi\gamma^\mu\partial_\mu\psi + M\bar\psi\psi) \quad (22.25)$$

The equivalence follows from the fact that in the two theories the perturbation expansions in M coincide. Here we encounter the situation mentioned in the introduction to Part IV: a model which is nonlinear in the bosonic formulation is a disguised theory of free massive fermions!

Since chiral sectors in the Gaussian model have independent dynamics, one may wonder whether it is possible to write down independent effective actions for the fields ϕ, $\bar\phi$. Indeed, it is possible. One just has to notice that the propagator of the Φ-field can be written as a sum of the chiral parts:

$$\langle\langle\Phi(-\omega, -q)\Phi(\omega, q)\rangle\rangle = \frac{1}{q^2 + \omega^2} = \frac{1}{2q}\left(\frac{1}{q + i\omega} + \frac{1}{q - i\omega}\right) \quad (22.26)$$

or, in real space,

$$\ln(z\bar z) = \ln z + \ln\bar z$$

This decomposition is reproduced if we adopt the following chiral actions for fields φ and $\bar{\varphi}$ (Jackiw and Woo, 1975; Moon *et al.*, 1993):

$$S_R = \int d\tau dx \, \partial_x \varphi (\partial_x \varphi - i\partial_\tau \varphi)$$

$$S_L = \int d\tau dx \, \partial_x \bar{\varphi} (\partial_x \bar{\varphi} + i\partial_\tau \bar{\varphi})$$

(22.27)

This representation is especially convenient when one needs to calculate correlation functions of chiral exponents (like correlation functions of fermion fields). Below I shall use it to calculate the single-fermion correlation function in the Tomonaga–Luttinger model.

References

Coleman, S. (1975). *Phys. Rev. D*, **11**, 2088.
Jackiw, R. and Woo, G. (1975). *Phys. Rev. D*, **12**, 1643.
Mandelstam, S. (1975). *Phys. Rev. D*, **11**, 3026.
Moon, K. Yi, H., Kane, C. L., Girvin, S. M. and Fisher, M. P. A. (1993). *Phys. Rev. Lett.*, **71**, 4381.

23

Relevant and irrelevant fields

In this chapter we continue to study the massless scalar field described by the action (22.1). We have seen that this model has a gapless excitation spectrum and the correlation functions of bosonic exponents follow power laws. This behaviour implies that the correlation length is infinite and the system is in its critical phase. It is certainly very important to know how stable this critical point is with respect to perturbations. Any model is only an idealization; when one derives it certain interactions are neglected. How does one decide which interactions are important and which are not? Obviously, weak interactions are those whose influence on the correlation functions is small. The trouble is, however, that usually the correlation functions are affected differently on different scales, the long distance asymptotics being affected the most. Therefore it can happen that a certain perturbation causes only tiny changes at short distances, but changes the large distance behaviour profoundly. In the renormalization group picture this is observed as a growth of the coupling constant associated with the perturbing operator. Such growth is a frequent phenomenon in critical theories; a slow decay of their correlation functions gives rise to divergences in their diagram series.

Operators whose influence grows on large scales (small momenta) are called 'relevant'. The problem of relevancy of perturbations can be formulated and solved in a general form. Suppose we have a system at criticality (i.e. all its correlation functions decay as a power law at large distances). Let it be described at the critical point by the action S_0 and let the physical field $A_d(\mathbf{r})$ be a field with scaling dimension d, that is

$$\langle A_d(\mathbf{r})A_d^+(\mathbf{r}')\rangle \sim |\mathbf{r} - \mathbf{r}'|^{-2d}$$

Now let us consider the perturbed action

$$S = S_0 + g \int d^D r\, A_d(\mathbf{r})$$

Does it describe the same critical point, i.e. does the perturbation affect the scaling dimensions of the correlation functions?

Theorem. A perturbation with *zero conformal spin* and scaling dimension d is relevant if

$$d < D$$

and irrelevant if

$$d > D$$

The case $d = D$ is a marginal one and the answer depends on the sign of g. For two-dimensional theories we have the following remarkable result due to Polyakov (1972). He calculated the Gell-Mann–Low function of a general conformal theory perturbed by several marginal operators

$$S = S_0 + \sum_n g_n \int d^2 r\, A^n(\mathbf{r}) \tag{23.1}$$

Suppose that the operators A^n are normalized in such a way that

$$\langle A^n(r_1) A^m(r_2) \rangle = \delta_{nm} \frac{1}{|r_1 - r_2|^4}$$

Then the renormalization group equations for the coupling constants g_n have the form

$$\frac{dg_n}{d\ln(\Lambda/k)} = -2\pi C_n^{pq} g_p g_q + \mathcal{O}(g^3) \tag{23.2}$$

where C_n^{pq} are the coefficients of the third-order correlation functions

$$\langle A^p(r_1) A^q(r_2) A^l(r_3) \rangle = \frac{C_n^{pq}}{|r_1 - r_2|^2 |r_1 - r_3|^2 |r_2 - r_3|^2} \tag{23.3}$$

It is very instructive to discuss the consequences of the above theorem in the framework of our bosonic theory. This will provide a nice illustration and at the same time give us an opportunity to appreciate the importance of the previous discussions. Let us consider some perturbation of our bosonic theory. Since we have the operator basis of bosonic exponents we can expand all perturbations local in Φ and Θ. Since only perturbations with small scaling dimensions are really important, it is not necessary to consider fields containing derivatives of the bosonic exponents; all such fields have scaling dimensions larger than two and are irrelevant. Therefore it is sufficient to study the problem of relevance for the following simple perturbation:

$$g \int d^2 x \cos(\beta\Phi)$$

where β belongs to the content of conformal dimensions of our theory and the scaling dimension of the perturbation is equal to

$$d = \beta^2 / 4\pi$$

The perturbed model has the following action:

$$S = \int d^2 x \left[\frac{1}{2}(\nabla\Phi)^2 + g\cos(\beta\Phi) \right] \tag{23.4}$$

and is called the sine-Gordon model. In fact, this model is one of the most important models of $(1 + 1)$-dimensional physics. Minimal perturbations to criticality arise very frequently in different physical contexts and therefore we shall encounter the sine-Gordon model more than once in future discussions (recall the discussion of the model of spin nematic in Chapter 18).

For the sake of simplicity I shall consider the perturbation expansion of the simplest two-point correlation function of bosonic exponents:

$$\langle \exp[i\beta_0 \Phi(1)] \exp[-i\beta_0 \Phi(2)] \rangle$$

where β_0 is one of the permitted β. The first nonvanishing correction to this correlation function is equal to

$$g^2 \int d^2 x_3 d^2 x_4 \langle \exp[i\beta_0 \Phi(1)] \exp[-i\beta_0 \Phi(2)] \exp[i\beta \Phi(3)] \exp[-i\beta \Phi(4)] \rangle$$

$$\sim \int d^2 z_3 d^2 z_4 |z_1 - z_2|^{-\beta_0^2/2\pi} \left(\frac{|z_1 - z_3||z_2 - z_4|}{|z_1 - z_4||z_2 - z_3|} \right)^{\beta_0\beta/2\pi} |z_3 - z_4|^{-\beta^2/2\pi} \quad (23.5)$$

The double integral in this expression converges at large distances when

$$2d \equiv \beta^2/2\pi > 4$$

i.e. if $d > 2$. This means that at $d > 2$ the perturbation expansion does not contain infrared singularities; so if the bare coupling constant is small, its effect will remain small and will not be amplified in the process of renormalization.

Let us consider what happens if $d < 2$, i.e. for the case when the perturbation is relevant. In general, there are two possibilities. A relevant perturbation can drive the system to another critical point.[1] Another possibility is that the conformal symmetry is totally lost and the system acquires a finite correlation length. For a $(1 + 1)$-dimensional system this means that elementary excitations become massive particles. The latter occurs in the sine-Gordon model. This model will be discussed below in great detail in Chapter 34. As we shall see, the mass spectrum of the sine-Gordon model is strongly β dependent. In fact, we already know the spectrum at one point, for $d = 1$ ($\beta = \sqrt{4\pi}$): according to the bosonization scheme the sine-Gordon model with this β is equivalent to the model of free massive fermions (see (22.25)) with mass $m \propto g$. It is also easy to solve the model (23.4) at $d \ll 1$, where one can expand the cosine term around $\Phi = \pi/\beta$:

$$\cos(\beta\Phi) \approx -1 + \frac{1}{2}\beta^2(\Phi - \pi/\beta)^2$$

As a result we get the theory of massive bosons:

$$S_{sg} = \frac{1}{2} \int d^2 x [(\nabla \Phi')^2 + g\beta^2 \Phi'^2] \qquad \Phi' = \Phi - \pi/\beta \qquad (23.6)$$

---

[1] According to Zamolodchikov's theorem (Zamolodchikov, 1986), this new critical point always has a smaller central charge than the nonperturbed one. The definition of central charge is given later in Chapters 25 and 26.

with the following excitation spectrum and free energy:

$$\omega(p) = \sqrt{p^2 + m^2} \qquad m^2 = g\beta^2 \tag{23.7}$$

$$F/L = T \int \frac{dp}{(2\pi)} \ln\{1 - \exp[-\omega(p)/T]\} \tag{23.8}$$

Though the latter calculation does not describe the entire spectrum (it leaves behind the solitons), it illustrates the main point: all excitations are massive. For $\beta^2 \ll 1$ the spectrum has a gap as in the $\beta^2 = 4\pi$ case. The elementary excitations, however, are different: they are not fermions, but bosons. It is not difficult to calculate correlation functions of the bosonic exponents in this approximation. They are still given by (22.7), but with

$$G(\xi_1, \xi_2) = \frac{1}{(2\pi)^2} \int d^2k \frac{e^{ik\xi_{12}}}{k^2 + m^2} = \frac{K_0(m|\xi_{12}|)}{\pi}$$

For the pair correlation function, for instance, we have

$$D_{12} = \langle \exp[i\beta_0 \Phi(1)] \exp[-i\beta_0 \Phi(2)] \rangle - |\langle \exp[i\beta_0 \Phi(1)] \rangle|^2$$
$$\approx (ma)^{8\Delta}\{\exp[-4\Delta K_0(m|z_{12}|)] - 1\}$$

where $\Delta = \beta_0^2/8\pi$. At small distances $|z_{12}| \ll m^{-1}$ where $K_0(x) \approx \frac{1}{2}\ln(1/x)$ one recovers the unperturbed expression (22.8). At large distances $|z_{12}| \gg m^{-1}$ where $K_0(x) \approx \sqrt{\pi/2x}\exp(-x)$ we have

$$D_{12} = -8\Delta(ma)^{4\Delta}\exp(-m|z_{12}|)\sqrt{\pi/2|z_{12}|} + \cdots$$

The obtained results suggest that the apparently simple sine-Gordon model is rich in properties. A more detailed discussion of this is given in Chapter 34.

Let us now consider a case when the perturbation contains two operators with *integer* conformal spins $\pm S$, such as

$$V = g \int dz d\bar{z} \cos(\beta \Phi) \cos(\bar{\beta} \Theta) = \frac{g}{2} \int dz d\bar{z} \{\cos[(\beta + \bar{\beta})\phi + (\beta - \bar{\beta})\bar{\phi}]$$
$$+ \cos[(\beta - \bar{\beta})\phi + (\beta + \bar{\beta})\bar{\phi}]\} \tag{23.9}$$

where $\beta, \bar{\beta}$ belong to the set (22.22). Since $\beta\bar{\beta}/\pi = 2S(\text{even})$, we get for the scaling dimensions of both operators:

$$d = \frac{1}{4\pi}(\beta^2 + \bar{\beta}^2) = \frac{1}{4\pi}(\beta^2 + 4S^2\pi^2/\beta^2) \geq S \tag{23.10}$$

The criterion $d < 2$ suggests that the operator V is irrelevant (marginal) for $S > 2$ ($S = 2$) and for $S = 1$ restricts the area of its relevance by

$$\frac{\beta^2}{2\pi} < 2 - \sqrt{3} \tag{23.11}$$

(since the action in this case is invariant under the duality transformation $\Phi \rightarrow \Theta$, $\beta \rightarrow \bar{\beta}$, we shall consider $\beta^2 < 2\pi$ only). We shall see in a moment that these statements are not correct and that the criterion of relevance must be modified.

To establish a new criterion, let us consider the perturbation series for the free energy. The first divergence in the expansion of the partition function appears in the fourth order in g:

$$Z^{(4)} \sim g^4 \int d^2x_1 \cdots d^2x_4 \langle \exp[i(\beta + \bar{\beta})\phi(1)] \exp[i(\beta - \bar{\beta})\phi(2)] \exp[i(-\beta + \bar{\beta})\phi(3)]$$
$$\times \exp[i(-\beta - \bar{\beta})\phi(4)]\rangle \langle \exp[i(\beta - \bar{\beta})\bar{\phi}(1)] \exp[i(\beta + \bar{\beta})\bar{\phi}(2)]$$
$$\times \exp[i(-\beta - \bar{\beta})\bar{\phi}(3)] \exp[i(-\beta + \bar{\beta})\bar{\phi}(4)]\rangle$$
$$+ (\beta \to \bar{\beta}) \tag{23.12}$$

Suppose that

$$(\bar{\beta}^2 - \beta^2)/4\pi > 1 \tag{23.13}$$

which corresponds to

$$\frac{\beta^2}{2\pi} < -1 + \sqrt{S^2 + 1} \tag{23.14}$$

Then the integrals in x_{12} and x_{34} converge at small distances. Calculating these integrals first we can rewrite approximately

$$\int dz_2 \exp[i(\beta + \bar{\beta})\phi(1)] \exp[i(\beta - \bar{\beta})\phi(2)] \sim \exp[2i\beta\phi(1)]a^{1+(\beta^2 - \bar{\beta}^2)/4\pi} \tag{23.15}$$

The remaining integral in x_1 and x_3 is

$$Z^{(4)} \sim g_1^2 \int d^2x_1 d^2x_3 \langle \cos[2\beta\Phi(1)] \cos[2\beta\Phi(3)] \rangle \qquad g_1 = g^2 a^{2+(\beta^2 - \bar{\beta}^2)/2\pi} \tag{23.16}$$

In the opposite case $\beta > \bar{\beta}$ one can repeat the above derivation interchanging β and $\bar{\beta}$. The integrals convergent at small distances are now integrals in x_{13} and x_{24}. In this case the procedure generates the perturbation $g_2 \cos[2\bar{\beta}\Theta]$ where $g_2 = g^2 a^{2+(\bar{\beta}^2 - \beta^2)/2\pi}$. The criterion (23.13) is replaced by

$$(\beta^2 - \bar{\beta}^2)/4\pi > 1 \tag{23.17}$$

Thus the long distance behaviour of the correlation functions is the same as if we had the perturbations

$$V_1 = g_1 \cos(2\beta\Phi) \qquad (\beta < \bar{\beta})$$
$$V_2 = g_2 \cos(2\bar{\beta}\Theta) \qquad (\beta > \bar{\beta}) \tag{23.18}$$

Thus we see that the original perturbation with nonzero conformal spin generates the perturbations (23.18) with zero conformal spin. For their relevance we have the standard criteria supplemented by (23.13) for V_1 or by (23.17) for V_2. This mechanism of generation of relevant perturbations was first described by Brazovsky and Yakovenko (1985) and later by Kusmartsev et al. (1992).

References

Brazovsky, S. A. and Yakovenko, V. M. (1985). *J. Phys. Lett. (Paris)*, **69**, 46; *Sov. Phys. JETP*, **62**, 1340 (1985).

Kusmartsev, F. V., Luther, A. and Nersesyan, A. A. (1992). *JETP Lett.*, **55**, 692; *Phys. Lett. A*, **176**, 363 (1993).

Polyakov, A. M. (1972). *Zh. Eksp. Teor. Fiz.*, **63**, 24.

Zamolodchikov, A. B. (1986). *JETP Lett.*, **43**, 730. See also in the book *Conformal Invariance and Applications to Statistical Mechanics*, ed. C. Itzykson, H. Saleur and J.-B. Zubert. World Scientific, Singapore, 1988.

24

Kosterlitz–Thouless transition

In the previous chapter we considered the sine-Gordon model, a model of free massless bosons perturbed by the $\cos(\beta\Phi)$ term. As we know, the complete basis of operators in our model also includes exponents of the dual field $\Theta(z, \bar{z})$. These exponents are also fields with zero conformal spin and are therefore subject to the criteria of relevance discussed in the previous chapter. According to these criteria, the $\cos(\tilde{\beta}\Theta)$ perturbation is relevant for $\tilde{\beta}^2/4\pi < 1$. What is really interesting to discuss is the possible mechanism of appearance of such perturbations. The O(2)-symmetric nonlinear sigma model, which we have already discussed in another context in Part II, provides an excellent opportunity to discuss a mechanism of generation of dual exponents. Let us recall its effective action:

$$S = \frac{M_0^2}{2T} \int d^2x (\partial_\mu \mathbf{n})^2 \qquad \mathbf{n}^2 = 1 \tag{24.1}$$

where $\mathbf{n}$ is a unit vector with two components and M_0 is the average 'radial' component of the vector $\mathbf{\Phi}$: $\langle \mathbf{\Phi}^2 \rangle = M_0^2$. The two-component unit vector field $\mathbf{n}$ can be parametrized as follows:

$$\mathbf{n} = [\cos(\Phi\sqrt{T}/M_0), \sin(\Phi\sqrt{T}/M_0)]$$

The action (24.1) in this parametrization apparently coincides with the action of the free scalar field (22.1). The fact that operators of the original model (24.1), being local functionals of $\mathbf{n}$, are periodic functionals of Φ with a period

$$T_1 = 2\pi M_0/\sqrt{T} \tag{24.2}$$

specifies the set of scaling dimensions. They are given by (22.22) with $K = T/\pi M_0^2$. The dual exponents have the period

$$T_2 = 4\pi/T_1 = 2\sqrt{T}/M_0$$

The pair correlation function of $\mathbf{n}$-fields decays as a power law:

$$\langle \mathbf{n}(\xi_1)\mathbf{n}(\xi_2) \rangle = M_0^2 \left(\frac{a}{|\xi_{12}|} \right)^{T/2\pi M_0^2}$$

Now I am going to show that this simple-minded derivation contains a very important mistake. The careful derivation gives additional terms, i.e. the effective action for the model

Figure 24.1. The vector field looks like a plane flow current with a source at $\mathbf{r}_0$.

(24.1) is described not by the action (22.1), but by this action perturbed by the dual exponents:

$$S = \frac{1}{2} \int d^2x \left[(\nabla \Phi)^2 + \sum_{k=1} A_k \cos \left(\frac{2\pi k M_0 \Theta}{\sqrt{T}} \right) \right] \qquad (24.3)$$

where A_k are some coefficients.[1] These perturbations are relevant at large temperatures $T > \pi M_0^2/2$ where a finite correlation length $\xi(T)$ arises and the correlation functions decay exponentially at distances larger than ξ. In order to see this, let us recall the path integral procedure. In a path integral one integrates over continuous functions, in the present case over $\Phi(x)$. It seems that all singular functions are automatically excluded since, giving an infinitely large contribution to the exponent, they give a zero contribution to the path integral. This is not correct, however, because we have regularized our theory and all singularities are smoothed out over a distance $\sim a$. Therefore contributions from singular configurations to the action are never zero, rather they contain some negative powers of a. In particular, there are configurations of fields which give logarithmically divergent contributions $S \propto \ln a$. The simplest configuration is

$$\mathbf{n} = (\cos \alpha, \sin \alpha) = \frac{\mathbf{r} - \mathbf{r}_0}{|\mathbf{r} - \mathbf{r}_0|} \qquad (24.4)$$

where α is the angle between the $\hat{x}$-axis and a radius vector drawn from some point $\mathbf{r}_0$ (see Fig. 24.1). This configuration is singular because the vector field has an indefinite direction at $\mathbf{r} = \mathbf{r}_0$. Graphically the vector field looks like a plane flow current with a source at $\mathbf{r}_0$.

The configuration (24.4) gives the following contribution to the action:

$$S_0 = \frac{M_0^2}{2T} \int \frac{d^2r}{r^2} \approx \frac{\pi M_0^2}{T} \ln(R/a)$$

This contribution is infinite for an infinite system. A finite contribution comes from configurations containing a source and a 'drain' (Fig. 24.2). For this configuration we have $M_0 \Phi/\sqrt{T} = \alpha_1 - \alpha_2$ (the angles are defined in Fig. 24.2) and the action is estimated as:

$$S_{\text{dip}} \approx \frac{2\pi M_0^2}{T} \ln(l/a) \qquad (24.5)$$

where l is the distance between the source and the drain.

It is easy to see that all singular configurations of the $\Phi(z, \bar{z})$-field can be represented by branch cut singularities of a logarithmic function. One can write down a general function

---

[1] Note that these perturbations contain dual exponents with *even* numbers only.

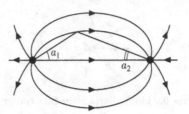

Figure 24.2. A field configuration containing a source and a 'drain'.

$\Phi(z, \bar{z})$ as a sum of singular and regular pieces:

$$\Phi = i \sum_i e_i \ln \left(\frac{z - z_i}{\bar{z} - \bar{z}_i} \right) + \Phi_{\text{reg}} \tag{24.6}$$

where

$$e_i = \pm \frac{M_0}{2\sqrt{T}}$$

Exercise

Check that the configuration in Fig. 24.1 corresponds to

$$\Phi_1 = -ie \ln \left(\frac{z - z_0}{\bar{z} - \bar{z}_0} \right)$$

and the configuration in Fig. 24.2 corresponds to

$$\Phi_2 = ie \ln \left(\frac{z - z_1}{\bar{z} - \bar{z}_1} \right) - ie \ln \left(\frac{z - z_2}{\bar{z} - \bar{z}_2} \right)$$

Substituting (24.6) into (22.1) we get

$$\partial \Phi \bar{\partial} \Phi = \left(i \sum_i e_i \frac{1}{z - z_i} + \partial \Phi_{\text{reg}} \right) \left(-i \sum_i e_i \frac{1}{\bar{z} - \bar{z}_i} + \bar{\partial} \Phi_{\text{reg}} \right)$$

To proceed further we first replace the derivative of Φ_{reg} by the derivative of Θ using the definition of the Θ-field (22.14) and then integrate the obtained expressions by parts using the identity

$$\partial_z \frac{1}{\bar{z}} + \partial_{\bar{z}} \frac{1}{z} = 2\pi \delta^{(2)}(x, y) \tag{24.7}$$

The resulting expression is given by

$$S_N \equiv 2 \int dz d\bar{z} \partial \Phi \bar{\partial} \Phi = S_{\text{cl}} + 4\pi i \sum_i e_i \Theta_{\text{reg}}(z_i, \bar{z}_i) + 2 \int dz d\bar{z} \partial \Phi_{\text{reg}} \bar{\partial} \Phi_{\text{reg}}$$

$$S_{\text{cl}} = 2 \int dz d\bar{z} \sum_{i,j} \frac{e_i e_j}{(z - z_i)(\bar{z} - \bar{z}_j)} = 2\pi \sum_{i,j} e_i e_j \ln \left(\frac{R^2}{|z_i - z_j|^2 + a^2} \right) \tag{24.8}$$

Thus the entire partition function has the following form:

$$Z = \sum_{N=0}^{\infty} \frac{1}{N!} \int \prod_{i=0}^{N} dz_i d\bar{z}_i \int D\Phi_{\text{reg}} \exp(-S_N)$$

$$= Z_0 + Z_0 \sum_{N=1}^{\infty} \frac{1}{N!} \prod_{i=1}^{N} \int dz_i d\bar{z}_i e^{(-S_{\text{cl}})} \langle \exp[4\pi i e_1 \Theta(1)] \cdots \exp[4\pi i e_N \Theta(N)] \rangle \quad (24.9)$$

Where Z_0 is the partition function without singularities which includes integration over Φ_{reg}.

Let us first consider the classical term S_{cl}, which appears in the exponent in front of the correlation function. For the infinite system S_{cl} is finite only for neutral configurations containing an equal number of sources and drains, which implies the condition $\sum_i e_i = 0$. According to the analysis given in Chapter 22, the same condition is necessary for the correlation function of the exponents to be nonzero. As is obvious from (24.8), S_{cl} represents the classical energy of a two-dimensional plasma of electric charges. This energy is finite for electrically neutral configurations and its magnitude is given by:

$$S_{\text{cl}} \approx N \frac{\pi M_0^2}{T} \ln(l/a) \quad (24.10)$$

where $l \sim R/\sqrt{N}$ is the average distance between the charges. Since the classical energy for each even N (electroneutrality!) is finite, the relevance of the singular configurations depends on convergence of the integrals of the correlation functions of the bosonic exponents. It is not difficult to figure out that correlation functions of the dual and the direct exponents are given by the same expression (22.10):

$$\langle \exp[4\pi i e_1 \Theta(1)] \cdots \exp[4\pi i e_N \Theta(N)] \rangle = \prod_{i>j} (|z_i - z_j|/a)^{8\pi e_i e_j} \quad (24.11)$$

The first two nonvanishing correlators are equal to:

$$\langle \exp[4\pi i e \Theta(1)] \exp[-4\pi i e \Theta(2)] \rangle = (|z_1 - z_2|/a)^{-8\pi e^2}$$

$$\langle \exp[4\pi i e \Theta(1)] \exp[-4\pi i e \Theta(2)] \exp[4\pi i e \Theta(3)] \exp[-4\pi i e \Theta(4)] \rangle \quad (24.12)$$

$$\sim \left(\frac{|z_1 - z_3||z_2 - z_4|}{|z_1 - z_2||z_1 - z_4||z_2 - z_3||z_3 - z_4|} \right)^{8\pi e^2}$$

Now by induction we can figure out the leading divergence of the correlation function at large distances. For this purpose we multiply all $z_i, \bar{z}_i$ in the Nth term in the expansion of the partition function by λ. Since the entire expression acquires a factor

$$\lambda^{N(2-4\pi e^2)}$$

we conclude that the integral converges only if the power of λ is negative, i.e.

$$2 - 4\pi e^2 < 0 \rightarrow \frac{\pi M_0^2}{2T} > 1 \quad (24.13)$$

Thus at temperatures lower than

$$T_{KT} = \frac{\pi M_0^2}{2}$$

the singular configurations can be taken into account perturbatively (KT is an abbreviation of the names Kosterlitz and Thouless who discovered the described phenomena). In contrast, at $T > T_{KT}$ when the integrals diverge, the singular configurations make an essential contribution and must be taken into account. It can be shown that at $T > T_{KT}$ the correlation functions decay exponentially at distances larger then the correlation length

$$\xi \sim \exp\left[\frac{\text{const}}{\sqrt{(T - T_{KT})}}\right] \tag{24.14}$$

At $T = T_{KT}$ the correlation length becomes infinite and the correlation functions of the **n**-field abruptly change their asymptotic behaviour from exponential to power law:

$$\langle \mathbf{n}(x)\mathbf{n}(y)\rangle|_{T=T_{KT}-0} \sim \frac{1}{|x - y|^{T_{KT}/2\pi M_0^2}} \tag{24.15}$$

$$T_{KT}/2\pi M_0^2 = 1/4$$

Let us summarize the physics. The O(2) nonlinear sigma model (24.1) has been derived as the long-wavelength action of the O(2) vector model in its low temperature phase. The action (24.1) includes the quantity M_0^2 which is equal to the average value of $|\Phi|^2$. Thus we expect this model to be valid in the temperature range where $|\Phi|$ fluctuates weakly. The quantity M_0^2 is the bare stiffness of the system. A simple analysis of dimensions shows that the characteristic temperature at which radial fluctuations of the order parameter are smaller than the transverse ones is $T^* \sim M_0^2$. Therefore at $T < T^*$ one can describe the system in terms of the sigma model. As follows from the previous discussion, there is a temperature interval $T_{KT} < T < T^*$ where the radial fluctuations are weak, but the angular ones are so strong that the system at large distances has zero stiffness. The latter statement is equivalent to saying that the correlation length is finite: parts of the system separated by distances larger than ξ are effectively disjointed and do not sense each other. At $T = T_{KT}$ the stiffness rises discontinuously. Therefore this point is a point of phase transition, the Kosterlitz–Thouless transition. Since at $T < T_{KT}$ the system remains critical all the way down to $T = 0$, T_{KT} marks the upper end of this critical line.

It is straightforward to generalize the results obtained in this chapter for the single bosonic field to the case of several bosonic fields, as discussed at the end of Chapter 18.

How the nonlinear sigma model on a torus becomes a linear one

As the conclusion to this chapter I would like to consider a model which illustrates rather nicely several important points about field theory in general and field theory in two dimensions in particular. Recall the nonlinear sigma models described in Chapter 9. In general their action is represented by (9.9). It has been mentioned in Chapter 9, that nonlinear sigma models on tori can be critical. Since this is an interesting statement, let us consider a sigma model on a two-dimensional torus (see Fig. 24.3).

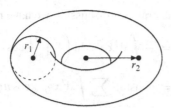

Figure 24.3.

The interval on the torus is given by

$$ds^2 = (r_2 + r_1 \cos \phi)^2 d\psi^2 + r_1^2 d\phi^2 \tag{24.16}$$

which defines the metric tensor $G_{\psi\psi} = (r_2 + r_1 \cos \phi)^2$, $G_{\phi\phi} = r_1^2$. Therefore the action is given by

$$S = \frac{1}{2} \int d^2x \left[(r_2 + r_1 \cos \phi)^2 (\partial_\mu \psi)^2 + r_1^2 (\partial_\mu \phi)^2 \right] \tag{24.17}$$

From what has been said in Chapter 9 we can expect that the nonlinear terms become irrelevant and the nonlinear sigma model becomes a linear one. Let us therefore separate the nonlinear terms and consider their scaling dimensions:

$$\delta S = \int d^2x \, A$$

$$A = \left(r_1 r_2 \cos \phi + \frac{r_1^2}{4} \cos 2\phi \right) (\partial_\mu \psi)^2 = \text{const} + \left(r_1 r_2 : \cos \phi : + \frac{r_1^2}{4} : \cos 2\phi : \right) : (\partial_\mu \psi)^2 :$$

$$+ \left(r_1 r_2 : \cos \phi : + \frac{r_1^2}{4} : \cos 2\phi : \right) \langle (\partial_\mu \psi)^2 \rangle \tag{24.18}$$

Obviously, the scaling dimensions of the first terms are larger than 2 and these terms are irrelevant. The last term, however, contains $: \cos \phi :, : \cos 2\phi :$ operators with scaling dimensions $d_1 = 1/2\pi r_1^2$ and $d_2 = 2/\pi r_1^2$ which can be smaller than 2. Thus something is wrong with our arguments. What is it which must be taken into account? I hope that in Chapter 24 we are not far enough from the earlier chapters to forget about measures of integration. In the above discussion I have deliberately said nothing about the measure, and the measure makes a lot of difference. For a nonlinear sigma model the measure is determined by the same metric tensor G_{ab}. That is, the path integral is defined as a limit of the multi-dimensional integral

$$d\mu = \prod_x \left[\sqrt{G(x)} dX^1(x) \cdots dX^N(x) \right] = \prod_x r_1[r_2 + r_1 \cos \phi(x)] d\psi(x) d\phi(x) \tag{24.19}$$

One can rewrite the measure as follows:

$$d\mu \sim \prod_x d\psi(x) d\phi(x) \exp \left\{ \sum_x \ln \left[1 + \frac{r_1}{r_2} \cos \phi(x) \right] \right\} \tag{24.20}$$

which makes it clear that in the previous discussion we have missed the term

$$\delta S = -\int d^2 x a^{-2} \ln\left(1 + \frac{r_1}{r_2} \cos\phi\right)$$

$$= \text{const} + \sum_{n=1}^{\infty} \int d^2 x g_n : \cos n\phi :$$

(24.21)

The latter expansion should be combined with the last terms in (24.18). One can check that the signs are correct and we get a cancellation. Therefore the nonlinear terms in the metric tensor are indeed irrelevant and the only relevant perturbations can come from vortex configurations of the fields ϕ and ψ, i.e. from the dual exponents. There are two dual fields in this case, Θ_1 for ϕ and Θ_2 for ψ. The dual exponents with the smallest scaling dimensions are $\exp(\pm 2 i \pi r_1 \Theta_1)$, their scaling dimension is $2\pi r_1^2$ and they are irrelevant at $S_1 = \pi r_1^2 > 1$, and $\exp(\pm 2 i \pi \tilde{R}\Theta_2)$, where $\tilde{R}^2 = \langle (r_2 + r_1 \cos\phi)^2 \rangle \approx r_2^2 + r_1^2/2$. The operator with the latter exponent is irrelevant at $\pi \tilde{R}^2 > 1$. Quantities πr_1^2 and $\pi \tilde{R}^2$ have a clear geometrical meaning (see Fig. 24.3). Thus the sigma model on a torus is critical when the torus itself is sufficiently large.

25
Conformal symmetry

As I have already demonstrated, despite its humble appearance the model of the free massless bosonic field is not such a trivial thing. In fact, we have not yet exhausted its wonders. The next wonderful property of this model is the presence of a special hidden symmetry: the conformal symmetry. This symmetry becomes manifest when one transforms the area A on which the field Φ is defined. Let A be an arbitrary area of the complex plane. As I have said before, the expression (22.10) for multi-point correlation functions of the bosonic exponents is valid for any A:

$$\langle \exp[i\beta_1 \Phi(\xi_1)] \cdots \exp[i\beta_N \Phi(\xi_N)] \rangle$$

$$= \exp\left[-\sum_{i>j} \beta_i \beta_j G(\xi_i; \xi_j) \right] \exp\left[-\frac{1}{2} \sum_i \beta_i^2 G(\xi_i; \xi_i) \right] \qquad (25.1)$$

One should just treat G as the Green's function of the Laplace operator on A. As shown in the theory of the Laplace equation, the Green's function $G(\xi_i; \xi_j)$ can be written explicitly if one knows a transformation $z(\xi)$ which maps A onto the infinite plane. Then the Green's function of the Laplace operator on A is given by:

$$G(\xi_1, \xi_2) = -\frac{1}{2\pi} \ln|z(\xi_1) - z(\xi_2)| - \frac{1}{4\pi} \ln[|\partial_{\xi_1} z(\xi_1) \partial_{\xi_2} z(\xi_2)|] \qquad (25.2)$$

Let us consider the case of two exponents: $N = 2$. Recall that the correlation functions (25.1) are products of analytic and antianalytic parts. Substituting expression (25.2) into (25.1) with $N = 2$ and $\beta_1 = \beta$, $\beta_2 = -\beta$, we get the following expression for the analytic part of the pair correlation function:

$$D(\xi_1; \xi_2) = \frac{1}{[z(\xi_1) - z(\xi_2)]^{2\Delta}} [\partial_{\xi_1} z(\xi_1) \partial_{\xi_2} z(\xi_2)]^{\Delta} \qquad (25.3)$$

where $\Delta = \beta^2/8\pi$.

As we see, the correlation functions transform locally under analytic coordinate transformations. Therefore one can consider these transformation properties as properties of the corresponding operators: the bosonic exponents $A_\Delta(z) \equiv \exp[i\beta\phi(z)]$. As is clear from (25.3), these operators transform as tensors of rank $(\Delta, 0)$:

$$A_\Delta(\xi) = A_\Delta[z(\xi)](dz/d\xi)^{\Delta} \qquad (25.4)$$

Correspondingly, the antianalytical exponent $\bar{A}_{\bar{\Delta}}(\bar{z})$ transforms as a tensor of rank $(0, \bar{\Delta})$.

There is one transformation which is particularly important for applications:

$$z(\xi) = \exp(2\pi\xi/L) \tag{25.5}$$

This transforms a strip of width L into a plane. For $(1 + 1)$-dimensional systems this transformation (i) relates correlation functions of finite quantum chains to those of infinite systems, and (ii) relates correlation functions at $T = 0$ to correlation functions at finite temperatures (in the latter case $T = i/L$). Substituting $z(\xi)$ into (25.3) we get

$$D(\xi_1; \xi_2) = \langle A_\Delta(z_1)A_\Delta^+(z_2)\rangle = \left\{\frac{\pi/L}{\sinh[\pi(\xi - \xi')/L]}\right\}^{2\Delta}$$
$$\tag{25.6}$$
$$\bar{D}(\xi_1; \xi_2) = \langle \bar{A}_{\bar{\Delta}}(\bar{z}_1)\bar{A}_{\bar{\Delta}}^+(\bar{z}_2)\rangle = \left\{\frac{\pi/L}{\sinh[\pi(\bar{\xi} - \bar{\xi}')/L]}\right\}^{2\bar{\Delta}}$$

Below I continue to treat our system as a quantum $(1 + 1)$-dimensional one at $T = 0$. Let $T = 0$ and the system be a circle of a finite length L. In this case we have

$$\xi = \tau + ix \qquad -\infty < \tau < \infty \qquad 0 < x < L$$

Let us expand the expressions for $D(1, 2)$, $\bar{D}(1, 2)$ at large τ_{12}. The result is

$$D(\xi_1; \xi_2)\bar{D}(\xi_1; \xi_2) = \sum_{n,m=0}^{\infty} C_{nm}(\pi/L)^d \exp\left[-\frac{2\pi}{L}(d + n)\tau_{12}\right] \exp\left[-ix_{12}\frac{2\pi}{L}(S + m)\right]$$
$$\tag{25.7}$$

This correlation function can be expanded in the Lehmann series:

$$D(\tau_{12}, x_{12})\bar{D}(\tau_{12}, x_{12}) = \sum_q |\langle 0| \exp[i\beta\phi(0) + i\bar{\beta}\bar{\phi}(0)]|q\rangle|^2 \exp[-E_q\tau_{12} - iP_q x_{12}]$$
$$\tag{25.8}$$

where q denote eigenstates of the Hamiltonian and E_q, P_q are eigenvalues of energy and momenta of the state q. Comparing these two expansions we find:

$$E_q = \frac{2\pi}{L}(d + n) \tag{25.9}$$

$$P_q = \frac{2\pi}{L}(S + m) \tag{25.10}$$

$$\left(\frac{\pi}{L}\right)^{(d+n)} C_{nm} = \sum_q |\langle 0| \exp[i\beta\phi(0) + i\bar{\beta}\bar{\phi}(0)]|q\rangle|^2$$
$$\times \delta\left[E_q - \frac{2\pi}{L}(d + n)\right]\delta\left[P_q - \frac{2\pi}{L}(S + m)\right] \tag{25.11}$$

All three expressions have a profound significance. The first two relate conformal dimensions of the correlation functions to the eigenvalues of energy and momentum operators. It is usually much easier to calculate energies than correlation functions. In the latter case the amount of computational work greatly increases since one must also calculate the matrix elements. Therefore it is a tremendous relief that for a quantum system with a gapless spectrum the relationships (25.9) and (25.10) allow us to avoid direct calculations of the correlation functions. Instead one can solve the problem of low-lying energy levels in a finite size system (which can be done numerically or even exactly) and then, using the relationships (25.6), (25.9) and (25.10), restore the correlation functions.

From (25.9) and (25.10) we see that the problem of conformal dimensions can be formulated as an eigenvalue problem for the following operators:

$$\Delta = \frac{L}{4\pi}(\hat{H} + \hat{P}) \equiv \hat{T}_0 \tag{25.12}$$

$$\bar{\Delta} = \frac{L}{4\pi}(\hat{H} - \hat{P}) \equiv \hat{\bar{T}}_0 \tag{25.13}$$

It can be shown that the operators $\hat{T}_0$, $\hat{\bar{T}}_0$ are related to analytic and antianalytic components of the stress energy tensor defined as

$$T_{ab}(x) = \frac{\delta S}{\delta g^{ab}(x)} \tag{25.14}$$

Namely,

$$\hat{T}_0 = \int_0^L dx\, T_{zz} \qquad \hat{\bar{T}}_0 = \int_0^L dx\, T_{\bar{z}\bar{z}}$$

It is not unnatural that the stress energy tensor appears in the present context. Indeed, the scaling dimensions are related to coordinate transformations and such transformations change the metric on the surface:

$$dz\, d\bar{z} \rightarrow (dz/d\xi)(d\bar{z}/d\bar{\xi}) d\xi\, d\bar{\xi}$$

The relations between scaling dimensions and the stress energy tensor have a general character and hold for all conformal theories. To make the new concept of the stress energy tensor more familiar, I calculate T_{ab} explicitly for the theory of free bosons. The action on a curved background is given by (recall Chapter 4)

$$S = \frac{1}{2} \int d^2x \sqrt{g}\, g^{ab} \partial_a \Phi \partial_b \Phi \tag{25.15}$$

where $g = \det g$ and $g^{ab} = (g^{-1})_{ab}$. In the vicinity of a flat space where the metric tensor is equal to $g_{ab} = \delta_{ab}$ we get:

$$T_{ab} = \frac{1}{2} : \left[\partial_a \Phi \partial_b \Phi - \frac{1}{2} \delta_{ab} (\partial_c \Phi)^2 \right] : \tag{25.16}$$

(the dots denote the normal ordering, i.e. it is assumed that the vacuum average of the operator is subtracted, $\langle T_{ab} \rangle = 0$).

Gaussian model in the Hamiltonian formulation

Since we need the eigenvalues of $\hat{T}_0$, $\hat{\bar{T}}_0$, it is convenient to express the Hamiltonian and the momentum operator $\hat{P}$ in terms of the creation and the annihilation operators. Thus from this moment we shall work in the Hamiltonian formalism. Our free bosonic theory (22.1) has the following Hamiltonian:

$$\hat{H}_0 = \frac{v}{2} \int dx [\hat{\pi}^2 + (\partial_x \Phi)^2]$$

$$[\pi(x), \Phi(y)] = -i\delta(x - y)$$

(25.17)

We shall assume that the velocity $v = 1$ as before. Using the fact that $\partial_x \Theta = \hat{\pi}$ we can rewrite the Hamiltonian in a slightly different form:

$$\hat{H}_0 = \frac{v}{2} \int dx [(\partial_x \Theta)^2 + (\partial_x \Phi)^2]$$

$$[\Theta(x), \Phi(y)] = -i\theta_H(x - y)$$

(25.18)

where $\theta_H(x)$ is the step function. From the latter commutation relations one can derive useful commutation relations for the holomorphic and antiholomorphic fields φ and $\bar{\varphi}$:

$$[\varphi(x), \bar{\varphi}(y)] = \frac{1}{4}[\Phi(x) + \Theta(x), \Phi(y) - \Theta(y)] = \frac{1}{4}\{[\Phi(x), \Theta(y)] - [\Theta(x), \Phi(y)]\} = -\frac{i}{4}$$

(25.19)

Thus these fields are not quite independent; however, this property influences correlation functions only at the distances of the order of the ultraviolet cut-off. Using this commutation relation and the Hausdorff formula

$$e^{\hat{A}} e^{\hat{B}} = e^{\hat{A}+\hat{B}} e^{\frac{1}{2}[\hat{A},\hat{B}]}$$

valid when the commutator $[\hat{A}, \hat{B}]$ is a c-number, we arrive at the following identity:

$$\exp[i\beta\varphi + i\bar{\beta}\bar{\varphi}] = \exp(i\beta\varphi) \exp(i\bar{\beta}\bar{\varphi}) \exp\left(\frac{i}{8}\beta\bar{\beta}\right)$$

(25.20)

The Hamiltonian (25.17) describes a set of coupled oscillators (a string). In order to introduce creation and annihilation operators, we expand the field Φ into the normal modes:

$$\Phi(\tau, x) = \Phi_0 + \sqrt{\pi}[iJ\tau + Qx]/L + \sum_{q \neq 0} \frac{1}{\sqrt{2|q|}} (\hat{a}_q^+ e^{|q|\tau - iqx} + \hat{a}_q e^{-|q|\tau + iqx})$$

(25.21)

where $q = 2\pi k/L$ (k is an integer). Then the momentum operator $\hat{\pi} = i\partial_\tau \Phi$ is

$$\hat{\pi}(\tau, x) = -\sqrt{\pi}J/L + i\sum_{q \neq 0} \sqrt{|q|/2}(\hat{a}_q^+ e^{|q|\tau - iqx} - \hat{a}_q e^{-|q|\tau + iqx})$$

(25.22)

As follows from the expressions for currents given in Table 22.1, the quantities Q and J are the total charge of the system and the total current through it.

It is straightforward to extract from the operator representation (25.21) the analytic and the antianalytic components of the field $\Phi(z, \bar{z})$:

$$\Phi(\tau, x) = \varphi(\tau + ix) + \bar{\varphi}(\tau - ix)$$
$$\Theta(\tau, x) = \varphi(\tau + ix) - \bar{\varphi}(\tau - ix)$$

$$\varphi(z) = \varphi_0 + \frac{i\sqrt{\pi}}{2}(J - Q)z/L + \sum_{q>0} \frac{1}{\sqrt{2q}}(e^{-qz}\hat{a}_{-q} + e^{qz}\hat{a}_{-q}^+) \qquad (25.23)$$

$$\bar{\varphi}(\bar{z}) = \bar{\varphi}_0 + \frac{i\sqrt{\pi}}{2}(J + Q)\bar{z}/L + \sum_{q>0} \frac{1}{\sqrt{2q}}(e^{q\bar{z}}\hat{a}_q^+ + e^{-q\bar{z}}\hat{a}_q) \qquad (25.24)$$

where $[\varphi_0, \bar{\varphi}_0] = -i/4$. Let us show that J and Q are related to the scaling dimensions. Indeed, we have postulated that we consider only the operators periodic in Φ and Θ with periods T_1 and $4\pi/T_1$, respectively. The fields Φ, Θ are defined on the circle with circumference L, but they are not periodic functions of x: as follows from (25.23) and (25.24), $\varphi(z + iL) = \varphi(z) - \sqrt{\pi}(J - Q)/2$, $\bar{\varphi}(\bar{z} - iL) = \bar{\varphi}(\bar{z}) + \sqrt{\pi}(J + Q)/2$. Therefore in order to maintain the periodicity of the operators $\exp[2i\pi(\varphi + \bar{\varphi})n/T_1 + i(\varphi - \bar{\varphi})T_1 m/2]$ one needs to satisfy the following conditions:

$$Q = T_1 m/2\sqrt{\pi} = \frac{m}{\sqrt{K}} \qquad J = 2n\sqrt{\pi}/T_1 = n\sqrt{K} \qquad (25.25)$$

Substituting expressions (25.23) and (25.24) for Φ into the Hamiltonian (25.17) we get the *Tomonaga–Luttinger* Hamiltonian:

$$\hat{H} = \frac{\pi v}{2L}\left(n^2 K + \frac{m^2}{K}\right) + \sum_q v|q|a_q^+ a_q \qquad (25.26)$$

(I have restored the velocity in this expression). The last term has eigenvalues $2\pi v/L \times$ integer so that the entire expression reproduces correctly those eigenvalues which follow from the correlation functions. The first two terms of the Hamiltonian (25.26) describe the motion of the centre of mass of the string which is quantized due to the imposed periodicity requirements. It is instructive also to have expressions for T_0 and $\bar{T}_0$. Since the stress energy tensor is defined with the normal ordering, we must subtract the infinite nonuniversal part of the ground state energy keeping, however, its finite L-dependent universal part:

$$T_0 = -\frac{1}{24L} + \frac{1}{8}\left(n\sqrt{K} + \frac{m}{\sqrt{K}}\right)^2 + \frac{L}{2\pi}\sum_{q>0} qa_q^+ a_q$$
$$\bar{T}_0 = -\frac{1}{24L} + \frac{1}{8}\left(n\sqrt{K} - \frac{m}{\sqrt{K}}\right)^2 - \frac{L}{2\pi}\sum_{q<0} qa_q^+ a_q \qquad (25.27)$$

Naturally, the motion of the centre of mass does not contribute to the bulk free energy in the limit $L \to \infty$. This free energy is determined exclusively by the third term of the Hamiltonian (25.26). It is remarkable, however, that the terms of the Hamiltonian which have a negligible effect on the thermodynamics determine the asymptotic behaviour of the

Figure 25.1. F. D. M. Haldane.

correlation functions! The bulk free energy, being the free energy of free bosons, is equal to

$$\frac{F_b}{L} = T \int_{-\infty}^{\infty} \frac{dp}{2\pi} \ln \left(1 - e^{-v|k|/T}\right) \tag{25.28}$$

It is easy to check, using explicit expressions for the corresponding table integrals, that

$$\frac{F_b}{L} = \text{const} - T \int_{-\infty}^{\infty} \frac{dp}{\pi} \ln \left(1 + e^{-v|k|/T}\right) = \text{const} - \frac{\pi}{6v} T^2 \tag{25.29}$$

i.e. this free energy coincides with the free energy of free spinless fermions (bosonization!).[1] This coincidence is just one extra demonstration of the fact that there is a one-to-one correspondence between the eigenstates of both theories. It follows from (25.29) that the product of the linear coefficient in the specific heat C_v by the velocity is a number:

$$C = \frac{3 C_v v}{\pi T} = 1 \tag{25.30}$$

This number C is called the central charge and plays an enormous role in conformal field theory. In the above example $C = 1$, but this is not always the case and C varies from theory to theory. The relationship between thermodynamics and conformal properties was discovered by Blote *et al.* (1986) and Affleck (1986). I shall return to the concept of central charge later in Chapter 26.

For critical theories with U(1) symmetry, conformal dimensions are functions of the interaction. As we shall see in a moment, there is a remarkable relationship between conformal dimensions and susceptibilities. This follows from the fact that the first term in (25.26) is the ground state energy of the system with total charge Q and therefore can be written

---

[1] Here it is more convenient to keep the velocity in explicit form, i.e. not putting $v = 1$.

as $Q^2/2\chi L$, where χ is the charge susceptibility. This leads to the following remarkable identity (Efetov and Larkin, 1975; Haldane, 1981):

$$K = \pi v \chi \tag{25.31}$$

References

Affleck, I. (1986). *Phys. Rev. Lett.*, **56**, 746.
Blote, H. W. J., Cardy, J. L. and Nightingale, M. P. (1986). *Phys. Rev. Lett.*, **56**, 742.
Efetov, K. B. and Larkin, A. I. (1975). *Sov. Phys. JETP*, **42**, 390.
Haldane, F. D. M. (1981). *Phys. Rev. Lett.*, **47**, 1840; *J. Phys. C*, **14**, 2585 (1981).

26

Virasoro algebra

In this chapter I continue to study the group of conformal transformations of the complex plane. The exposition of this and the following chapter is based on the pioneering paper by Belavin *et al.* (1984). Naturally, such studies are hardly necessary for the Gaussian model where one can calculate the correlation functions directly. However, these general considerations become indispensable in more complicated cases.

In two dimensions the group of conformal transformations is isomorphic to the group of analytic transformations of the complex plane. Since the number of analytic functions is infinite, this group is infinite-dimensional, i.e. has an infinite number of generators. An infinite number of generators generates an infinite number of Ward identities for correlation functions. These identities serve as differential equations on correlation functions. No matter how many operators you have in your correlation function, you always have enough Ward identities to specify it. Thus in conformally invariant theories (for brevity we shall call them 'conformal' theories) one can (at least in principle) calculate all multi-point correlation functions. We have already calculated multi-point correlation functions of the bosonic exponents for the Gaussian model; below we shall see less trivial examples.

Let us take in good faith that there are models besides the Gaussian one whose spectrum is linear, whose correlation functions factorize into products of analytic and antianalytic parts and which have operators transforming under analytic (antianalytic) transformations like (25.4). Such operators will be called *primary* fields. This is almost a definition of what conformal theory is. To make this definition rigorous we have to add one more property, namely, to postulate that the three-point correlation functions of the primary fields have the following form:

$$\langle A_{\Delta_1}(z_1)A_{\Delta_2}(z_2)A_{\Delta_3}(z_3)\rangle = \frac{C_{123}}{(z_{12})^{\Delta_1+\Delta_2-\Delta_3}(z_{13})^{\Delta_1+\Delta_3-\Delta_2}(z_{23})^{\Delta_2+\Delta_3-\Delta_1}} \qquad (26.1)$$

where C_{123} are some constants.

Since the eigenvalues of the Hamiltonian and momentum are related to the zeroth components of the stress energy tensor, it is logical to suggest that the generators of transformations (25.4) are Fourier components of the stress energy tensor. In order to find commutation relations between these components and the primary fields we have to consider the infinitely small version of the transformation (25.4):

$$z' = z + \epsilon(z) \qquad (26.2)$$

According to (25.4) this transformation changes the primary field $A_\Delta(z)$:

$$\delta A_\Delta(z) = A_\Delta(z+\epsilon)(1+\partial\epsilon)^\Delta - A_\Delta(z) = \epsilon(z)\partial A_\Delta(z) + \partial\epsilon(z)\Delta A_\Delta(z) + \mathcal{O}(\epsilon^2) \quad (26.3)$$

Let us restrict our correlation functions to the line $\tau = 0$ where $z = ix$. This eliminates the time dependence and therefore we can use the operator language of quantum mechanics. As we know from quantum mechanics, if an operator changes under some transformation, this change can be expressed as an action of some other operator, a generator of this transformation. For our case it means that

$$\delta\hat{A}_\Delta(x) = [\hat{Q}_\epsilon, \hat{A}_\Delta(x)] \quad (26.4)$$

where $\hat{Q}_\epsilon$ is an element of the group of conformal transformations corresponding to the infinitesimal transformation (26.2). For a general Lie group G with generators $\hat{t}_i$ an infinitesimal transformation is given by $\epsilon_i\hat{t}_i$, where ϵ_i are infinitely small parameters, coordinates of the transformation. Since conformal transformations in two dimensions are characterized not by a finite set of parameters, but by an entire function $\epsilon(z)$, the conformal algebra is infinite-dimensional. Therefore a general infinitesimal transformation can be written as an integral:[1]

$$\hat{Q}_\epsilon = \frac{1}{2\pi}\int dy\,\epsilon(y)\hat{T}(y)$$

Therefore we have

$$\frac{i}{2\pi}\int dy\,\epsilon(y)[\hat{T}(y), \hat{A}_\Delta(x)] = \epsilon(x)\partial_x A_\Delta(x) + \partial_x\epsilon(x)\Delta A_\Delta(x) \quad (26.5)$$

which corresponds to the following local commutation relations:

$$\frac{i}{2\pi}[\hat{T}(y), \hat{A}_\Delta(x)] = \delta(x-y)\partial_x A_\Delta(x) - \partial_y\delta(y-x)\Delta A_\Delta(x) \quad (26.6)$$

These commutation relations must be satisfied in any conformal theory.

According to the general rules explained in Chapter 14 (see the discussion around (14.28)), there is a one-to-one correspondence between commutation relations and operator product expansion (OPE). In this particular case *under the* $\langle\cdots\rangle$ *sign* at $z \to z_1$ one can replace the product of the operators T and A_Δ by the following expansion:

$$T(z)A_\Delta(z_1) = \frac{\Delta}{(z-z_1)^2}A_\Delta(z_1) + \frac{1}{(z-z_1)}\partial_{z_1}A_\Delta(z_1) + \cdots \quad (26.7)$$

where dots stand for terms nonsingular at $(z-z_1) \to 0$.

Exercise

Using (14.28) check that (26.6) does follow from (26.7).

Let us now study the algebraic properties of the stress energy tensor $T(z)$. For this purpose we shall study its correlation functions using the Gaussian model stress energy

---

[1] The factor 2π is introduced to conform to the accepted notation.

tensor (25.16) as an example. In order to simplify the results, we change the notation and introduce the new components[2]

$$T_{zz} = -\pi(T_{11} - T_{22} - 2iT_{12}) \equiv T = -2\pi : (\partial\varphi)^2 :$$
$$T_{\bar{z}\bar{z}} = -\pi(T_{11} - T_{22} + 2iT_{12}) \equiv \bar{T} = -2\pi : (\bar{\partial}\bar{\varphi})^2 : \tag{26.8}$$
$$\mathrm{Tr}T = T_{11} + T_{22} = 0$$

Now it is obvious that correlation functions of operators T and $\bar{T}$ depend on z or $\bar{z}$ only. Since the two-point correlation function is the simplest one, let us calculate it first. A straightforward calculation yields:

$$\langle T(1)T(2)\rangle = \frac{C}{2(z_1 - z_2)^4}$$
$$\tag{26.9}$$
$$\langle \bar{T}(1)\bar{T}(2)\rangle = \frac{C}{2(\bar{z}_1 - \bar{z}_2)^4}$$

where $C = 1$ in the given case. We keep C in the relations (26.9) because in this form they hold for all conformal theories. As far as the correlation functions of $\mathrm{Tr}T$ are concerned, it seems that (26.8) dictates them to be identically zero. The miracle is that this is *wrong*! Here we encounter another example of an anomaly which is called the *conformal* anomaly.

Let us consider the matter more carefully. Since the stress energy tensor is a conserved quantity, i.e.

$$\partial_a T_{ab} = 0$$

which reflects the conservation of energy and momentum, the two-point correlation function must satisfy the identity

$$q_a\langle T_{ab}(-q)T_{cd}(q)\rangle = 0$$

This suggests that the Fourier transformation of (26.8) must be modified:

$$\langle T_{ab}(-q)T_{cd}(q)\rangle = \frac{Cq^2}{48\pi}\left(\delta_{ab} - \frac{q_a q_b}{q^2}\right)\left(\delta_{cd} - \frac{q_c q_d}{q^2}\right) \tag{26.10}$$

which gives the nonvanishing correlation function of the traces:

$$\langle \mathrm{Tr}T(-q)\mathrm{Tr}T(q)\rangle = \frac{Cq^2}{48\pi} \tag{26.11}$$

In real space the pair correlation function of the traces is ultralocal:

$$\langle \mathrm{Tr}T(x)\mathrm{Tr}T(y)\rangle = \frac{C}{48\pi}\nabla^2\delta^{(2)}(x - y) \tag{26.12}$$

Since this correlation function is short range, it is generated at high energies. It is no wonder, therefore, that we have missed it in the straightforward calculations.

---

[2] The factor 4π is introduced in order to make the present notation conform to the standard one.

As we have seen in the previous chapter, the central charge appears in the expression for the specific heat. Later we shall learn that it also determines the conformal dimensions of the correlation functions. The fact that the specific heat of a conformal theory is proportional to its central charge means that the latter is related to the number of states. Since the conformal dimensions are eigenvalues of the stress energy tensor components, the relation between C and the correlation functions suggests that C also influences the scaling dimensions. For what follows it is necessary to know the fusion rules for the stress energy tensor itself. We can find these commutation relations using general properties of the stress energy tensor. At first, the stress energy tensor is really a tensor, and as such it has definite transformation properties. From this fact it follows that its conformal dimensions are $(2, 0)$, which is consistent with (26.9). Therefore the fusion rules for $T(z)T(z_1)$ must include the right-hand side of (26.7) with $\Delta = 2$. This is not all, however; the expansion must contain a term with the identity operator, since otherwise the pair correlation function $\langle T(z)T(z_1)\rangle$ would vanish. This term can be deduced from (26.9); collecting all these terms together we get the following OPE (the Virasoro fusion rules):

$$T(z)T(z_1) = \frac{C}{2(z - z_1)^4} + \frac{2}{(z - z_1)^2}T(z_1) + \frac{1}{(z - z_1)}\partial_{z_1}T(z_1) + \cdots \quad (26.13)$$

A comparison of (26.13) with (26.7) shows that the stress energy tensor *is not a primary field*, that is, despite its name, the stress energy tensor does not transform as a *tensor* under conformal transformations. It can be shown that instead of the tensorial law (25.4) the stress energy tensor transforms under finite conformal transformations as follows:

$$T(\xi) = T[z(\xi)](\mathrm{d}z/\mathrm{d}\xi)^2 + \frac{C}{12}\{z, \xi\}$$

$$\{z, \xi\} = \frac{z'''}{z'} - \frac{3}{2}\left(\frac{z''}{z'}\right)^2 \quad (26.14)$$

In particular, it follows from this expression that the stress energy tensor on a strip of width L in the x-direction ($z = \exp(2\pi\xi/L)$) is equal to

$$T_{\text{strip}}(\xi) = (2\pi/L)^2 \left[z^2 T_{\text{plane}}(z) - C/24\right] \quad (26.15)$$

Since $\langle T_{\text{plane}}(z)\rangle = 0$, we find

$$\langle T_{\text{strip}}(z)\rangle = -\frac{C}{24}(2\pi/L)^2 \quad (26.16)$$

Now recalling the relationship between the stress energy tensor and the Hamiltonian (25.13) we reproduce from the latter equation the $1/L$ correction to the stress energy tensor eigenvalues (25.27).

It is also convenient to have the Virasoro fusion rules (26.13) in operator form with commutators:

$$\frac{1}{2i\pi}[\hat{T}(x), \hat{T}(y)] = \delta(x - y)\partial_x T(x) - 2\partial_x\delta(x - y)T(x) + \frac{C}{6}\partial_x^3\delta(x - y) \quad (26.17)$$

Since the commutation relations (26.17) are local, they do not depend on the global geometry of the problem.

Exercise

Using (14.28) check that (26.17) does follow from (26.13).

It is customary to write the Virasoro fusion rules using the Laurent components of the stress energy tensor and the operators defined on the infinite complex plane:

$$\hat{T}(z) = \sum_{-\infty}^{\infty} \frac{L_n}{z^{n+2}} \qquad \hat{A}_\Delta(z) = \sum_{-\infty}^{\infty} \frac{A_{\Delta,n}}{z^{n+2}}$$

The transformation laws (26.5) and (26.14) establish a connection between this expansion and the Fourier expansion which we use in a strip geometry:

$$A_\Delta^{\text{strip}}(x) = \left(\frac{2\pi}{L}\right)^\Delta \sum_n A_{\Delta,n} e^{-2\pi i x(n-\Delta)/L}$$

$$T^{\text{strip}}(x) = \left(\frac{2\pi}{L}\right)^2 \left(\sum_n L_n e^{-2\pi i n x/L} - C/24\right)$$

$$(26.18)$$

Substituting these expansions into the fusion rules (26.13) and (26.5) we get the following commutation relations for the Laurent components:

$$[L_n, L_m] = (n - m)L_{n+m} + \frac{C}{12}n(n^2 - 1)\delta_{n+m,0} \quad (26.19)$$

and

$$[L_n, A_{\Delta,m}] = [\Delta(1 - n) + m + n]A_{\Delta,m+n} \quad (26.20)$$

From (26.19) we see that the set of operators L_n, I (I is the identity operator) is closed with respect to the operation of commutation. Therefore components of the stress energy tensor in conformal theories together with the identity operator form an algebra (Virasoro algebra). As we have said, this fact opens up the possibility of formal studies of conformal theories as representations of the Virasoro algebra.

Ward identities

The identity (26.5) can be used to derive Ward identities for correlation functions of the stress energy tensor with primary fields. Let us insert this identity into a correlation function:

$$\frac{1}{2\pi i}\int_C dz\epsilon(z)\langle T(z)A_{\Delta_1}(1)\cdots A_{\Delta_N}(N)\rangle = \sum_{i=1}^{N}[\epsilon(z_i)\partial_{z_i}\langle A_{\Delta_1}(1)\cdots A_{\Delta_N}(N)\rangle$$

$$+ \partial_{z_i}\epsilon(z_i)\Delta_i\langle A_{\Delta_1}(1)\cdots A_{\Delta_N}(N)\rangle] \quad (26.21)$$

and assume that $\epsilon(z)$ is analytic in the domain enclosing the points z_i and the contour C encircles these points. Then (26.21) suggests that as a function of z the integrand of the left-hand side has simple and double poles at $z = z_i$:

$$\langle T(z)A_{\Delta_1}(1)\cdots A_{\Delta_N}(N)\rangle = \sum_{i=1}^{N}\left[\frac{\Delta_i}{(z-z_i)^2} + \frac{1}{z-z_i}\partial_{z_i}\right]\langle A_{\Delta_1}(1)\cdots A_{\Delta_N}(N)\rangle \quad (26.22)$$

This is the Ward identity for $T(z)$ we had in mind.

Subalgebra sl(2)

It follows from (26.19) that three operators L_0, $L_{\pm 1}$ compose a subalgebra of the Virasoro algebra isomorphic to the algebra sl(2):

$$[L_0, L_{\pm 1}] = \mp 2L_{\pm 1} \qquad [L_1, L_{-1}] = 2L_0 \qquad (26.23)$$

The corresponding group is the group of all rational transformations of the complex plane

$$w(z) = \frac{az+b}{cz+d} \qquad ad-bc = 1 \qquad (26.24)$$

Let us show that all correlation functions in critical two-dimensional theories are invariant with respect to the SL(2) transformations (26.24). This fact follows from the Ward identity (26.22). Indeed, since $\langle T(z)\rangle = 0$ on the infinite plane, the correlation function in the left-hand side of (26.22) must decay at infinity as

$$\langle T(z)A_{\Delta_1}(1)\cdots A_{\Delta_N}(N)\rangle \sim z^{-4}D(z_1, \cdots z_N) \qquad (26.25)$$

Expanding the right-hand side of (26.21) in powers of z^{-1} we find that the terms containing z^{-1}, z^{-2} and z^{-3} must vanish, which gives rise to the following identities:

$$\sum_i \partial_{z_i}\langle A_{\Delta_1}(1)\cdots A_{\Delta_N}(N)\rangle = 0 \qquad (26.26)$$

$$\sum_i (z_i\partial_{z_i} + \Delta_i)\langle A_{\Delta_1}(1)\cdots A_{\Delta_N}(N)\rangle = 0 \qquad (26.27)$$

$$\sum_i \left(z_i^2\partial_{z_i} + 2\Delta_i z_i\right)\langle A_{\Delta_1}(1)\cdots A_{\Delta_N}(N)\rangle = 0 \qquad (26.28)$$

It is easy to see that the operators

$$L_{-1} = \partial_z \qquad L_0 = (z\partial_z + \Delta) \qquad L_1 = (z^2\partial_z + 2\Delta z) \qquad (26.29)$$

satisfy the commutation relations (26.23) thus composing a representation of the sl(2) algebra.

Exercise

Show that the SL(2) invariance dictates that a four-point correlation function of four primary fields with the same conformal dimension has the following general form:

$$\langle A(1) \cdots A(4) \rangle = \left(\left| \frac{z_{13} z_{24}}{z_{12} z_{14} z_{23} z_{34}} \right| \right)^{4\Delta} G(x, \bar{x}) \tag{26.30}$$

where

$$x = \frac{z_{12} z_{34}}{z_{13} z_{24}}$$

Reference

Belavin, A. B., Zamolodchikov, A. A. and Polyakov, A. M. (1984). *Nucl. Phys. B*, **241**, 333; *Conformal Invariance and Applications to Statistical Mechanics*, ed. C. Itzykson, H. Saleur and J.-B. Zubert. World Scientific, Singapore, 1988.

27

Differential equations for the correlation functions

In this chapter I shall generate eigenvectors of the Hilbert space of conformal theories using the Virasoro operators. One may wonder why we use Virasoro operators whose commutation relations are so awkward. The answer is that we want to make sure that the conformal symmetry is preserved, and the best way to guarantee this is to use as creation and annihilation operators the generators of the conformal group.

In fact, the procedure I am going to follow is a standard one in group theory, where it is used for constructing representations. In order to feel ourselves on familiar ground, let us recall how one constructs representations of spin operators in quantum mechanics. Consider a three-dimensional rotator whose Hamiltonian is the sum of squares of generators of the SU(2) group, components of the spin:

$$\hat{H} = \sum_{i=x,y,z} \hat{S}_i^2 = \hat{S}_z^2 + \frac{1}{2}(\hat{S}_+\hat{S}_- + \hat{S}_-\hat{S}_+)$$

Since $[\hat{H}, \hat{S}^z] = 0$, $\hat{S}^z$ is diagonal in the basis of eigenfunctions of $\hat{H}$. The key point is to write down the operator algebra in the proper form:

$$[\hat{S}_\pm, \hat{S}_z] = \mp\hat{S}_\pm \quad [\hat{S}_+, \hat{S}_-] = 2\hat{S}_z$$

In doing so we separate the diagonal generator $\hat{S}_z$ from the raising and lowering operators. As follows from the commutation relations, the operators $\hat{S}_+$ ($\hat{S}_-$) increase (decrease) an eigenvalue of $\hat{S}_z$ by one. Let us suppose that there exists an eigenvector of $\hat{S}^z$ which is annihilated by $\hat{S}_-$:

$$\hat{S}_- |-S\rangle = 0 \quad \hat{S}_z |-S\rangle = -S |-S\rangle$$

Then $(\hat{S}_+)^j |-S\rangle$ is also an eigenvector of $\hat{S}_z$ with the eigenvalue $j - S$. As we know, it can also be shown that if $2S$ is a positive integer, the state

$$|\chi\rangle = (\hat{S}_+)^{(2S+1)} |-S\rangle \tag{27.1}$$

is a null vector, that is representations of the SU(2) group with integer $(2S + 1)$ are finite-dimensional.

Let us show that in the case of the Virasoro algebra the state similar to $|-S\rangle$ is

$$|\Delta, \bar{\Delta}\rangle = A_{\Delta,0} A_{\bar{\Delta},0} |0\rangle$$

According to (25.7), $\langle n|A_\Delta(z)|0\rangle = 0$ for all negative n, which according to the definition of $A_{\Delta,n}$ is equivalent to

$$A_{\Delta,n}|0\rangle = 0 \qquad n > 0 \tag{27.2}$$

Similar considerations for the two-point correlation function of the stress energy tensor lead to

$$L_n|0\rangle = 0 \qquad n > 0 \tag{27.3}$$

Now we can prove that operators L_n ($n > 0$) annihilate the state $|\Delta, \bar{\Delta}\rangle$. Indeed, according to (26.20)

$$L_n A_{\Delta,0}|0\rangle = [\Delta(1-n)+n]A_{\Delta,n}|0\rangle + A_{\Delta,0}L_n|0\rangle = 0 \tag{27.4}$$

Therefore we can use L_n with positive n as lowering operators and L_n with negative n as raising operators. Thus we look for the eigenvectors in the following form:

$$|\bar{n}_1, \bar{n}_2, \ldots, \bar{n}_M; n_1, n_2, \ldots, n_M; \Delta, \bar{\Delta}\rangle = \bar{L}_{-\bar{n}_1}\bar{L}_{-\bar{n}_2}\cdots\bar{L}_{-\bar{n}_M}L_{-n_1}L_{-n_2}\cdots L_{-n_M}|\Delta, \bar{\Delta}\rangle \tag{27.5}$$

We see that a tower of states is built on each primary field, including the trivial identity operator. It follows from the Virasoro commutation relations (26.19), that these vectors are eigenstates of the operators L_0 and $\bar{L}_0$ with eigenvalues

$$\Delta_n = \Delta + \sum_j n_j$$
$$\bar{\Delta}_n = \bar{\Delta} + \sum_j \bar{n}_j \tag{27.6}$$

Operators obtained from primary fields by action of the Virasoro generators L_{-n} are called *descendants* of the primary fields. It can be shown that for the Gaussian model all primary fields and their descendants are linearly independent and their Hilbert space is isomorphic to the Hilbert space built up by the bosonic creation operators. This, however, is only one particular representation of the Virasoro algebra. There are many others, and different methods suggested for finding nontrivial representations of the Virasoro algebra. Some representations can be constructed by truncation of the Gaussian theory Hilbert space, i.e. by postulating that there are linearly dependent states among the states (27.5).

In this chapter I consider only the simplest case when the truncation occurs on the second level. Suppose that the states

$$L_{-2}|\Delta\rangle \qquad L_{-1}^2|\Delta\rangle$$

are linearly dependent, i.e. there is a number α such that

$$|\chi\rangle = \left(L_{-2} + \alpha L_{-1}^2\right)|\Delta\rangle = 0 \tag{27.7}$$

that is $|\chi\rangle$ is a null vector similar to (27.1).

Since all other Virasoro generators L_{-n} with $n > 2$ are generated by these two via the commutation relations (26.20), the fact that L_{-2} is proportional to L_{-1}^2 means that the only linearly independent eigenvectors from the set (27.5) are

$$|n, \Delta\rangle = (L_{-1})^n |\Delta\rangle$$

Let us consider conditions for existence of the null state $|\chi\rangle$ (27.7). Due to their mutual independence the left and right degrees of freedom can be considered separately. Since we want to preserve conformal invariance of the theory, condition (27.7) must survive conformal transformations, i.e.

$$L_n|\chi\rangle = 0 \qquad n > 0 \tag{27.8}$$

Using the Virasoro commutation relations (26.20) and the properties

$$L_0|\Delta\rangle = \Delta|\Delta\rangle \qquad L_n|\Delta\rangle = 0 \qquad n > 0$$

we get from (27.8) the following two equations:

$$3 + 2\alpha + 4\alpha\Delta = 0 \tag{27.9}$$

$$4\Delta(2 + 3\alpha) + C = 0 \tag{27.10}$$

whose solution is

$$\alpha = -\frac{3}{2(1 + 2\Delta)} \qquad \Delta = \frac{1}{16}(5 - C \pm \sqrt{(5 - C)^2 - 16C}) \tag{27.11}$$

Since Δ must be real, we conclude that $C \leq 1$ or $C \geq 25$.

At this point I would like to stress that we have already left the clear waters of the Gaussian model. We have truncated its Hilbert space, that is thrown away some of its states, which is consistent with the condition $C \leq 1$ (notice however that there is a solution $C > 25$!). At this point we do not even know what is a path integral representation for such a procedure, that is we do not know what is a Hamiltonian or an action for our truncated Gaussian model. However, as we shall see, we can calculate its correlation functions even without the action. For this purpose I shall use the Ward identities for the stress energy tensor.

As the first step let us write down the complete OPE for the stress energy tensor with primary field $A_\Delta(z_1)$ which is a generalization of (26.7):

$$T(z)A_\Delta(z_1) = \sum_{n=0}^{\infty}(z - z_1)^{n-2}L_{-n}A_\Delta(z_1) \tag{27.12}$$

The absence in this expansion of the Virasoro generators with positive index follows from the Ward identity (26.22). Comparing this OPE with (26.7) or (26.22) we obtain the following identities:

$$L_0 A_\Delta(z_1) = \Delta A_\Delta(z_1) \qquad L_{-1}A_\Delta(z_1) = \partial_{z_1}A_\Delta(z_1) \tag{27.13}$$

The next step is to use these identities to extract the action of L_{-2}. This we do by subtracting the singular part of the stress energy tensor at $z \to z_1$ and using (26.22):

$$\langle L_{-2} A_{\Delta_1}(1) \cdots A_{\Delta_N}(N) \rangle = \left\langle \left[T(z) - \frac{L_0}{(z - z_1)^2} - \frac{L_{-1}}{z - z_1} \right] A_{\Delta_1}(1) \cdots A_{\Delta_N}(N) \right\rangle \Bigg|_{z \to z_1}$$

$$= \sum_{j \neq 1} \left[\frac{\Delta_j}{(z_1 - z_j)^2} + \frac{1}{(z_1 - z_j)} \partial_j \right] \langle A_{\Delta_1}(1) \cdots A_{\Delta_N}(N) \rangle \tag{27.14}$$

According to (27.7), L_{-1}^2 is identified with $-\alpha \partial^2$, and we obtain from (27.14) the linear differential equation for a multi-point correlation function of primary fields $A_j(z_j, \bar{z}_j)$ with analytic conformal dimension Δ_j:

$$\left\{ \frac{3}{2(1 + 2\Delta_1)} \partial_1^2 - \sum_{j \neq 1} \left[\frac{\Delta_j}{(z_1 - z_j)^2} + \frac{1}{(z_1 - z_j)} \partial_j \right] \right\} \langle A_1(z_1) \cdots A_N(z_N) \rangle = 0 \tag{27.15}$$

I emphasize that this equation holds only for the simplest representation of the Virasoro algebra, namely the one defined by (27.7).

For the four-point correlation function the presence of the SL(2) symmetry helps to reduce (27.15) to an ordinary differential equation. In the case when all conformal dimensions are equal the general expression for a four-point correlation function (26.30) acquires the following form:

$$\langle A(1) \cdots A(4) \rangle = \left(\left| \frac{z_{13} z_{24}}{z_{12} z_{14} z_{23} z_{34}} \right| \right)^{4\Delta} G(x, \bar{x}) \tag{27.16}$$

where

$$x = \frac{z_{12} z_{34}}{z_{13} z_{24}} \qquad 1 - x = \frac{z_{14} z_{23}}{z_{13} z_{24}} \tag{27.17}$$

are the so-called anharmonic ratios.

Substituting (27.16) into (27.15) we obtain the following conventional differential equation for $G(x, \bar{x})$:

$$x(1 - x)G'' + \frac{2}{3}(1 - 4\Delta)(1 - 2x)G' + \frac{4\Delta}{3}(1 - 4\Delta)G = 0 \tag{27.18}$$

(and the same equation for $\bar{x}$).

Equations of this type appear frequently in conformal field theory and therefore we shall discuss the solution in detail. Equation (27.18) is a particular case of the hypergeometric equation. Its special property is invariance with respect to the transformation $x \to 1 - x$ which reflects the invariance of the correlation function with respect to permutation of the coordinates ($2 \to 4$, for instance). Thus if $\mathcal{F}(x)$ is a solution, $\mathcal{F}(1 - x)$ is also a solution. If $\frac{2}{3}(1 - 4\Delta)$ is not an integer number, two linearly independent solutions of (27.18) nonsingular in the vicinity of $x = 0$ are given by the hypergeometric functions:

$$\mathcal{F}^{(0)}(x) = F(a, b, c; x)$$
$$\mathcal{F}^{(1)}(x) = x^{1-c} F(a - c + 1, b - c + 1, 2 - c; x) \qquad (27.19)$$
$$a = \frac{1}{3}(1 - 4\Delta) \qquad b = -4\Delta \qquad c = \frac{2}{3}(1 - 4\Delta)$$

When c is an integer, the second solution is not a hypergeometric function. I shall discuss this case later. A general solution for $G(x, \bar{x})$ is

$$G(x, \bar{x}) = W_{ab} \mathcal{F}^{(a)}(x) \mathcal{F}^{(b)}(\bar{x}) \qquad (27.20)$$

This solution must satisfy two requirements. The first is that it must be invariant with respect to interchange of $x \to 1 - x$ (*crossing* symmetry) which reflects the fact that the entire correlation function does not change when any two coordinates are interchanged. The second requirement is that the correlation function must be single valued on the $(x, \bar{x})$ plane. These requirements fix the matrix W (up to a factor).

If c is not an integer, we use the known identities for hypergeometric functions and the fact that $a + b = 2c - 1$, to find

$$\mathcal{F}^{(a)}(1 - x) = A_{ab} \mathcal{F}^{(b)}(x)$$
$$A_{00} = \frac{\Gamma(c)\Gamma(1 - c)}{\Gamma(c - a)\Gamma(1 + a - c)} \qquad A_{01} = \frac{\Gamma(c)\Gamma(c - 1)}{\Gamma(a)\Gamma(b)} \qquad (27.21)$$
$$A_{10} = \left(1 - A_{00}^2\right)/A_{01} \qquad A_{11} = -A_{00}$$

The crossing symmetry condition gives

$$W_{ab} A_{ac} A_{bd} = W_{cd} \qquad (27.22)$$

The monodromy matrix $\hat{A}$ has eigenvalues $\lambda = \pm 1$. Therefore a general solution of (27.22) is

$$W_{ab} = C_{+} e_{a}^{(+)} e_{b}^{(+)} + C_{-} e_{a}^{(-)} e_{b}^{(-)} \qquad (27.23)$$

where $e^{(\pm)}$ are the eigenvectors of $\hat{A}^{t}$. Substituting the explicit expressions for these eigenvectors into the expression for $G(x, \bar{x})$, we find

$$G(x, \bar{x}) = C_{+} \left[A_{01} \mathcal{F}^{(1)}(x) + (1 + A_{00}) \mathcal{F}^{(0)}(x) \right] \left[A_{01} \mathcal{F}^{(1)}(\bar{x}) + (1 + A_{00}) \mathcal{F}^{(0)}(\bar{x}) \right]$$
$$+ C_{-} \left[A_{01} \mathcal{F}^{(1)}(x) + (A_{00} - 1) \mathcal{F}^{(0)}(x) \right] \left[A_{01} \mathcal{F}^{(1)}(\bar{x}) + (A_{00} - 1) \mathcal{F}^{(1)}(\bar{x}) \right]$$
$$(27.24)$$

To determine $C_{\pm}$ we recall that the correlation function must be a single-valued function. From (27.19) we see that $\mathcal{F}^{(0)}(x)$ is analytic in the vicinity of $x = 0$, but $\mathcal{F}^{(1)}(x)$ has a branch cut. Therefore there should be no terms containing cross products of these two solutions. Such cross products vanish if

$$C_{+}(1 + A_{00}) = C_{-}(1 - A_{00}) \qquad (27.25)$$

Finally we get

$$G(x, \bar{x}) \sim A_{01}^2 \mathcal{F}^{(1)}(x)\mathcal{F}^{(1)}(\bar{x}) + \left(1 - A_{00}^2\right) \mathcal{F}^{(0)}(x)\mathcal{F}^{(0)}(\bar{x}) \tag{27.26}$$

For future purposes it is convenient to express the solution in terms of $\mathcal{F}^{(0)}$. Using (27.21), we get

$$\begin{aligned}
G(x, \bar{x}) \sim \ & \left[\mathcal{F}^{(0)}(x)\mathcal{F}^{(0)}(1 - \bar{x}) + \mathcal{F}^{(0)}(\bar{x})\mathcal{F}^{(0)}(1 - x)\right] \\
& - A_{00}^{-1}\left[\mathcal{F}^{(0)}(1 - x)\mathcal{F}^{(0)}(1 - \bar{x}) + \mathcal{F}^{(0)}(x)\mathcal{F}^{(0)}(\bar{x})\right]
\end{aligned} \tag{27.27}$$

(here we have changed the normalization factor). The purpose of the latter exercise is to obtain a limit of integer c. In this case, according to (27.21), $A_{00}^{-1} \to 0$ and the cross-symmetric and single-valued solution is

$$G(x, \bar{x}) = \mathcal{F}(x)\mathcal{F}(1 - \bar{x}) + \mathcal{F}(\bar{x})\mathcal{F}(1 - x) \tag{27.28}$$

Now let us discuss some particular cases.

Ising model. One interesting case is $C = 1/2$; it corresponds to the critical Ising model. From (27.11) we get two solutions

$$\Delta_- = 1/16 \qquad \Delta_+ = 1/2 \tag{27.29}$$

which indeed correspond to conformal dimensions of the Ising model primary fields (see Chapter 28 about the Ising model for details).

Fields with conformal dimensions $(1/16, 1/16)$ are called order (disorder) parameter fields and are denoted as $\sigma(z, \bar{z})$ $(\mu(z, \bar{z}))$. These two fields are mutually nonlocal and one can choose either of them for the basis of primary fields.

Exercise

Check that the following correlation function satisfies (27.18) with $\Delta = 1/16$ and $C = 1/2$:

$$\begin{aligned}
G(x, \bar{x}) = \ & \left\{\left[1 + (1 - x)^{1/2}\right]\left[1 + (1 - \bar{x})^{1/2}\right]\right\}^{1/2} \\
& + \left\{\left[1 - (1 - x)^{1/2}\right]\left[1 - (1 - \bar{x})^{1/2}\right]\right\}^{1/2}
\end{aligned} \tag{27.30}$$

$C = 0$ *model.* As is easy to guess, this theory is nonunitary and therefore cannot appear as a quantum field theory. However, it appears in applications as a statistical field theory. In particular, it describes classical percolation. There are two primary fields with dimensions $\Delta_+ = 5/8$ and $\Delta_- = 0$. The latter field has the same conformal dimension as the unity operator which leads to the degenerate situation described above. Namely, one has to use for the four-point function of the $\Delta = 5/8$ operator, (27.28) with

$$\mathcal{F}(x) = x^2 F(-1/2, 3/2, 3; x)$$

$C = -2$ *model.* This model is also nonunitary. It frequently appears in applications describing, for example, dense polymers (Rozansky and Saleur, 1992). The conformal dimensions are $\Delta_- = -1/8$ and $\Delta_+ = 1$. Here again we have a degenerate situation where for the $\Delta = -1/8$ operator we need to use (27.28) with

$$\mathcal{F}(x) = F(1/2, 1/2, 1; \; x)$$

Equation (27.16) represents the most general form of four-point correlation function in conformal field theory. In the simplest case of a Gaussian model such a correlation function (as all others) factorizes into a product of analytic and antianalytic functions (22.10). For the Gaussian model, this factorization implies a factorization of the corresponding primary fields: the bosonic exponents are represented as products of chiral exponents (22.12). However, the calculations of this section demonstrate that such straightforward decomposition can no longer be performed even for the simplest nontrivial conformal field theory. Thus the N-point correlation function satisfies linear differential equations in the holomorphic and antiholomorphic sectors (see (27.16)); such equations have several solutions (conformal blocks) and the multi-point function is a linear combination of these solutions

$$\langle A(1) \dots A(N) \rangle = \sum_{p,q} W_{pq} \mathcal{F}^p(z_1, \dots, z_N) \bar{\mathcal{F}}^q(\bar{z}_1, \dots, \bar{z}_N) \qquad (27.31)$$

where the matrix W_{pq} is chosen in such a way that the resulting expression is a single-valued function on the $(z, \bar{z})$ plane. This equation certainly excludes any simple factorization of primary fields into products of chiral ones. So one cannot write down a primary field as a product of two operators, one holomorphic and one antiholomorphic. Worse still, in general one *cannot* even write down a primary field as a sum of products of chiral operators, that is as

$$A_\Delta(z, \bar{z}) = \sum_{a=1}^{l} U^a(z) \bar{U}^a(\bar{z}) R_{ab} \qquad (27.32)$$

where $U^a(z)$ and $\bar{U}^a(\bar{z})$ would be fields with conformal dimensions $(\Delta, 0)$ and $(0, \Delta)$ and R_{ab} is a constant matrix. Even if one could manage to find such a decomposition for some correlation function, it would not work for others. The reasons for this will become more transparent in the next section.

Coulomb gas construction for the minimal models

The path integral representation for the above truncation procedure was suggested by Dotsenko and Fateev (1984). It is based on the observation that the Gaussian model can be viewed as a model of a two-dimensional classical Coulomb gas where the bosonic exponents generate charges. Then the simplest way to modify the theory is to add an extra charge at infinity (or spread it over the background). Such modification does not change the spectrum in the bulk and therefore will not break the conformal symmetry.

For the sake of brevity we shall denote

$$V_\alpha = \exp[i\sqrt{4\pi}\,\alpha\Phi]$$

With the extra charge $-Q$ added to the system we shall adopt a new definition for correlation functions:

$$\langle V_{\alpha_1}(1)\cdots V_{\alpha_N}(N)\rangle_Q = \lim_{R\to\infty} R^{2Q^2}\langle V_{\alpha_1}(1)\cdots V_{\alpha_N}(N)V_{-Q}(R)\rangle \qquad (27.33)$$

In particular, the only nonvanishing two-point correlation function is now given by

$$\langle V_\alpha(1)V_{(Q-\alpha)}(2)\rangle_Q \equiv \lim_{R\to\infty} R^{2Q^2}\langle\exp[i\sqrt{4\pi}\,\alpha\Phi(1)]$$

$$\times \exp[i\sqrt{4\pi}(Q-\alpha)\Phi(2)]\exp[-i\sqrt{4\pi}\,Q\Phi(R)]\rangle = (|z_{12}|)^{-4\Delta} \qquad (27.34)$$

$$\Delta = \frac{1}{2}\alpha(\alpha - Q) \qquad (27.35)$$

So, as expected, the correlation function still decays as a power law, but the exponents are modified thus indicating that the stress energy tensor is changed. Notice that the bosonic exponents having the same conformal dimension are now V_α and $V_{Q-\alpha}$. If our deformation generates a realistic critical theory, this theory must have nonvanishing four-point correlation functions of these exponents. It is obvious, however, that one cannot combine four exponents with charges α and $Q - \alpha$ to satisfy the electroneutrality condition

$$\sum_i \alpha_i = Q \qquad (27.36)$$

Figure 27.1. Vladimir Fateev.

Therefore such a simple modification of the Gaussian model is not sufficient to obtain a nontrivial theory. The modification procedure is not complete and we have to add some other terms to the action to obtain a nontrivial conformal theory. The only operators one can add without breaking the conformal symmetry are truly marginal operators, that is operators with scaling dimension 2. Then the modified action is

$$S = \int d^2x \left[\frac{1}{2}(\partial_\mu \Phi)^2 + i\sqrt{4\pi} Q R^{(2)} \Phi + \sum_{\sigma=\pm} \mu_\sigma \exp(i\sqrt{4\pi}\alpha_\sigma \Phi) \right] \quad (27.37)$$

where the 'charges' $\alpha_\pm$ are chosen in such a way that the dimensions of the corresponding exponents (*screening* operators) given by (27.35) are equal to one:

$$\alpha_\pm(\alpha_\pm - Q) = 2$$

or

$$\alpha_\pm = Q/2 \pm \sqrt{Q^2/4 + 2} \quad (27.38)$$

The quantity $R^{(2)}$ represents the Riemann curvature of the surface. In this representation the extra charge Q is spread over the entire surface. Such representation makes it easier to calculate the stress energy tensor, but is not very convenient for calculations of the correlation functions.

The action (27.37) may appear odd, because the presence of $i = \sqrt{-1}$ in the action raises doubts whether the theory is unitary. We shall prove, however, that for certain values of Q the unitarity is preserved. It may also appear that by adding nonlinear terms to the action we make the theory intractable. This, however, is not the case. As we shall see, the multi-particle correlation functions of $V_{\alpha_\pm}$ operators vanish and thus the perturbation expansion in $V_{\alpha_\pm}$ contains only a finite number of terms.

Let us make sure that model (27.37) indeed possesses nontrivial multi-point correlation functions. We know that for $\mu_\sigma = 0$ only those correlation functions do not vanish whose exponents satisfy the 'electroneutrality' condition (27.36). As we have already seen, this condition is easily satisfied for two operators and we have concluded that exponents with the charges $-\alpha$ and $\alpha - Q$ have the same conformal dimensions. Now we want to have nontrivial multi-point correlation functions of such operators. Without screening charges such functions will always vanish. However, if α belongs to the set

$$\alpha_{n,m} = -\frac{1}{2}(n-1)\alpha_- - \frac{1}{2}(m-1)\alpha_+ \quad (27.39)$$

the four-point correlation function is not zero:

$$\langle V_{\alpha_{n,m}}(1) V_{\alpha_{n,m}}(2) V_{\alpha_{n,m}}(3) V_{Q-\alpha_{n,m}}(4) \rangle_Q = \lim_{R \to \infty} R^{2Q^2} \mu^{(n+m-2)}$$

$$\times \int d^2\xi_1 \cdots d^2\xi_{n+m-2} \langle V_{\alpha_{n,m}}(1) V_{\alpha_{n,m}}(2) V_{\alpha_{n,m}}(3) V_{Q-\alpha_{n,m}}(4) V_{\alpha_+}(\xi_1) \cdots$$

$$\times V_{\alpha_+}(\xi_{n-1}) V_{\alpha_-}(\xi_n) \cdots V_{\alpha_-}(\xi_{n+m-2}) V_Q(R) \rangle \quad (27.40)$$

I obtained this expression using the perturbation expansion in powers of $\mu_\pm$. The remarkable fact making this calculation possible is that it contains *just one term*, unless $\alpha_+ p + \alpha_- q = 0$ for some integer p, q. According to (27.39) the latter would amount to the existence of a primary field with zero conformal dimension (this occurs, for instance, in the theory with $C = 0$ mentioned above).

If there are no such operators, however, the model (27.37) remains tractable despite being nonlinear.

The quantization condition (27.39) determines the spectrum of conformal dimensions of the theory (27.37). Substituting (27.39) into (27.35) we get the formula for the permitted conformal dimensions:

$$\Delta_{n,m} = \frac{1}{2}[(\alpha_- n - \alpha_+ m)^2 - (\alpha_- + \alpha_+)^2] \tag{27.41}$$

Now we have to calculate the central charge. The stress energy tensor can be calculated straightforwardly, differentiating the action (27.37) with respect to the metric. The answer is

$$T(z) = -2\pi : \partial \Phi \partial \Phi : + i\sqrt{4\pi} Q \partial^2 \Phi \tag{27.42}$$

Calculating the correlation function $\langle T(z)T(0)\rangle = C/2z^4$ we find the value of the central charge:

$$C = 1 - 3Q^2 \tag{27.43}$$

Exercise

We leave it to the reader to check that thus defined, $T(z)$ has the correct OPE (26.13) with the bosonic exponent V_α, which reproduces the conformal dimension (27.35).

The family of models described by the action (27.37) contains a special subset of unitary models with

$$C = 1 - \frac{6}{p(p+1)} \qquad \Delta_{n,m} = \frac{[pn - (p+1)m]^2 - 1}{4p(p+1)} \tag{27.44}$$

where $p = 3, 4 \ldots$. These models are called *minimal*. The number of primary fields in these models is finite. The minimal model with $p = 3$ has $C = 1/2$ and is equivalent to the critical Ising model; the model with $p = 4$ has $C = 7/10$ and is equivalent to the Ising model at the tricritical point; and the model with $p = 5$ has $C = 4/5$ and is equivalent to the Z_3 Potts model at criticality.

In order to get a better insight into how the Dotsenko–Fateev scheme works, let us consider a four-point correlation function of $O_{1,2}$ fields. This field can be represented by the exponents $V_{-\alpha_+/2}$ and $V_{Q+\alpha_+/2}$. The nonvanishing four-point correlation function is obtained if one adds one screening operator:

$$\lim_{R \to \infty} R^{2Q^2} \langle V_{-\alpha_+/2}(1)V_{-\alpha_+/2}(2)V_{-\alpha_+/2}(3)V_{Q+\alpha_+/2}(4)V_{\alpha_+}(\xi)V_{-Q}(R)\rangle$$

Again, one can check that there are no contributions from other screening operators. As a result we have (from now on we shall drop the subscript Q in notation of correlation functions)

$$
\langle A(1) \cdots A(4) \rangle \equiv \langle O_{1,2}(1) O_{1,2}(2) O_{1,2}(3) O_{1,2}(4) \rangle
$$

$$
= \mu \frac{|z_{12} z_{13} z_{23}|^{\alpha_+^2/2}}{|z_{14} z_{24} z_{34}|^{\alpha_+(\alpha_+ + 2Q)/2}} \int d^2\xi \frac{|z_4 - \xi|^{\alpha_+(\alpha_+ + 2Q)}}{|(z_1 - \xi)(z_2 - \xi)(z_3 - \xi)|^{\alpha_+^2}} \quad (27.45)
$$

Let us make sure that the obtained expression is conformally invariant, that is, it can be written in the canonical form (27.16). To achieve this we perform a transformation of the variable in the integral:

$$
\xi = \frac{z_{13} z_4 \eta + z_{34} z_1}{z_{13} \eta + z_{34}} \quad (27.46)
$$

This transformation maps the points $\xi = z_1, z_2, z_3, z_4$ onto $0, x, 1$ and ∞ respectively. Substituting this expression into (27.45) and taking into account the fact that $\alpha_+(\alpha_+ - Q) = 2$ we obtain

$$
\langle A(1) \cdots A(4) \rangle = \frac{1}{|z_{13} z_{24}|^{4\Delta}} |x(1-x)|^{\alpha_+^2/2} \int d^2\eta \, [|\eta(1-\eta)(x-\eta)|]^{-\alpha_+^2/2} \quad (27.47)
$$

where $\Delta = \alpha_+(\alpha_+ + 2Q)/4$. The latter expression coincides with (27.16) with

$$
G(x, \bar{x}) = |x(1-x)|^{\alpha_+^2/2 + 4\Delta} \int d^2\eta \, [|\eta(1-\eta)(x-\eta)|]^{-\alpha_+^2/2} \quad (27.48)
$$

The integral over η may diverge at certain α. In this case it should be treated as an analytic continuation from the area of α where it is convergent. The problem of calculation of such integrals was solved by Dotsenko and Fateev, who have obtained the following expression for a general four-point correlation function:

$$
\langle O_{n_1,m_1}(1) O_{n_2,m_2}(2) O_{n_3,m_3}(3) O_{n_4,m_4}(4) \rangle
$$

$$
= \frac{|z_{13}|^{2[\Delta(\alpha_1 + \alpha_3 + \alpha_+) - \Delta_1 - \Delta_3 + \alpha_+ \alpha_2]} |z_{24}|^{2[\Delta(\alpha_2 + \alpha_4) - \Delta_2 - \Delta_4 + \alpha_+ \alpha_2]}}{|z_{12}|^{2[\Delta_1 + \Delta_2 - \Delta(\alpha_1 + \alpha_2)]} |z_{23}|^{2[\Delta_2 + \Delta_3 - \Delta(\alpha_2 + \alpha_3)]}}
$$

$$
\times |z_{34}|^{-2[\Delta(\alpha_3 + \alpha_4 + \alpha_+) - \Delta_3 - \Delta_4]} |z_{14}|^{-2[\Delta(\alpha_1 + \alpha_4 + \alpha_+) - \Delta_1 - \Delta_4]}
$$

$$
\times \{\sin[\pi\alpha_-(\alpha_1 + \alpha_2 + \alpha_3)] \sin[\pi\alpha_-\alpha_2] |I_1(x)|^2
$$

$$
+ \sin[\pi\alpha_-\alpha_1] \sin[\pi\alpha_-\alpha_3] |I_2(x)|^2 \} \quad (27.49)
$$

where $\Delta(\alpha)$ is determined by the formula (27.35) and

$$
I_1(x) = \int_1^\infty dt \, t^{\alpha-\alpha_1} (t-1)^{\alpha-\alpha_2} (t-x)^{\alpha-\alpha_3}
$$

$$
= \frac{\Gamma[-1 - \alpha_-(\alpha_1 + \alpha_2 + \alpha_3)] \Gamma(\alpha_-\alpha_2)}{\Gamma[-\alpha_-(\alpha_1 + \alpha_3)]}
$$

$$
\times F[-\alpha_-\alpha_3, -1 - \alpha_-(\alpha_1 + \alpha_2 + \alpha_3), -\alpha_-(\alpha_1 + \alpha_3); x] I_2(x)
$$

$$= \int_0^x dt \, t^{\alpha-\alpha_1}(t-1)^{\alpha-\alpha_2}(t-x)^{\alpha-\alpha_3}$$

$$= x^{1+\alpha_-(\alpha_1+\alpha_3)} \frac{\Gamma(1+\alpha_-\alpha_1)\Gamma(1+\alpha_-\alpha_3)}{\Gamma[2+\alpha_-(\alpha_1+\alpha_3)]} F[-\alpha_-\alpha_2, 1+\alpha_-\alpha_1, 2+\alpha_-(\alpha_1+\alpha_3); x]$$

References

Dotsenko, V. S. and Fateev, V. A. (1984). *Nucl. Phys. B*, **240**, 312.

Rozansky, L. and Saleur, H. (1992). *Nucl. Phys. B*, **376**, 461.

Ising model

This chapter is fully devoted to two-dimensional classical and equivalent one-dimensional quantum Ising models. Sometimes I will not distinguish between the two. In what follows I will not mention the number of dimensions and the reader should take the meaning of term 'Ising model' in the sense explained above.

In the previous chapter we have briefly discussed the so-called minimal conformal theories obtained via truncation of the Hilbert space of the theory of free bosons. This truncation is really a sophisticated elimination of certain eigenstates of the original bosonic theory (22.1) without violation of the conformal symmetry. In this chapter I discuss the simplest 'nontrivial' model of this sort, namely, the model with conformal charge $C = 1/2$ ($p = 3$). It is known that this theory describes the two-dimensional Ising model at the point of second-order phase transition. I shall derive a relation between the Ising model and the theory of free real (Majorana) fermions and prove this statement.

The importance of the Ising model for the theory of strong interactions is difficult to overemphasize. As we shall see further, it appears in such important applications as spin liquids. The Ising model has many equivalent representations. Since we are interested in field theory, I describe only continuous limits of these representations.

The most straightforward representation of the Ising model is the Ginzburg–Landau theory of a real field φ (the so-called φ^4 model):

$$S = \int d^2x \left(\frac{1}{2}(\partial_\mu \varphi)^2 + \frac{\tau}{2}\varphi^2 + \frac{g}{4}\varphi^4 \right) \tag{28.1}$$

where $g > 0$.

At the critical point the Ising model can be represented as the minimal model (27.39) with $Q^2 = 1/6$.

Another equivalent form of the action is fermionic (see (28.10) and the discussion below).

This triadic equivalence is a very powerful tool in analysis of strongly correlated systems related to the Ising model.

Ising model as a minimal model

To consider the minimal model with $C = 1/2$ is especially instructive because in this case it is easy to demonstrate how the reduction of the Hilbert space occurs. Let us construct a conformally invariant model with $C = 1/2$ *ab initio* using a model of free spinless fermions with a linear spectrum as the starting point. The Hamiltonian is equal to

$$\hat{H} = \int dx[-i\psi_R^+\partial_x\psi_R + i\psi_L^+\partial_x\psi_L] = \sum_q[q\psi_R^+(q)\psi_R(q) - q\psi_L^+(q)\psi_L(q)] \quad (28.2)$$

and the free energy is given by (25.29). Now let us separate the eigenvalues with positive and negative energies:

$$\hat{H} = \hat{H}_+ + \hat{H}_-$$

$$\hat{H}_+ = \sum_{q>0} q[\psi_R^+(q)\psi_R(q) + \psi_L^+(-q)\psi_L(-q)] \quad (28.3)$$

$$\hat{H}_- = -\sum_{q>0} q[\psi_R^+(-q)\psi_R(-q) + \psi_L^+(q)\psi_L(q)]$$

The free energy associated with the $+$ part is equal to

$$\frac{F_+}{L} = -T \int_{-\infty}^{\infty} \frac{dp}{2\pi} \ln(1 + e^{-|p|/T}) = -\frac{\pi}{12}T^2 \quad (28.4)$$

i.e. according to (25.30), it corresponds to $C = 1/2$ as required. More than that, the excitation spectrum of the truncated theory remains linear, $\omega(q) = |q|$, which guarantees separation of left and right movers and factorization of the correlation functions into analytic and antianalytic parts. Now I want to make sure that we can write this truncated theory as a local quantum field theory of some field whose stress energy tensor satisfies the Virasoro algebra (26.13). Let us introduce the *following fields*:[1]

$$\chi_R(z) = \sum_{q>0}[\psi_R^+(q)e^{-qz} + \psi_R(q)e^{qz}]$$

$$\chi_L(\bar{z}) = \sum_{q>0}[\psi_L^+(-q)e^{-q\bar{z}} + \psi_L(-q)e^{q\bar{z}}] \quad (28.5)$$

At $\tau = 0$ these operators are real: $\chi^+(x) = \chi(x)$. They satisfy the following anticommutation relations:

$$\{\chi_a(x), \chi_b(y)\} = \delta_{ab}\delta(x - y)$$

Therefore we can call them real fermions (they are also called Majorana fermions). It is easy to check that one can rewrite the truncated Hamiltonian in terms of the Majorana fields:

$$H_+ = \frac{i}{2}\int dx[-\chi_R\partial_x\chi_R + \chi_L\partial_x\chi_L] \quad (28.6)$$

The correlation function of the right-moving components is equal to

$$\langle\langle\chi_R(z)\chi_R(0)\rangle\rangle \equiv \theta(\tau)\langle\chi_R(x, \tau)\chi_R(0)\rangle - \theta(-\tau)\langle\chi_R(0, 0)\chi_R(x, \tau)\rangle$$

$$= \sum_{q>0}\langle\psi_R(q)\psi_R^+(q)\rangle[e^{-iqx-q\tau}\theta(\tau) - e^{iqx+q\tau}\theta(-\tau)] = \frac{1}{2\pi(\tau + ix)} \equiv \frac{1}{2\pi z}$$

$$(28.7)$$

---

[1] They do not include the eliminated states!

For the left-moving components we get

$$\langle\langle \chi_L(\bar{z})\chi_L(0)\rangle\rangle = \frac{1}{2\pi\bar{z}} \tag{28.8}$$

Thus the Majorana fermions χ_R, χ_L represent primary fields with conformal dimensions $(1/2, 0)$ and $(0, 1/2)$, respectively. A field

$$\epsilon(z, \bar{z}) = i\chi_R(z)\chi_L(\bar{z})$$

has conformal dimensions $(1/2, 1/2)$.

The stress energy tensor is also local in Majorana fermion fields. It is straightforward to check that the following operators

$$T(z) = -2\pi \chi_R \partial_z \chi_R$$
$$\bar{T}(\bar{z}) = -2\pi \chi_L \partial_{\bar{z}} \chi_L \tag{28.9}$$

satisfy the Virasoro algebra (26.13) with $C = 1/2$, thus being the components of the stress energy tensor for the Ising model.

Exercise

Show that operators (28.9) are true stress energy tensors, that is they satisfy (26.13), and (26.17). *Hint*: consider two- and three-point correlation functions of operators (28.9) and compare them with the canonical ones.

As will be shown further in this chapter, the field $\epsilon(z, \bar{z})$ is equivalent to the energy density field in the Ising model. It will also be shown (see also Dotsenko and Dotsenko (1982) and McCoy and Wu (1973)) that in the vicinity of the transition the Ising model is described by the following Hamiltonian:

$$\hat{H} = \int dx \left[-\frac{i}{2}\chi_R \partial_x \chi_R + \frac{i}{2}\chi_L \partial_x \chi_L + im\chi_R\chi_L \right]$$

$$= \sum_{q>0} \{q[\psi_R^+(q)\psi_R(q) + \psi_L^+(-q)\psi_L(-q)] + m[\psi_R^+(q)\psi_L(-q) + \text{H.c.}]\} \tag{28.10}$$

where $m = b(\theta/\theta_c - 1)$ and θ, θ_c are the temperature and the critical temperature, respectively, and b is a numerical constant.[2] Do not confuse them with T, the 'quantum' temperature, which in the present context is related to the width of the strip on which the two-dimensional Ising model is defined. It is remarkable that the Ising model remains exactly solvable even at $\theta \neq \theta_c$.

It will not lead us too far off the track if we discuss the specific heat of the *classical* Ising model. The partition function of the Ising model is equal to

$$Z = \int D\chi_R D\chi_L \exp\left[-S_0 - b(\theta/\theta_c - 1)\int d^2x \epsilon(x)\right] \tag{28.11}$$

---

[2] For the Ising model on a square lattice $b^2 = 32(\sqrt{2}+1)$.

where $x_1 = \tau$, $x_2 = x$ and

$$S_0 = \int d^2x \left[\frac{1}{2} \chi_R (\partial_\tau - i\partial_x) \chi_R + \frac{1}{2} \chi_L (\partial_\tau + i\partial_x) \chi_L \right]$$

The entropy is equal to

$$S \approx -\theta_c \frac{\partial \ln Z}{\partial \theta} = b \int d^2x \langle \epsilon(x) \rangle \qquad (28.12)$$

and the specific heat per unit area is given by

$$C \approx b^2 \int d^2x \langle\langle \epsilon(x)\epsilon(0) \rangle\rangle$$

$$\approx b^2 \int_{|z|<(\theta/\theta_c-1)^{-1}} \frac{dz d\bar{z}}{4\pi^2 |z|^2} \approx -\frac{b^2}{8\pi} \ln(\theta/\theta_c - 1) \qquad (28.13)$$

i.e. exactly as for the Ising model.

Quantum Ising model

The Ising model in its original formulation is a model of a magnet and is formulated in terms of spins. We have not yet discussed the relationship between the spins and Majorana fermions. To do this we have to return to microscopic derivation. In that context it is easier to discuss not the classical two-dimensional Ising model, but its one-dimensional quantum equivalent: the so-called quantum Ising model or Ising model in a transverse magnetic field. I will not derive this equivalence here, referring the reader to the original papers (Ferrell, 1973; Fradkin and Susskind, 1978).

The quantum Ising model frequently appears in applications. Its Hamiltonian has the following form:

$$H = -\sum_n \left(J\sigma_n^x \sigma_{n+1}^x + h\sigma_n^z \right) \qquad (28.14)$$

where σ^a are Pauli matrices. This Hamiltonian can be diagonalized exactly using the Jordan–Wigner transformation discussed in Chapter 19 (Pfeuty, 1970). In fact, we can do a better job and consider a more general model

$$H = -\sum_n \left(J_x \sigma_n^x \sigma_{n+1}^x + J_y \sigma_n^y \sigma_{n+1}^y + h\sigma_n^z \right) \qquad (28.15)$$

which can be solved by the same method.

Using formulas (19.4) we find

$$\sigma_n^x \sigma_{n+1}^x = (\psi_n^+ - \psi_n)(\psi_{n+1}^+ + \psi_{n+1}) \qquad \sigma_n^y \sigma_{n+1}^y = -(\psi_n^+ + \psi_n)(\psi_{n+1}^+ - \psi_{n+1})$$

Substituting these expressions together with $\sigma^z = 2\psi^+\psi - 1$ into (28.15) and passing to momentum space I obtain the following Hamiltonian:

$$H = \sum_{k>0} \{ [(J_x + J_y)\cos k - 2h](\psi_k^+ \psi_k + \psi_{-k}^+ \psi_{-k}) + (J_x - J_y)\sin k(\psi_k^+ \psi_{-k}^+ + \psi_{-k}\psi_k) \}$$

$$(28.16)$$

which looks like a Hamiltonian of a superconductor with $\epsilon_k = [(J_x + J_y)\cos k - 2h]$ and $\Delta_k = (J_x - J_y)\sin k$. It can be diagonalized by the Bogolyubov transformation:

$$\psi_k = \cos\theta_k \, A_k + \sin\theta_k \, B_k^+ \qquad \psi_{-k} = -\sin\theta_k \, A_k + \cos\theta_k \, B_k^+ \qquad (28.17)$$

where $\tan(2\theta_k) = \Delta_k/\epsilon_k$, with the result

$$H = \sum_{k>0} E(k)(A_k^+ A_k + B_k^+ B_k)$$

$$E(k) = \{[(J_x + J_y)\cos k - 2h]^2 + (J_x - J_y)^2 \sin^2 k\}^{1/2} \qquad (28.18)$$

The spectrum becomes gapless at $2h = J_x + J_y$; in the vicinity of this point the spectrum has a relativistic form

$$E^2(k) \approx m^2 + v^2 k^2 \qquad m = (J_x + J_y) - 2h \qquad v = (J_x - J_y) \qquad (28.19)$$

The continuous limit of the quantum Ising model Hamiltonian is given by (28.10). There are two conclusions one can draw from the above discussion.

Majorana fermions do not represent anything mysterious, they are just Bogolyubov quasi-particles,

The J_y-term in the Ising model Hamiltonian just shifts the critical value of h and renormalizes the velocity v. Unless $J_y = J_x$, its presence is irrelevant. Therefore from now on I will put $J_y = 0$.

Exercise

Show that the $J_z \sigma_n^z \sigma_{n+1}^z$ term has a similar effect and therefore is also irrelevant at small J_z.

Order and disorder operators

The quantum Ising model (28.14) has the most curious property of *self-duality*. Namely, its Hamiltonian remains invariant, being rewritten in terms of new operators μ^x, μ^z, which are nonlocal with respect to σ^z, σ^x, but possess the same commutation relations. These dual operators are defined on sites of a *dual lattice*, that is on half-integer sites $n + 1/2$:

$$\mu_{n+1/2}^x = \prod_{j=1}^{n} \sigma_j^z \qquad \mu_{n-1/2}^x \mu_{n+1/2}^x = \sigma_n^z$$

$$\mu_{n+1/2}^z = \sigma_n^x \sigma_{n+1}^x \qquad \sigma_n^x = \prod_{j=1}^{n} \mu_{j+1/2}^z \qquad (28.20)$$

In terms of the new variables the Hamiltonian preserves its form, though J and h are interchanged:

$$H = -\sum_n \left(h\mu_{n-1/2}^x \mu_{n+1/2}^x + J\mu_{n+1/2}^z \right) \qquad (28.21)$$

This corresponds to a change of sign of the mass term in the fermionic Hamiltonian (28.10). Naturally, the spectrum

$$E(k) = \sqrt{J^2 + h^2 - 2Jh\cos k} \qquad (28.22)$$

remains oblivious to such change.

Though the spectrum does not change, the correlation functions of σ^x and μ^x experience the most dramatic transformation. To get a qualitative understanding of this let us consider the limits of large h and large J. If $J \gg h$ there is a finite average $\langle\sigma^x\rangle$. In what follows I shall call $\sigma \equiv \sigma^x$ the *order parameter* field. On the contrary, if $h \gg J$ this average is absent, but there is a finite average $\langle\mu^x\rangle$. I shall call $\mu \equiv \mu^x$ the *disorder parameter* field. The change of regime occurs at the critical point, as might be expected.

There is an interesting relationship between correlation functions of σ and μ and the correlation functions of the sine-Gordon model with $\beta^2 = 4\pi$ (the free fermionic point). This relationship comes from the fact that the sum of two Ising models is equivalent to the model of a massive noninteracting Dirac fermion. This identification follows immediately from the fact that one can make a Dirac fermion from two Majoranas:

$$\psi = \chi_1 + i\chi_2 \qquad \psi^+ = \chi_1 - i\chi_2 \qquad (28.23)$$

However, bosonizing the model of free massive Dirac fermions we obtain the sine-Gordon model with $\beta^2 = 4\pi$ (see the discussion around (22.25)). It turns out (I will not derive this, referring the reader to the original papers by Zuber and Itzykson (1977) and Schroer and Truong (1978)) that order and disorder parameter fields of Ising models 1 and 2 are related to the trigonometric functions of the bosonic field Φ and its dual field Θ. The following formulas are written for the case $J < 2h$ ($m > 0$) when the order parameter field has zero average:

$$\sigma_1(x)\sigma_2(x) = \frac{(ma_0)^{1/4}}{\sqrt{\pi}} \; : \sin\sqrt{\pi}\,\Phi(x) :$$

$$\mu_1(x)\mu_2(x) = \frac{(ma_0)^{1/4}}{\sqrt{\pi}} \; : \cos\sqrt{\pi}\,\Phi(x) :$$

$$(28.24)$$

$$\sigma_1(x)\mu_2(x) = \frac{(ma_0)^{1/4}}{\sqrt{\pi}} \; : \cos\pi\,\Theta(x) :$$

$$\mu_1(x)\sigma_2(x) = \frac{(ma_0)^{1/4}}{\sqrt{\pi}} \; : \sin\pi\,\Theta(x) :$$

To obtain similar expressions for $J > 2h$ one has to interchange σ and μ. The dots correspond to normal ordering.

Since the Ising models do not interact, we can write down the following relatioships:

$$\langle\sigma(1)\cdots\sigma(N)\rangle^2 = \left[\frac{(ma_0)^{1/4}}{\sqrt{\pi}}\right]^N \langle:\sin\pi\,\Phi(1): \cdots :\sin\pi\,\Phi(N):\rangle$$

$$(28.25)$$

$$\langle\mu(1)\cdots\mu(N)\rangle^2 = \left[\frac{(ma_0)^{1/4}}{\sqrt{\pi}}\right]^N \langle:\cos\pi\,\Phi(1): \cdots :\cos\pi\,\Phi(N):\rangle$$

These expressions are valid both at and outside the critical point. At the critical point, however, the situation is greatly simplified since correlation functions of bosonic exponents are simple.

Exercise

Using (28.24) show that at the critical point the four-point correlation function of spin fields in the Ising model is equal to

$$\langle \sigma(1) \cdots \sigma(4) \rangle^2 = \left(\frac{r_{14}r_{32}}{r_{12}r_{34}r_{13}r_{42}} \right)^{1/2} + \left(\frac{r_{13}r_{42}}{r_{12}r_{34}r_{14}r_{32}} \right)^{1/2} + \left(\frac{r_{12}r_{34}}{r_{13}r_{42}r_{14}r_{32}} \right)^{1/2} \tag{28.26}$$

where $r = |z|$. Consult the book by Itzykson and Drouffe (1989) if necessary.

Correlation functions outside the critical point

In this section I give (without derivation) asymptotic expansions of two-point correlation functions for the order and disorder parameters of the *off-critical* two-dimensional Ising model in the limits $\tilde{r} \ll 1$ and $\tilde{r} \gg 1$, where $\tilde{r} = mr$. (Details of calculations can be found in Wu *et al.* (1976).) Assuming that $m > 0$ (i.e. $T > T_c$), the long-distance asymptotics ($\tilde{r} \gg 1$) are given by

$$\langle \sigma(r)\sigma(0) \rangle \equiv G_\sigma(\tilde{r}) = \frac{A_1}{\pi} K_0(\tilde{r}) + \mathcal{O}(e^{-3\tilde{r}}) \tag{28.27}$$

$$\langle \mu(r)\mu(0) \rangle \equiv G_\mu(\tilde{r})$$

$$= A_1 \left\{ 1 + \frac{1}{\pi^2} \left[\tilde{r}^2 \left(K_1^2(\tilde{r}) - K_0^2(\tilde{r}) \right) - \tilde{r} K_0(\tilde{r})K_1(\tilde{r}) + \frac{1}{2} K_0^2(\tilde{r}) \right] \right\} + \mathcal{O}(e^{-4\tilde{r}}) \tag{28.28}$$

where A_1 is a nonuniversal parameter, and $K_n(\tilde{r})$ are the Bessel functions of imaginary argument.

In the short-distance limit, $\tilde{r} \ll 1$, the leading asymptotics of the correlation functions are of power law form:

$$G_\sigma(\tilde{r}) = G_\mu(\tilde{r}) = A_2 \tilde{r}^{-1/4} \tag{28.29}$$

The ratio of the constants A_1 and A_2 is a universal quantity involving Glaisher's constant A:

$$\frac{A_2}{A_1} = 2^{-1/6} A^{-3} e^{1/4} \tag{28.30}$$

$$A = 1.282\,427\,129 \ldots$$

One can find information about Ising model correlation functions at finite temperatures in the paper by Leclair *et al.* (1996). Expressions for n-point correlation functions were derived by Abraham (1978).

Deformations of the Ising model

Two coupled Ising models

This subsection is based on the results obtained in the papers by Delfino and Mussardo (1998) and Fabrizio *et al.* (2000).

Let us consider two coupled quantum Ising models

$$H = H_1 + H_2 - J' \sum_n \sigma_1^x \sigma_2^x \tag{28.31}$$

where $H_{1,2}$ are given by (28.14).

Close to the critical point we can write down the continuous version of the effective action using (28.24):

$$S = \int d\tau dx \mathcal{L} \tag{28.32}$$

$$\mathcal{L} = \frac{1}{2}(\partial_\mu \Phi)^2 - m \cos(2\sqrt{\pi}\Phi) + J' \sin(\sqrt{\pi}\Phi) \tag{28.33}$$

At $J' \neq 0$, model (28.33) is solvable only at $m = 0$ where it is equivalent to the sine-Gordon model with $\beta^2 = \pi$. The spectrum at this point is massive and contains kink, antikink and six breathers (see the subsequent chapters about the sine-Gordon model). At $m \neq 0$ the model is no longer integrable. Naively one might expect that at large m it is possible to develop a perturbation theory in $J'/m^{7/4}$ taking the basis of free Dirac fermions as a starting point. Let us consider in detail whether such expectations are justified.

Quantum numbers (topological charges)

$$Q = \frac{1}{2\sqrt{\pi}} \int dx \partial_x \Phi \tag{28.34}$$

are determined by a distance between the minima of the effective potential. In a soliton configuration field Φ approaches different minima at $x = \pm\infty$ giving rise to nonzero Q. At $J' = 0$, a soliton (antisoliton) has topological charge $+1/2$ $(-1/2)$.

At $J' \neq 0$ the situation depends on the sign of m. At $m < 0$, which corresponds to the case when the individual Ising models are in the ordered phase $\langle \sigma_i \rangle \neq 0$, the perturbation lifts a degeneracy between the minima of the potential

$$V(\Phi) = -m \cos(2\sqrt{\pi}\Phi) + J' \sin(\sqrt{\pi}\Phi) \tag{28.35}$$

Now the minima $\sqrt{\pi}\Phi = \pi/2 + 2\pi n$ and $\sqrt{\pi}\Phi = 3\pi/2 + 2\pi n$ have different energy. Only one set corresponds to the vacuum state (for $J' > 0$ it is the second one). The distance between the minima is doubled and (anti)solitons carry topological charge ± 1. This is a situation of *confinement*: particles with topological charges $\pm 1/2$ are altogether removed from the spectrum and only their bound states remain. These bound states are adiabatically connected to the kinks of the $\beta^2 = \pi$ sine-Gordon model. With such dramatic restructuring of the spectrum it is obvious that perturbation theory in J' fails at low energies. At $m > 0$ the positions of the minima in (28.35) are shifted in the opposite directions, but their energies remain the same. This means that there are two kinds of solitons with different topological charges $Q_\pm = 1/2 \pm \alpha J'$ where α is some coefficient. The change is smooth which makes it possible to develop a perturbation theory approach.

The increase in J' in the ordered state leads to a phase transition. This transition belongs to the Ising model universality class. Indeed, let us expand potential (28.35) around its minimum:

$$V(\Phi) = V(3\sqrt{\pi}/2) + \frac{1}{2}V''(3\sqrt{\pi}/2)(\Phi - 3\sqrt{\pi}/2)^2$$
$$+ \frac{1}{4!}V^{(4)}(3\sqrt{\pi}/2)(\Phi - 3\sqrt{\pi}/2)^4 + \cdots \qquad (28.36)$$

where

$$V''(3\sqrt{\pi}/2) = 4|m| - J'$$

Thus at some positive value of J' (quantum corrections strongly renormalize its classical value $4|m|$) the quadratic term vanishes. The remaining Lagrangian density

$$\mathcal{L}_{\text{eff}} = \frac{1}{2}(\partial_\mu \Phi')^2 + \gamma(\Phi')^4 \qquad \Phi' = \Phi - 3\sqrt{\pi}/2 \qquad (28.37)$$

describes the critical Ising model.

Ising model in a magnetic field

$$H = H_0 + \mathcal{H}\sum_n \sigma_n^x \qquad (28.38)$$

where H_0 describes the Ising model at the critical point. The magnetic field is a relevant perturbation and generates spectral gaps. In the continuous limit the model is integrable (Zamolodchikov, 1988, 1989) and its spectrum and certain correlation functions are known (Delfino and Mussardo, 1995). The excitation spectrum contains eight massive particles with masses

$$m_1 \qquad m_2 = 2m_1 \cos(\pi/5) \qquad m_3 = 2m_1 \cos(\pi/30) \qquad m_4 = 2m_2 \cos(7\pi/30)$$
$$m_5 = 2m_2 \cos(2\pi/15) \qquad m_6 = 2m_2 \cos(\pi/30) \qquad m_7 = 4m_2 \cos(\pi/5)\cos(7\pi/30)$$
$$m_8 = 4m_2 \cos(\pi/5)\cos(2\pi/15)$$

All these particles contribute to the two-point correlation function of the order parameter operator, but with unequal weight:

$$\langle\langle \sigma(\omega, q)\sigma(-\omega, -q) \rangle\rangle = \sum_{j=1}^{8} \frac{Z_j}{\omega^2 - q^2 - m_j^2} \qquad (28.39)$$

The highest contributions come from the first, second, third and fourth particles with relative weights $Z_1:Z_2:Z_3:Z_4 = 1:0.21:0.09:0.05$.

References

Abraham, D. (1978). *Commun. Math. Phys.*, **60**, 205.

Belavin, A. B., Zamolodchikov, A. A. and Polyakov, A. M. (1984). *Nucl. Phys. B*, **241**, 333; *Conformal Invariance and Applications to Statistical Mechanics*, ed. C. Itzykson, H. Saleur and J.-B. Zubert. World Scientific, Singapore, 1988.

Delfino, G. and Mussardo, G. (1995). *Nucl. Phys. B*, **455**, 724.

Delfino, G. and Mussardo, G. (1998). *Nucl. Phys. B*, **516**, 675.

Dotsenko, Vik. S. and Dotsenko, V. S. (1982). *J. Phys. C*, **15**, 495.

Fabrizio, M. Gogolin, A. O. and Nersesyan, A. A. (2000). *Nucl. Phys. B*, **580**, 647.

Ferrell, R. A. (1973). *J. Stat. Phys.*, **8**, 265.

Fradkin, E. and Susskind, L. (1978). *Phys. Rev. D*, **17**, 2637.

Itzykson, C. and Drouffe, J.-M. (1989). *Statistical Field Theory*, Vol. 1, Chapter 2, Vol. 2, Chapter 9. Cambridge University Press, Cambridge.

Leclair, A., Lesage, F., Sachdev, S. and Saleur, H. (1996). *Nucl. Phys. B*, **482**, 579.

McCoy, B. M. and Wu, T. T. (1973). *The Two-Dimensional Ising Model*. Harvard University Press.

Pfeuty, P. (1970). *Ann. Phys.* (NY), **57**, 79.

Schroer, B. and Truong, T. T. (1978). *Nucl. Phys. B*, **144**, 80.

Wu, T. T., McCoy, B. M., Tracy, C. A. and Barouch, E. (1976). *Phys. Rev. B*, **13**, 316.

Zamolodchikov, A. B. (1988). *Int. J. Mod. Phys. A*, **3**, 743; *Adv. Stud. Pure Math.*, **19**, 641 (1989).

Zuber, B. and Itzykson, C. (1977). *Phys. Rev. D*, **15**, 2875.

One-dimensional spinless fermions:
Tomonaga–Luttinger liquid

Let us consider spinless fermions on a lattice with a short-range interaction. Such a problem may arise in the context of one-dimensional spin $S = 1/2$ Heisenberg chains which, as we know from Chapter 19, can be reformulated as a model of spinless fermions. Assuming that the interaction is much smaller than the bandwidth, let us write the continuous limit of the Lagrangian. From Chapter 14 we know we can linearize the spectrum in the vicinity of the Fermi points and decompose the Fermi fields into the slow components:

$$\psi(x) = \exp(-ip_F x)R(x) + \exp(ip_F x)L(x) \tag{29.1}$$

In the continuous limit the most general form of the Lagrangian we can get is

$$L = \int dx[R^+(\partial_\tau - iv\partial_x)R + L^+(\partial_\tau + iv\partial_x)L + gR^+RL^+L] \tag{29.2}$$

where the velocity v and the coupling constant g depend on the bare parameters.

Using the bosonization rules formulated in Table 22.1, we can write down the bosonized version of model (29.2):

$$L = \frac{(1 + g/\pi v)}{2} \int dx[(v^{-1}(\partial_\tau \Phi)^2 + v(\partial_x \Phi)^2] \tag{29.3}$$

To reduce this Lagrangian to the canonical form of the Gaussian model (22.1), one needs to rescale the fields:

$$\Phi \to K^{1/2}\Phi \qquad \Theta \to K^{-1/2}\Theta \qquad K = (1 + g/\pi v)^{-1} \tag{29.4}$$

One highly nontrivial consequence of this rescaling is that the fermionic fields cease to be chiral:

$$R = \frac{1}{\sqrt{2\pi a_0}} \exp\{i\sqrt{\pi}[(K^{1/2} + K^{-1/2})\varphi + (K^{1/2} - K^{-1/2})\bar{\varphi}]\}$$

$$L = \frac{1}{\sqrt{2\pi a_0}} \exp\{-i\sqrt{\pi}[(K^{1/2} + K^{-1/2})\bar{\varphi} + (K^{1/2} - K^{-1/2})\varphi]\} \tag{29.5}$$

which leads to the following nontrivial change in the single-fermion Green's function (it is assumed that $T = 0$):

$$G(\tau, x) = \frac{1}{2\pi} \left(\frac{e^{-ip_F x}}{v\tau + ix} + \frac{e^{ip_F x}}{v\tau - ix} \right) \left(\frac{a_0^2}{v^2\tau^2 + x^2} \right)^{\theta/2} \tag{29.6}$$

where a_0 is the ultraviolet cut-off and

$$\theta = \frac{1}{2}(K^{1/2} - K^{-1/2})^2 \tag{29.7}$$

In frequency momentum space the function has a branch cut.

Single-electron correlator in the presence of Coulomb interaction

A problem one may encounter in applications is a problem of one-dimensional electrons interacting via the Coulomb interaction. In this section I consider two situations: (a) the electrons occupy a single chain or wire, (b) the chains are assembled into a lattice. In reality one has to take into account the fact that electrons have spin, which leads to minor complications. So I will neglect spin in the first approximation. Since I will be particularly interested in the single-fermion correlation function, I will approach the problem using a chiral decomposition of the Gaussian model given by (22.27).

Let us consider first the problem of the wire. The contribution of the Coulomb interaction to the bosonized action is given by

$$V = \frac{1}{2}\sum_{q,\omega} \rho(q,\omega)V(q)\rho(-q,-\omega) = \frac{1}{2\pi}\sum_{q,\omega} q^2 V(q)(\varphi + \bar\varphi)_{(q,\omega)}(\varphi + \bar\varphi)_{(-q,-\omega)} \tag{29.8}$$

where

$$V(q) = -\frac{e^2}{\pi}\ln|qd|$$

with d being the thickness of the wire. Putting together (29.8) and (22.27) I get the following effective action:

$$S = \frac{1}{2}\sum_{q,\omega}(\varphi,\bar\varphi)_{-q,-\omega}\begin{pmatrix} 2q(vq + i\omega) + q^2 V(q) & q^2 V(q) \\ q^2 V(q) & 2q(vq - i\omega) + q^2 V(q) \end{pmatrix}\begin{pmatrix}\varphi \\ \bar\varphi\end{pmatrix}_{q,\omega}$$

The correlation functions are

$$\begin{pmatrix} \langle\langle\varphi\varphi\rangle\rangle & \langle\langle\varphi\bar\varphi\rangle\rangle \\ \langle\langle\bar\varphi\varphi\rangle\rangle & \langle\langle\bar\varphi\bar\varphi\rangle\rangle \end{pmatrix} = \frac{1}{4q\{\omega^2 + v^2 q^2[1 + \frac{e^2}{\pi v}\ln(1/|q|d)]\}}$$

$$\times \begin{pmatrix} 2(vq - i\omega) + \frac{e^2 q}{\pi}\ln(1/|q|d) & -\frac{e^2 q}{\pi}\ln(1/|q|d) \\ -\frac{e^2 q}{\pi}\ln(1/|q|d) & 2(vq + i\omega) + \frac{e^2 q}{\pi}\ln(1/|q|d) \end{pmatrix} \tag{29.9}$$

From here we can extract the single-electron Green's function:

$$\langle\langle R(\tau, x)R^+(0,0)\rangle\rangle = \exp[-\mathcal{K}(\tau, x)] \tag{29.10}$$

$$\mathcal{K}(\tau, x) = 2\pi[\langle\langle\varphi(\tau, x)\varphi(0,0)\rangle\rangle - \langle\langle\varphi(0,0)\varphi(0,0)\rangle\rangle]$$

$$= \int \frac{d\omega dq}{(2\pi)^2}[e^{-i\omega\tau - iqx} - 1]\langle\langle\varphi(-\omega, -q)\varphi(\omega, q)\rangle\rangle$$

$$= \int \frac{d\omega dq}{2\pi}[e^{-i\omega\tau - iqx} - 1]\frac{2(vq - i\omega) + \frac{e^2 q}{\pi}\ln(1/|q|d)}{4q\{\omega^2 + v^2 q^2[1 + \frac{e^2}{\pi v}\ln(1/|q|d)]\}} \tag{29.11}$$

I shall estimate this integral only for $x = 0$. With logarithmic accuracy we have

$$G_R(\tau, x = 0) = \text{sign}\tau \, \exp[-f(\tau) + f(\tau_0)]$$

$$f(\tau) = \frac{4}{3\xi} \left[1 + \frac{\xi}{4} \ln(\tau v_\tau / d)\right] \sqrt{1 + \xi \ln(\tau v_\tau / d)} \qquad (29.12)$$

$$v_\tau^2 = v^2[1 + \xi \ln(v\tau / d)] \qquad \xi = \frac{e^2}{\pi v} \qquad \tau_0 v \sim \epsilon_F$$

The distinct feature of this function is that at large times $\tau > \tau^* \sim v^{-1} d \exp(1/\xi)$ it decays faster then any power law, but still slower than an exponential:

$$G_R(\tau, x = 0) \sim \text{sign}\tau \, (1/|\tau|)^{\frac{1}{3}\sqrt{\xi \ln(\tau/\tau^*)}} \qquad (29.13)$$

This behaviour gives rise to a pseudogap in the tunnelling density of states.

I suggest the reader consider the problem of many chains coupled by the Coulomb interaction as an exercise. Now the fields $\varphi, \bar{\varphi}$ depend on the chain index n. The Coulomb interaction also depends on the distance between the chains. This leads to the following modification of (29.8):

$$V = \frac{1}{2\pi} \sum_{q,\omega} q^2 V(q, q_\perp)(\varphi + \bar{\varphi})_{(q,q_\perp,\omega)}(\varphi + \bar{\varphi})_{(-q,-q_\perp,-\omega)} \qquad (29.14)$$

where

$$V(q, q_\perp) = \sum_G V_0(\mathbf{q} + \mathbf{G}) \qquad V_0(\mathbf{q}) = \frac{4\pi e^2}{|\mathbf{q}|^2} \qquad (29.15)$$

where the sum goes over inverse lattice vectors.

Spin $S = 1/2$ Heisenberg chain

As I have said, the model of spinless fermions has a direct connection to the $S = 1/2$ Heisenberg chain. The latter model very frequently appears in applications since there are quite a few magnetic systems which can be described as compositions of weakly interacting magnetic chains (see the review articles by Landee (1987) and Regnault et al. (1989)). Some of the best examples of spin $S = 1/2$ quasi-one-dimensional antiferromagnets are $KCuF_3$ and $CuCl_2 \cdot 2N(C_5D_5)$, where magnetic copper ions belong to chains well separated from each other in space. The recent neutron measurements on $KCuF_3$ are in excellent agreement with the theory presented below (Cowley et al., 1993). A large magnitude of the inchain exchange integral ($J \sim 1000$ K) exists in Sr_2CuO_3 compound in combination with a tiny ordering temperature $T_N \sim 5$ K (see Keren et al., 1993).

Historically, the spin $S = 1/2$ Heisenberg chain was among the first models treated by bosonization (Luther and Peschel, 1975).

Let us start our consideration with the case when the exchange integral in the z-direction is much smaller than the exchange intergrals $J_x = J_y = J$. In this case the interaction in the fermion model is weak and we can apply directly the results of the previous discussion. This means that only states close to the Fermi points $\pm k_F = \pm\pi/2a$ (a being the lattice

Figure 29.1. Alan Luther.

constant) are important, and one can linearize the bare-particle spectrum $\epsilon(k) = J \cos ka$ in the vicinity of these points, as we have done above:

$$\epsilon(k) \simeq \mp v(k \mp k_F) \qquad |k \mp k_F| \ll k_F \tag{29.16}$$

where $v = Ja$ is the Fermi velocity of the Jordan–Wigner fermions. As follows from representation (19.4), the z-component of the spin density operator of the XXZ chain coincides with the normal ordered density operator for the Jordan–Wigner fermions,

$$S_j^z =: \psi_j^\dagger \psi_j := \psi_j^\dagger \psi_j - 1/2$$

Since in one-dimensional Fermi systems the Fermi 'surface' is reduced to two points $\pm k_F$, only Fourier components of the operator $\rho_q = \sum_k \psi_k^\dagger \psi_{k+q}$ with momenta q close to 0 and $\pm 2k_F$ describe low energy density fluctuations. Consequently, in the continuum limit, the spin density operator $S^z(x)$ is given by a sum of slow and rapidly oscillating (staggered) contributions:

$$S_j^z \to a S^z(x)$$
$$S^z(x) = \rho(x) + (-1)^j M(x) \tag{29.17}$$

Here

$$\rho(x) = :R^\dagger(x)R(x): + :L^\dagger(x)L(x): = \bar{J}(x) + J(x) \tag{29.18}$$

where $J, \bar{J}$ are the left and right Abelian currents satisfying the U(1) Kac–Moody algebra (see Chapter 14), and

$$M(x) = R^\dagger(x)L(x) + L^\dagger(x)R(x) \tag{29.19}$$

Using decompositions (29.17) and (29.18), we obtain the continuum version of the XXZ model

$$H = H_0 + H_{int}$$

$$H_0 = -iv \int dx (R^\dagger \partial_x R - L^\dagger \partial_x L)$$

(29.20)

$$H_{int} = v\Delta \int dx [:\rho(x)\rho(x+a): - M(x)M(x+a)]$$

(29.21)

Let us now bosonize this Hamiltonian. As we already know from Chapter 19, a model of free massless fermions is equivalent to the model of a free massless Bose field. Therefore

$$H_0 = \frac{v}{2} \int dx [\hat{\pi}^2 + (\partial_x \Phi)^2]$$

(29.22)

where $\hat{\pi}(x)$ is the momentum conjugate to the field $\Phi(x)$, satisfying the canonical commutation relation

$$[\Phi(x), \hat{\pi}(x')] = i\delta(x - x')$$

(29.23)

When bosonizing H_{int}, the limit $a \to 0$ can be safely taken in the first, $\rho\rho$ term of (29.21), and the smooth part of the density operator can be replaced by

$$\rho(x) = \frac{1}{\sqrt{\pi}} \partial_x \Phi(x)$$

(29.24)

However, the second term in (29.21) should be treated with care. Representing $M(x)$ as

$$M(x) \simeq -\frac{1}{\pi a} \sin \sqrt{4\pi} \Phi(x)$$

(29.25)

one can derive the following operator product expansion:

$$\lim_{a \to 0} \left\{ \frac{1}{\pi a} \sin \sqrt{4\pi} \Phi(x) \right\} \left\{ \frac{1}{\pi a} \sin \sqrt{4\pi} \Phi(x + a) \right\}$$

$$= -\frac{1}{(\pi a)^2} \cos \sqrt{16\pi} \Phi(x) - \frac{1}{\pi} (\partial_x \Phi)^2 + const$$

(29.26)

Using this relation, one finds the bosonized version of the XXZ model to be of the form (see Lukyanov (1997) for more details)

$$H = \int dx \left\{ \frac{v}{2} [\hat{\pi}^2 + (1 + 4\Delta/\pi)(\partial_x \Phi)^2] + \frac{v\Delta}{(\pi a)^2} \cos \sqrt{16\pi} \Phi \right\}$$

(29.27)

The cosine term in (29.27) originates from Umklapp processes

$$R^\dagger(x)R^\dagger(x + a)L(x + a)L(x) + H.c.$$

'hidden' in the interaction of the staggered components of $S^z(x)$. These processes are allowed to appear because the band of the Jordan–Wigner fermions is half-filled, and $4k_F$ coincides with a reciprocal lattice vector (Haldane, 1980, 1981; Black and Emery, 1981;

den Nijs, 1981). The Umklapp scattering breaks the continuous chiral (γ^5) symmetry of the one-dimensional fermionic model (29.20) down to a discrete (Z_2) one. The latter can be broken spontaneously in two dimensions and, in fact, at $\Delta = 1$ the system undergoes an Ising-like transition from the gapless disordered phase to a gapful, long-range ordered Néel state. However, in the whole range $|\Delta| \leq 1$ the scaling dimension of the cosine perturbation is larger than 2. As was explained in Chapter 23, such a perturbation should be regarded as irrelevant and, therefore, can be taken into account perturbatively. This has been done by Eggert *et al.* (1994) and Lukyanov (1998) who established an excellent agreement between the field theory and the exact solution.

The remaining Hamiltonian can be rewritten as follows:

$$H = \frac{u}{2} \int dx \left[K\hat{\pi}^2 + \frac{1}{K}(\partial_x \Phi)^2 \right]$$

(29.28)

To transform it to the canonical form (29.22), one should rescale the field and momentum keeping the canonical commutator (29.23) preserved,

$$\Phi(x) \to \sqrt{K}\,\Phi(x) \qquad \hat{\pi}(x) \to \frac{1}{\sqrt{K}}\hat{\pi}(x)$$

(29.29)

and renormalize the velocity

$$v \to u = \frac{v}{K}$$

(29.30)

At small Δ

$$K \simeq 1 - \frac{2\Delta}{\pi} + 0(\Delta^2)$$

Thus, the low energy sector of the spin-1/2 XXZ chain is described by the Gaussian model with parameters u and K depending on the anisotropy Δ. The conformal dimensions of various operators are given by (22.22). One should take into account, however, that operators local in terms of spins have integer conformal spins, and operators local in terms of fermions may have half-integer spins. In fact, the Gaussian model description, which we have derived in the limit $\Delta \ll 1$, holds throughout the entire critical region $-1 < \Delta \leq 1$. The exact expressions for u and K can be extracted from the Bethe ansatz solution of the spin-1/2 XXZ chain (Johnson *et al.*, 1973):

$$K = \frac{\pi}{2(\pi - \mu)} \qquad u = \frac{\pi \sin \mu}{2\mu}$$

(29.31)

According to the Abelian bosonization rules given in Table 22.1, with transformations (29.29) taken into account, the spin density operators are represented as

$$S^z(x) = \sqrt{\frac{K}{\pi}}\partial_x \Phi(x) - \lambda_z(-1)^n \sin \sqrt{4\pi K}\,\Phi(x)$$

(29.32)

$$S^{\pm}(x) = \lambda_x(-1)^n \exp[\pm i\sqrt{\pi/K}\,\Theta(x)]$$

(29.33)

where λ_z, λ_x are given as (Lukyanov, 1998), for example:

$$\lambda_x^2 = \frac{1}{2}(\pi/\mu)^2 \left[\frac{\Gamma(\pi/2\mu - 1/2)}{2\sqrt{\pi}\Gamma(\pi/2\mu)} \right]^{(1-\mu/\pi)}$$

$$\times \exp\left\{ -\int_0^\infty \frac{dx}{x} \left[\frac{\sinh(1 - \pi/\mu)x}{\sinh x \cosh(\pi x/\mu)} - (\pi/\mu - 1)e^{-2x} \right] \right\} \quad (29.34)$$

I emphasize that the latter expression is valid only for the Heisenberg model with nearest-neighbour exchange interaction. In (29.33) I have retained the most singular, staggered part of the transverse magnetization.

Since the smooth part of $S^z(x)$ is proportional to the sum of the U(1) currents,

$$S_{\text{smooth}}^z(x) = \sqrt{K}[J(x) + \bar{J}(x)] \quad (29.35)$$

the decay of the corresponding correlation function is characterized by universal critical exponent 2:

$$\langle S_{\text{smooth}}^z(x, \tau)S_{\text{smooth}}^z(0, 0)\rangle = K[\langle J(x, \tau)J(0, 0)\rangle + \langle \bar{J}(x, \tau)\bar{J}(0, 0)\rangle] = \frac{K}{4\pi^2}\left(\frac{1}{z^2} + \frac{1}{\bar{z}^2}\right)$$

$$(29.36)$$

where $z = u\tau + ix$, $\bar{z} = u\tau - ix$. Notice that under Wick's back rotation to $(1 + 1)$-dimensional space-time, this correlator is not Lorentz invariant, because $S_{\text{smooth}}^z(x)$ is proportional to a fixed (temporal) component j^0 of the two-current $j^\mu \sim \epsilon^{\mu\nu}\partial_\nu\Phi$. However, the correlation functions of the staggered components of the magnetization are Lorentz invariant, with critical exponents continuously depending on the anisotropy parameter Δ (Luther and Peschel, 1975):

$$\langle S_{\text{stag}}^z(x, \tau)S_{\text{stag}}^z(0, 0)\rangle \sim \frac{1}{|z|^{\nu_z}} \quad (29.37)$$

$$\langle S_{\text{stag}}^+(x, \tau)S_{\text{stag}}^-(0, 0)\rangle \sim \frac{1}{|z|^{\nu_x}} \quad (29.38)$$

where

$$\nu_z = \frac{1}{\nu_x} = 2K \quad (29.39)$$

At the isotropic point, $\Delta = 1$, $K = 1/2$, the above correlation functions display manifestly SU(2)-invariant behavior, with $\nu_x = \nu_z = 1$. An alternative bosonization scheme for the Heisenberg (SU(2)-invariant) spin chain is discussed in Chapter 30.

Explicit expression for the dynamical magnetic susceptibility

The spin-spin correlation functions are directly measurable by inelastic neutron scattering. The neutron's differential cross-section at energy transfer ω and wave vector q is proportional to the imaginary part of the Fourier image of the dynamical spin-spin correlation function. The latter is related to the retarded spin-spin correlation function $\chi^{(R)}(\omega, q)$:

$$\frac{d\sigma(\omega, k)}{d\Omega} \propto \frac{1}{1 - e^{-\omega/T}} \Im m \chi^{(R)}(\omega, k) \tag{29.40}$$

Since the bosonization approach usually provides us with correlation functions in space-time representation, it is convenient to use the following explicit expression where a retarded correlation function is written as a Fourier transformation of the space-time correlation function:

$$\Im m D^{(R)}(\omega) = \frac{1}{2} \int_{-\infty}^{\infty} dt [D_-(it + \epsilon) - D_+(it + \epsilon)] e^{i\omega t} \tag{29.41}$$

where the functions $D_\pm(\tau)$ are determined from the decomposition of the original thermodynamic Green's function:

$$D(\tau) = \theta(\tau) D_+(\tau) \pm \theta(-\tau) D_-(\tau) \tag{29.42}$$

For bosonic operators $D_-(\tau) = D_+(-\tau)$.

It is not really a problem for us to get an explicit expression for the thermodynamic spin-spin correlation function. As was explained at the beginning of Chapter 25, one should make a conformal transformation

$$z(\xi) = \exp(2\pi i T \xi)$$

Taking into account the operator relations (29.38) for the transverse spin components, we get the following expression for the thermodynamic correlation function of their staggered components:

$$\langle\langle S^+(\tau, x) S^-(0, 0)\rangle\rangle \propto \left\{ \frac{(\pi T)^2}{\sinh[\pi T(x - i\tau)] \sinh[\pi T(x + i\tau)]} \right\}^{2\Delta} (-1)^{x/a}$$

with $\Delta = 1/8K = \frac{1}{4}(1 - \mu/\pi)$.[1] In order to find $D^+(\tau, q)$ and $D^-(\tau, q)$, it is convenient to calculate the Fourier transformation in x first ($q = \pi - k$, where k is a real wave vector):

$$D(\tau, q > 0) = \int_{-\infty}^{\infty} dx e^{iqx} \left\{ \frac{(\pi T)^2}{\sinh[\pi T(x - i\tau)] \sinh[\pi T(x + i\tau)]} \right\}^{2\Delta}$$

Bending the contour of integration to the upper plane where the integrand has a cut we get

$$2\sin(2\pi \Delta) \int_{|\tau|}^{\infty} dy e^{-qy} \left\{ \frac{(\pi T)^2}{\sinh[\pi T(iy - i\tau)] \sinh[\pi T(iy + i\tau)]} \right\}^{2\Delta}$$

From this we find

$$D_+(\tau, q > 0) - D^-(\tau, q > 0)$$

$$= -2\sin(2\pi \Delta) \int_{-\tau}^{\tau} dy e^{-qy} \left\{ \frac{(\pi T)^2}{\sinh[\pi T(iy - i\tau)] \sinh[\pi T(iy + i\tau)]} \right\}^{2\Delta} \tag{29.43}$$

[1] Here and below the renormalized velocity of magnons is assumed to be $u = 1$.

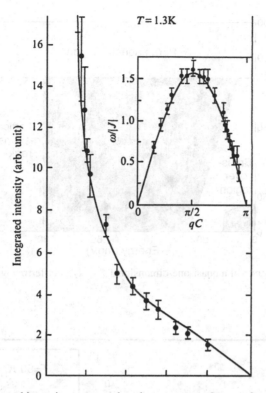

Figure 29.2. The integrated intensity versus reduced wave vector $Q = qa$ for $CuCl_2 \cdot 2N(C_5D_5)$ (from Endoh *et al.*, 1974). The solid line represents the theoretical prediction.

The latter expression allows the straightforward analytic continuation $\tau = it$. Finally we can write down the integral representation for $\Im m D(\omega, q)$:

$$-\Im m D(\omega, q > 0)$$

$$= \sin(2\pi \Delta) \int_{-\infty}^{\infty} dt e^{i\omega t} \int_{-|t|}^{|t|} dx e^{-iqx} \left\{ \frac{(\pi T)^2}{\sinh[\pi T(t - x)] \sinh[\pi T(x + t)]} \right\}^{2\Delta} \quad (29.44)$$

This integral is reduced to a table one by the substitution $z = x + t$, $\bar{z} = t - x$:

$$-\Im m D(\omega, q > 0) = \sin(2\pi \Delta)(\pi T)^{4\Delta} \Im m \int_0^\infty dz \frac{e^{iz(\omega - q)}}{[\sinh(\pi T z)]^{2\Delta}} \int_0^\infty d\bar{z} \frac{e^{i\bar{z}(\omega + q)}}{[\sinh(\pi T \bar{z})]^{2\Delta}}$$

$$\sim \sin(2\pi \Delta) \frac{1}{T^{2-4\Delta}} \Im m \left[\rho\left(\frac{\omega - q}{4\pi T}\right) \rho\left(\frac{\omega + q}{4\pi T}\right) \right] \quad (29.45)$$

where $q = |\pi - k|$ and

$$\rho(x) = \frac{\Gamma(\Delta - ix)}{\Gamma(1 - \Delta - ix)}$$

The above result was first obtained by Schulz and Bourbonnais (1983) and Schulz (1986).

Figure 29.2 represents the data for the imaginary part of the magnetic susceptibility integrated over all momenta $I(q) = \int d\omega \, \Im m \chi(\omega, q)$. For an isotropic Heisenberg chain

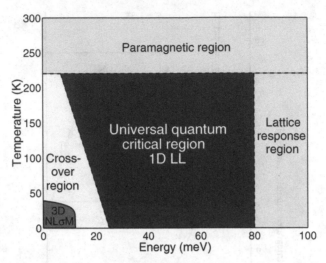

Figure 29.3. Phase diagram of a quasi-one-dimensional $S = 1/2$ antiferromagnet $KCuF_3$ (provided by A. Tennant).

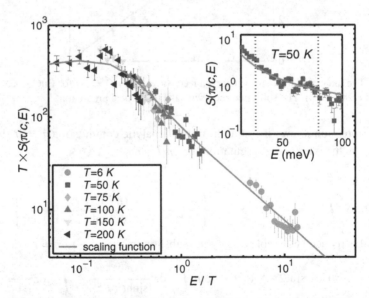

Figure 29.4. The scaling plot of the spectral function of $KCuF_3$ measured at the Néel wave vector. The black curve corresponds to (29.45) with $\Delta = 1/4$. The inset shows the same function as a function of energy at $T = 50$ K (provided by A. Tennant).

where $\Delta = 1/4$ in the limit of zero temperature we derive from (29.45) $I \sim -\ln(qa)$, which agrees well with the data at $|qa| \ll 1$.

There is another way to measure magnetic susceptibility, by nuclear magnetic resonance (NMR), which unfortunately does not measure the entire function $\chi(\omega, q)$ as neutron scattering does. NMR measures the relaxation time of a nuclear spin τ_1 which is related to the

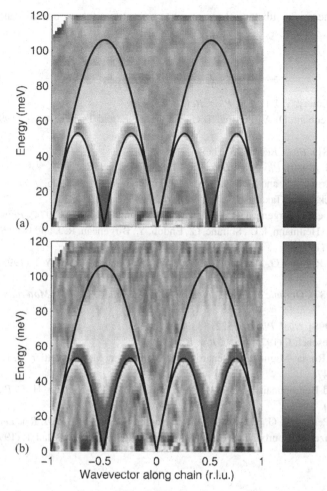

Figure 29.5. The neutron scattering intensity distribution for a quasi-one-dimensional $S = 1/2$ antiferromagnet KCuF$_3$ below (a) and above (b) the transition. The brighter shades correspond to larger values of $\chi''(\omega, q)$ (from Tennant *et al.*, 1993).

magnetic susceptibility of the medium:

$$\frac{1}{\tau_1 T} \propto \lim_{\omega \to 0} \int d^D k \frac{\Im m \chi^{(R)}(\omega, k)}{\omega} \tag{29.46}$$

It follows from (29.45) that the contribution from the region $|k| \approx \pi$ is of the order of

$$\frac{1}{\tau_1 T} \propto T^{-2+4\Delta} \tag{29.47}$$

NMR measurements of various quasi-one-dimensional materials do demonstrate this power law dependence (see Wzietek *et al.*, 1993). See Figs. 29.3, 29.4 and 29.5.

In Figs. 29.3, 29.4 and 29.5 I present several experimental pictures for a quasi-one-dimensional $S = 1/2$ antiferromagnet KCuF$_3$. This material consists of weakly coupled copper chains. Though it orders antiferromagnetically below 10 K (see the phase diagram in Fig. 29.3), there is a substantial region of temperatures where one-dimensional scaling is observed.

References

Black, J. L. and Emery, V. J. (1981). *Phys. Rev. B*, **23**, 429.

Cowley, R. A., Tennant, D. A., Perring, T. G., Nagler, S. E. and Tsvelik, A. M. (1993). *Physica A*, **194**, 280.

den Nijs, M. (1981). *Phys. Rev. B*, **23**, 6111.

Haldane, F. D. M. (1980). *Phys. Rev. Lett.*, **45**, 1358; **47**, 1840 (1981).

Johnson, J. D., Krinsky, S. and McCoy, B. M. (1973). *Phys. Rev. A*, **8**, 2526.

Eggert, S., Affleck, I. and Takahashi, M. (1994). *Phys. Rev. Lett.*, **73**, 332.

Endoh, Y., Shirane, G., Birgeneau, R. J., Richards, P. M. and Holt, S. L. (1974). *Phys. Rev. Lett.*, **32**, 171; see also Heilmann, I. U., Shirane, G., Endoh, Y., Birgeneau, R. J. and Holt, S. L., *Phys. Rev. B*, **18**, 3530 (1978).

Keren, A., Le, L. P., Luke, G. M., Sternlieb, B. J., Wu, W. D. and Uemura, Y. J. (1993). *Phys. Rev. B*, **48**, 12926.

Landee, C. P. (1987). *Organic and Inorganic Low-Dimensional Crystalline Materials*, p. 75. Plenum, New York.

Lukyanov, S. (1998). *Nucl. Phys. B*, **522**, 533.

Luther, A. and Peschel, I. (1975). *Phys. Rev. B*, **12**, 3908.

Regnault, L. P., Rossat-Mignod, J., Renard, J. P., Verdaguer, M. and Vettier, C. (1989). *Physica*, **156–157**, 247.

Schulz, H. J. and Bourbonnais, C. (1983). *Phys. Rev. B*, **27**, 5856; Schulz, H. J., *Phys. Rev. B*, **34**, 6372 (1986).

Tennant, D. A., Perring, T. G., Cowley, R. A. and Nagler, S. E. (1993). *Phys. Rev. Lett.*, **70**, 4003.

Wzietek, P., Creuzet, F., Bourbonnais, C., Jerome, D., Bechgaard, K. and Batail, P. (1993). *J. Physique* **3**, 171.

30

One-dimensional fermions with spin: spin-charge separation

The bosonization procedure works equally well for fermions with spin, thus providing us with an essentially nonperturbative approach to one-dimensional metals. The foundations of this approach were laid in the late 1970s and early 1980s by various authors (see the review article by Brazovsky and Kirova (1984) and references therein).

The bosonization approach is based on the fact that coherent excitations in one-dimensional interacting systems are not renormalized electrons, but collective excitations – bosons. For spinless fermions there is only one branch of collective excitations – charge density waves. For fermions with spin, another branch appears which represents spin density waves. Elementary excitations in the charge sector carry charge $\pm e$ and spin 0; excitations in the spin sector are neutral and carry spin $1/2$. Since the different branches have different symmetry properties, one can expect them to have quite different spectra.[1] This can even go to such extremes that one branch has a gap and the other does not. It is only natural under such circumstances that electrons, which carry quantum numbers from both the spin and charge sectors, cannot propagate coherently. Roughly speaking, the parts of the electron containing different degrees try to tear it in pieces; it is customary to call this phenomenon *spin-charge separation*. Empirically this loss of coherence is revealed as an absence of a quasi-particle pole in the single-electron Green's function, an effect whose existence is supported by experimental observations (see discussion at the end of the chapter).

All phenomenona described above (the spin-charge separation and existence of excitations with fractions of quantum numbers) can be described by the model considered below. Let us consider a model of one-dimensional fermions with an interaction which preserves the SU(2) symmetry:

$$H_{\text{int}} = \frac{1}{2} \int dx \, dx' \, \psi_\alpha^\dagger(x) \psi_\beta^\dagger(x') Q_{\alpha\beta\alpha'\beta'}(x - x') \psi_{\beta'}(x') \psi_{\alpha'}(x) \tag{30.1}$$

where

$$Q_{\alpha\beta\alpha'\beta'}(x - x') = U_1(x - x')\delta_{\alpha\alpha'}\delta_{\beta\beta'} - U_2(x - x')\delta_{\alpha\beta'}\delta_{\beta\alpha'} \tag{30.2}$$

In my derivation I will treat the interaction as weak, $U_i \rho(\epsilon_F) \ll 1$, so that only states close to the Fermi points are involved in interactions. In this case one can linearize the

---

[1] Charge excitations have the U(1) and spin excitations the SU(2) symmetry.

spectrum in the vicinity of the Fermi points and decompose the Fermi fields into the slow components:

$$\psi_\sigma(x) = \exp(-i p_F x) R_\sigma(x) + \exp(i p_F x) L_\sigma(x) \tag{30.3}$$

First we shall assume that $4k_F \neq 2\pi$ and drop the oscillatory terms containing $\exp(\pm 4 i k_F x)$. For a one-dimensional Fermi system on a lattice, this is the case when the band is not half-filled.

The linearized noninteracting Hamiltonian has the following form:

$$\hat{H}_0 = v \int dx [-i R_\sigma^+(x) \partial_x R_\sigma(x) + i L_\sigma^+(x) \partial_x L_\sigma(x)] \tag{30.4}$$

For what follows it will be convenient to introduce the so-called current operators:

$$J = \sum_\alpha L_\alpha^+ L_\alpha \qquad \bar{J} = \sum_\alpha R_\alpha^+ R_\alpha \tag{30.5}$$

and

$$J^a = \frac{1}{2} \sum_{\alpha\beta} L_\alpha^+ \sigma_{\alpha\beta}^a L_\beta$$

$$\bar{J}^a = \frac{1}{2} \sum_{\alpha\beta} R_\alpha^+ \sigma_{\alpha\beta}^a R_\beta \tag{30.6}$$

where σ^a are Pauli matrices. These operators represent the smooth parts of the charge and the spin density of the left- and right-moving fermions. The zero momentum Fourier components of the currents are generators of symmetry transformations for the groups U(1) and SU(2) respectively. The current operators will play the key role in all subsequent discussion and are the most important objects in the theory of strongly correlated systems.

The current operators possess the following important properties.

The charge and spin currents mutually commute.
The currents of different chirality (right or left) commute.
As we established in Chapter 14, the charge currents satisfy the following algebra:

$$[J(x), J(y)] = \frac{i}{\pi} \partial_x \delta(x - y)$$

$$[\bar{J}(x), \bar{J}(y)] = \frac{i}{\pi} \partial_x \delta(x - y) \tag{30.7}$$

The generalization of this algebra for the spin currents is

$$[J^a(x), J^b(y)] = \frac{i}{4\pi} \delta_{ab} \partial_x \delta(x - y) + i \epsilon^{abc} J^c(y) \delta(x - y)$$

$$[\bar{J}^a(x), \bar{J}^b(y)] = \frac{i}{4\pi} \delta_{ab} \partial_x \delta(x - y) + i \epsilon^{abc} \bar{J}^c(y) \delta(x - y) \tag{30.8}$$

The extra terms appearing in the latter expressions are not anomalous and come from the ordinary commutators. Equations (30.8) represent a particular case of *Kac–Moody* algebra. We shall discuss this algebra in greater detail in subsequent chapters.

Almost all results obtained for one-dimensional electrons are based on the following two facts.

The free Hamiltonian (30.4) can be recast in terms of the current operators:

$$H_0 = H[U(1)] + H[SU(2)_1] \tag{30.9}$$

$$H[U(1)] = 2\pi v \int dx [:J(x)^2: + :\bar{J}^2(x):] \tag{30.10}$$

$$H[SU(2)_1] = \frac{2\pi v}{3} \int dx [:\mathbf{J}(x)^2: + :\bar{\mathbf{J}}^2(x):] \tag{30.11}$$

where the dots denote the normal ordering.
Interaction (30.1) can also be written in terms of currents:

$$H_{\text{int}} = \int dx \, \{g_4(J^2 + \bar{J}^2) + g_c J \bar{J} + g_4'(\mathbf{J} \cdot \mathbf{J} + \bar{\mathbf{J}} \cdot \bar{\mathbf{J}}) + g_s \mathbf{J} \cdot \bar{\mathbf{J}}\} \tag{30.12}$$

where

$$g_4 = \frac{1}{4}[2V_1(0) - V_2(0)] \qquad g_4' = -V_2(0) \qquad g_s = -2[V_2(0) + V_1(2k_F)]$$

$$g_c = [V_1(0) + V_2(2k_F)] - \frac{1}{2}[V_2(0) + V_1(2k_F)]$$

$V_i(k)$ being the Fourier transforms of $U_i(x)$.[2]

To derive (30.12) I substituted (30.3) into (30.10) and used the following identities:

$$R_\sigma^+ L_\sigma L_{\sigma'}^+ R_{\sigma'} = -\frac{1}{2}J\bar{J} - 2\mathbf{J} \cdot \bar{\mathbf{J}} \tag{30.13}$$

$$\lim_{x \to x'} L_\sigma^+(x)L_{\sigma'}(x)L_{\sigma'}^+(x')L_\sigma(x') = \frac{1}{2}J J + 2\mathbf{J} \cdot \mathbf{J} \tag{30.14}$$

$$\lim_{x \to x'} R_\sigma^+(x)R_{\sigma'}(x)R_{\sigma'}^+(x')R_\sigma(x') = \frac{1}{2}\bar{J}\bar{J} + 2\bar{\mathbf{J}} \cdot \bar{\mathbf{J}} \tag{30.15}$$

Exercise

Show that for the Hubbard model

$$\hat{H} = -t \sum_{j,\sigma}(c_{j+1,\sigma}^+ c_{j,\sigma} + c_{j,\sigma}^+ c_{j+1,\sigma}) + U \sum_j n_{j\uparrow}n_{j\downarrow}$$

$$\tag{30.16}$$

$$g_4 = \frac{U}{4} \qquad g_4' = -\frac{U}{3} \qquad g_s = -2U \qquad g_c = \frac{U}{2}$$

and $v = 2t \sin k_F$.

---

[2] We suppose that these matrix elements are nonsingular.

Exercise

Using the formulas from the bosonization table (22.1), show that model (30.10) is equivalent to the Gaussian model (29.3). Derive a relationship between K_c and v_c, g_c.

From these two capital facts we derive that the Hamiltonian of one-dimensional interacting electrons splits into two mutually commuting parts:

$$H = H_c + H_s \equiv H_{U(1)} + H_{SU_1(2)} \tag{30.17}$$

where

$$H_c = \frac{\pi v_c}{2} \int dx \, (:JJ: + :\bar{J}\bar{J}:) + g_c \int dx \, J\bar{J} \tag{30.18}$$

$$H_s = \frac{2\pi v_s}{3} \int dx \, (:\mathbf{J} \cdot \mathbf{J}: + :\bar{\mathbf{J}} \cdot \bar{\mathbf{J}}:) + g_s \int dx \, \mathbf{J} \cdot \bar{\mathbf{J}} \tag{30.19}$$

with renormalized velocities

$$v_c = v_F + \frac{2g_4}{\pi} \qquad v_s = v_F + \frac{3g'_4}{2\pi} \tag{30.20}$$

and therefore can be studied separately. In other words, the spectrum of the interacting model is composed of two branches of independent excitations. This effect is called *spin-charge separation*. The Hamiltonian (30.18) is equivalent to the Hamiltonian of a spinless Tomonaga–Luttinger liquid (29.3) and hence to the Gaussian model. As we know from the previous discussion, this model is characterized by two parameters: the velocity v_c and constant K_c determining the scaling dimensions. The latter constant is related to g_c.

Relationship (30.20) between g_c, g_s, v_c, v_s and parameters of the bare Hamiltonian is not universal and holds only for weak interactions. The spin-charge separation itself is universal, which means that the low energy sector of the theory is always described by Hamiltonians (30.18) and (30.19). This description is corrected by irrelevant operators and is asymptotically exact at zero energy.

When the band is half-filled ($2k_F = \pi$), we have $\exp(4ik_F na) = 1$, and the Umklapp process

$$g_u(R^\dagger_\sigma L_\sigma R^\dagger_{\sigma'} L_{\sigma'} + \text{H.c.}) \qquad g_u = V_1(4k_F) + V_2(4k_F) \tag{30.21}$$

should be taken into account in the derivation of the continuum limit of the Hamiltonian. This extra term contributes to the charge part of the Hamiltonian modifying its simple Gaussian form. In fact, with the Umklapp term included, the charge Hamiltonian H_c can be written entirely in terms of new, the so-called *pseudospin*, vector currents $\mathbf{I}$ and $\bar{\mathbf{I}}$. The latter can be obtained from the usual spin vector currents $\mathbf{J}$, $\bar{\mathbf{J}}$ by a particle–hole transformation in one spin component of the Fermi field (Lieb and Wu, 1968; Yang, 1989; Yang and Zhang, 1990):

$$c_{j\downarrow} \to (-1)^j c^\dagger_{j\downarrow} : \qquad R_\downarrow \to R^\dagger_\downarrow \qquad L_\downarrow \to L^\dagger_\downarrow \tag{30.22}$$

which implies that

$$I^3 = \sum_\sigma :R_\sigma^\dagger R_\sigma : = \frac{1}{2}J \qquad I^+ = R_\uparrow^\dagger R_\downarrow^\dagger \qquad I^- = (I^+)^\dagger \qquad (30.23)$$

with similar expressions for the right chiral components $\bar{I}^a$. One can easily check that the pseudospin currents satisfy the $SU_1(2)$ Kac–Moody algebra.

Written in terms of $\mathbf{I}, \bar{\mathbf{I}}$, the Hamiltonian H_c takes the form

$$H_c = \frac{2\pi v_c}{3} \int dx \, (:\mathbf{I}\cdot\mathbf{I}: + :\bar{\mathbf{I}}\cdot\bar{\mathbf{I}}:)$$

$$+ \int dx \, [4g_c I^3 \bar{I}^3 + 2g_u(I^+\bar{I}^- + I^-\bar{I}^+)] \qquad (30.24)$$

Bosonic form of the $SU_1(2)$ Kac–Moody algebra

For futher analysis of the problem I employ the bosonization technique. Since we have two species of fermions with spin up and spin down, we have to introduce two bosonic fields Φ_α, $\alpha =\uparrow, \downarrow$, with the corresponding dual fields. The fields with different spin indices commute. Therefore one might think that to describe the situation one has to replicate the bosonization table 22.1. However, a slight complication arises due to the fact that fermion operators with different spin *anticommute*. To take care of this, new elements should be introduced into the scheme, namely the Klein factors. Klein factors are *coordinate independent* operators satisfying the anticommutation relations

$$\{\eta_\sigma, \eta_{\sigma'}\} = 2\delta_{\sigma\sigma'} \qquad \eta_\sigma^+ = \eta_\sigma \qquad (30.25)$$

They can be represented by the Pauli matrices:

$$\eta_\uparrow = \tau^x \qquad \eta_\downarrow = \tau^y \qquad \eta_\uparrow\eta_\downarrow = i\tau^z \qquad (30.26)$$

Since the Klein factors have no dynamics, we can always choose one of the two eigenvalues of τ^z, e.g. $+1$. So, in what follows, I shall assume that

$$\eta_\uparrow\eta_\downarrow = i$$

The modified bosonization formulas for fermions include these factors and are

$$R_\sigma(x) \simeq (2\pi a_0)^{-1/2}\eta_\sigma \exp(i\sqrt{4\pi}\,\varphi_\sigma(x))$$
$$L_\sigma(x) \simeq (2\pi a_0)^{-1/2}\eta_\sigma \exp(-i\sqrt{4\pi}\,\bar{\varphi}_\sigma(x)) \qquad (30.27)$$

For future convenience we introduce linear combinations of bosonic fields:

$$\varphi_c = \frac{\varphi_\uparrow + \varphi_\downarrow}{\sqrt{2}} \qquad \bar{\varphi}_c = \frac{\bar{\varphi}_\uparrow + \bar{\varphi}_\downarrow}{\sqrt{2}}$$

$$\varphi_s = \frac{\varphi_\uparrow - \varphi_\downarrow}{\sqrt{2}} \qquad \bar{\varphi}_s = \frac{\bar{\varphi}_\uparrow - \bar{\varphi}_\downarrow}{\sqrt{2}} \qquad (30.28)$$

The most essential fact is that since

$$(\partial_\mu \Phi_\uparrow)^2 + (\partial_\mu \Phi_\downarrow)^2 = (\partial_\mu \Phi_c)^2 + (\partial_\mu \Phi_s)^2 \qquad (30.29)$$

this transformation leaves invariant the action (or the Hamiltonian) of the noninteracting fermions (30.4).

With these definitions the $k = 1$ Kac–Moody currents (30.6) can be expressed solely in terms of the spin field:

$$J^+ = R^+_\uparrow R_\downarrow = \frac{i}{2\pi a_0} e^{-i\sqrt{8\pi}\varphi_s} \qquad \bar{J}^+ = L^+_\uparrow L_\downarrow = \frac{i}{2\pi a_0} e^{i\sqrt{8\pi}\bar{\varphi}_s}$$

$$J^- = R^+_\downarrow R_\uparrow = \frac{-i}{2\pi a_0} e^{i\sqrt{8\pi}\varphi_s} \qquad \bar{J}^- = L^+_\downarrow L_\uparrow = \frac{-i}{2\pi a_0} e^{-i\sqrt{8\pi}\bar{\varphi}_s} \qquad (30.30)$$

$$J^3 = \frac{1}{2}(R^+_\uparrow R_\uparrow - R^+_\downarrow R_\downarrow) = \frac{i}{\sqrt{2\pi}}\partial\varphi_s \qquad \bar{J}^3 = -\frac{i}{\sqrt{2\pi}}\bar{\partial}\varphi_s$$

Though definitions (30.30) contain the cut-off explicitly, the current-current correlation functions are cut-off independent and reveal the underlying SU(2) symmetry:

$$\langle\langle J^a(x)J^b(x')\rangle\rangle = -\frac{\delta^{ab}}{4\pi^2}\frac{1}{(x-x')^2} \qquad (30.31)$$

It is easy to generalize the above derivation for the charge Kac–Moody currents (30.23). The resulting expressions are the same with φ_s, $\bar{\varphi}_s$ being substituted by φ_c, $\bar{\varphi}_c$.

Besides the current operators there are other important fields, the most important among them being the charge and the spin density operators

$$\rho(x) \approx \psi^+_\sigma(x)\psi_\sigma(x) \approx R^+_\sigma R_\sigma + L^+_\sigma L_\sigma + [\exp(-2ik_Fx)R^+_\sigma L_\sigma + \text{H.c.}] + \cdots \quad (30.32)$$

$$S^a(x) = \frac{1}{2}\psi^+_\sigma(x)\sigma^a_{\sigma\sigma'}\psi_{\sigma'}(x)$$

$$\approx J^a + \bar{J}^a + \frac{1}{2}[\exp(-2ik_Fx)R^+_\sigma\sigma^a_{\sigma\sigma'}L_{\sigma'} + \text{H.c.}] + \cdots \qquad (30.33)$$

where the dots stand for operators with conformal dimensions higher than one. Using (30.27) one can derive the following bosonized expressions for the density operators:

$$\rho(x) = J + \bar{J} + \frac{a_c}{2\pi a_0}\sin(2k_Fx + \sqrt{2\pi}\,\Phi_c)\cos(\sqrt{2\pi}\,\Phi_s)$$

$$+ a_u\cos(4k_Fx + \sqrt{8\pi}\,\Phi_c) \qquad (30.34)$$

$$S(x) = J + \bar{J} + \frac{a_s}{2\pi a_0}\cos(2k_Fx + \sqrt{2\pi}\,\Phi_c)\,\mathbf{n}(x) \qquad (30.35)$$

where

$$\mathbf{n} = (-\sin(\sqrt{2\pi}\,\Phi_s), \cos(\sqrt{2\pi}\,\Theta_s), \sin(\sqrt{2\pi}\,\Theta_s)) \qquad (30.36)$$

Though derived for noninteracting electrons, formulas (30.34), (30.35) remain valid in the presence of interactions. Indeed, the interactions may affect only two things: (a) they can modify the dynamics of fields Φ_c, Φ_s, but this does alter the form of (30.36), (30.37); (b) they can change the values of the amplitudes a_c, a_s, a_u. It turns out that the amplitudes are not universal, that is they are determined not just by the low energy processes taken into account in the derivation of the low energy effective action, but also depend on high energy physics. The most important change introduced by the interactions is the generation of finite amplitudes a_u for the $4k_F$-oscillation of charge density.

Spin $S = 1/2$ Tomonaga–Luttinger liquid

Physics described by Hamiltonians (30.18) (or (30.24) if the band is half-filled) and (30.19) strongly depends on signs of the coupling constants and a value of the band filling. If $2k_F \neq \pi$ the Umklapp processes are irrelevant and the charge sector is described by the Gaussian effective action (30.18). This means that excitations in the charge sector are gapless and their spectrum is linear. The behaviour in the spin sector depends on the relevance of the current-current interaction. This interaction is marginal and its relevance depends on the sign of the coupling constant g_s. If $g_s < 0$ it renormalizes to zero. Then the spin sector is described by Hamiltonian (30.19) with $g_s = 0$

$$H_s = \frac{2\pi v_s}{3} \int dx \, (:\mathbf{J} \cdot \mathbf{J}: + :\bar{\mathbf{J}} \cdot \bar{\mathbf{J}}:) \tag{30.37}$$

This Hamiltonian represents a particular case of the Wess–Zumino–Novikov–Witten (WZNW) model. The WZNW model plays an extraodinarily important role in the theory of strong interactions and I will discuss it in greater detail in subsequent chapters. Here I would just like to point out that the version of the WZNW model represented by (30.37) has an equivalent representation in the form of the Gaussian model with $K_s = 1$.

The Luttinger parameter of the charge sector is not fixed and, according to (29.4), depends on the coupling constant g_c. As I have mentioned already, this relation is universal only for weak coupling. For stronger coupling one can extract it using the relationship between energy levels in a finite system and scaling dimensions of the operators explained in Chapter 25. The energy levels are calculated either numerically or using the exact solution (if it is available). The reader can find examples of such calculations in the papers by Frahm and Korepin (1990) and Kawakami and Yang (1991).

Thus we see that one part of the phase diagram of one-dimensional spin-1/2 fermions is occupied by a critical state, the *Tomonaga–Luttinger liquid*. This state is characterized by two velocities of gapless collective modes v_c and v_s and two Luttinger parameters K_c and K_s which determine conformal dimensions of all operators. In the presence of SU(2) symmetry $K_s = 1$.

From (30.34), (30.35) we can find the corresponding two-point correlation functions:

$$\langle \mathbf{S}(1)\mathbf{S}(2) \rangle = \frac{\chi_s}{8\pi v_s} \left[\frac{1}{(\tau + ix/v_s)^2} + \frac{1}{(\tau - ix/v_s)^2} \right] + \frac{a_s^2 \cos(2k_F x)}{|\tau + ix/v_c|^{K_c} |\tau + ix/v_s|} \tag{30.38}$$

$$\langle \rho(1)\rho(2) \rangle = \frac{\chi_c}{8\pi v_c} \left[\frac{1}{(\tau + ix/v_c)^2} + \frac{1}{(\tau - ix/v_c)^2} \right] + \frac{a_c^2 \cos(2k_F x)}{|\tau + ix/v_c|^{K_c} |\tau + ix/v_s|}$$

$$+ \frac{a_u^2 \cos(4k_F x)}{|\tau + ix/v_c|^{4K_c}} \tag{30.39}$$

Taking into account rescaling of the charge fields (29.4), one can derive from (30.27), (30.28) the bosonized form of the single-particle creation and annihilation operators (compare it with the spinless case (29.5)):

$$R_\sigma(x) \simeq (2\pi a_0)^{-1/2} \eta_\sigma \mathcal{O}^c \mathcal{O}^s_\sigma$$
$$L_\sigma(x) \simeq (2\pi a_0)^{-1/2} \eta_\sigma \bar{\mathcal{O}}^c \bar{\mathcal{O}}^s_\sigma$$

(30.40)

$$\mathcal{O}^c = \exp\left\{ i\sqrt{\pi/2}\left[(K_c^{1/2} + K_c^{-1/2})\varphi_c + (K_c^{1/2} - K_c^{-1/2})\bar{\varphi}_c \right] \right\}$$
$$\bar{\mathcal{O}}^c = \exp\left\{ -i\sqrt{\pi/2}\left[(K_c^{1/2} + K_c^{-1/2})\bar{\varphi}_c + (K_c^{1/2} - K_c^{-1/2})\varphi_c \right] \right\}$$

(30.41)

$$\mathcal{O}^s_\sigma = \exp(i\sigma\sqrt{2\pi}\,\varphi_s) \qquad \bar{\mathcal{O}}^s_\sigma = \exp(-i\sigma\sqrt{2\pi}\,\bar{\varphi}_s)$$

(30.42)

where $\sigma = \pm 1$. This leads to the following expression for the single-electron Green's function:

$$G(x,\tau) = \frac{1}{2\pi}\left[\frac{a_0^2}{v_c^2\tau^2 + x^2} \right]^{\theta/2} \left[\frac{\exp(ik_F x)}{\sqrt{(v_c\tau - ix)(v_s\tau - ix)}} + \frac{\exp(-ik_F x)}{\sqrt{(v_c\tau + ix)(v_s\tau + ix)}} \right]$$

(30.43)

where

$$\theta = \frac{1}{4}\left(K_c^{1/2} - K_c^{-1/2} \right)^2$$

The reader can see that the presence of spin serves as an additional source of incoherence: in addition to the first factor, which was present in the spinless case (see (29.6)), there is another branch cut related to a difference between spin and charge velocities.

Incommensurate charge density wave

Another scenario is realized when $g_s > 0$. In this case the current-current interaction in the spin sector is marginally relevant and flows to strong coupling. The spin sector acquires a gap, the charge sector remains gapless. This leads to exponential decay of all irreducible correlation functions containing the Φ_s-field and to the formation of a nonzero average

$$\langle \cos(\sqrt{2\pi}\,\Phi_s) \rangle \neq 0$$

To see this let us consider the bosonized form of Hamiltonian (30.19). The Hamiltonian density is

$$\mathcal{H}_s = \frac{1}{2}[(\partial_x\Theta_s)^2 + (\partial_x\Phi_s)^2] + \frac{g_s}{2\pi}\partial_x\varphi_s\partial_x\bar{\varphi}_s - \frac{g_s}{(2\pi a_0)^2}\cos(\sqrt{8\pi}\,\Phi_s)$$

(30.44)

The SU(2) symmetry of the original Hamiltonian is encoded in the robust structure of the last two terms parametrized by a single coupling constant g_s. The sign of the cosine term is very important since it determines the vacuum value of the Φ_s-field. With the minus sign the system chooses for its ground state one value from the set

$$\sqrt{8\pi}\,\Phi_s = 2\pi n \qquad n = \text{integer}$$

(30.45)

such that $\cos(\sqrt{2\pi}\,\Phi_s)$ acquires a nonzero vacuum expectation value and $\sin(\sqrt{2\pi}\,\Phi_s)$ does not. This is an example of spontaneous symmetry breaking. Such an event may occur at $T = 0$ in a $(1 + 1)$-dimensional system provided that the symmetry is discrete. The easiest way to determine the correct sign is to use the fermionic representation of the currents where the cosine term is given by

$$g_s[R_\uparrow^+ R_\downarrow L_\downarrow^+ L_\uparrow + R_\downarrow^+ R_\uparrow L_\uparrow^+ L_\downarrow] = -g_s[R_\uparrow^+ L_\uparrow L_\downarrow^+ R_\downarrow + L_\uparrow^+ R_\uparrow R_\downarrow^+ L_\downarrow]$$

and use the bosonization formula

$$R_\sigma^+ L_\sigma = \frac{i}{2\pi a_0} e^{-i\sqrt{4\pi}\,\Phi_\sigma}$$

The correlation functions of the Φ_c-field remain power law. The dynamics of this field is described by the Gaussian model characterized by the parameter K_c and the charge velocity v_c. Replacing the field $\cos(\sqrt{2\pi}\,\Phi_s)$ in (30.34) by its average we obtain the following expression for the strongly fluctuating part of the charge density:

$$\rho(x) = J + \bar{J} + A\cos(2k_F x + \sqrt{2\pi}\,\Phi_c) + a_u \cos(4k_F x + \sqrt{8\pi}\,\Phi_c) \quad (30.46)$$

where

$$A = \frac{a_c}{2\pi a_0} \langle \cos(\sqrt{2\pi}\,\Phi_s) \rangle$$

Expression (30.46) is valid at distances much larger than the spin correlation length. At these distances the $2k_F$ harmonics of the density-density correlation function behaves as

$$\chi_{2k_F}(\tau, x) \sim |\tau + ix/v_c|^{-K_c} \quad (30.47)$$

that is it decays slower than the corresponding correlation function in the gapless state (see (30.39)). Here we have a very important fact: due to the gap formation in the spin sector the $2k_F$ response in the charge sector becomes more singular. In other words, the system becomes more sensitive to perturbations, which includes the charge sector.

Half-filled band

If the band is half-filled, the proper description of the charge sector is given by (30.24). In the limiting case $g_u = g_c$ this model coincides with model (30.19). Using the bosonization rules for the currents (30.30) one can show that model (30.24) is equivalent to the sine-Gordon model:

$$S = \int d\tau dx \left[\frac{K}{2}(\partial_\mu \Phi_c)^2 + M\cos(\sqrt{8\pi}\,\Phi_c) \right],$$

$$K = 1 + g_c/2\pi v_c \qquad M = \frac{g_u}{(2\pi a_0)^2} \quad (30.48)$$

Here we have a situation similar to the charge density wave, but the sign at the cosine is positive. As we know from the previous subsection, the sign is important since it determines the vacuum state for the field Φ_c. Now it is $\sin(\sqrt{2\pi}\,\Phi_c)$ that acquires a nonzero vacuum expectation value. The easiest way to establish the correct sign is to use the fermionic form of the interaction (30.21). So the charge sector is described by the sine-Gordon model (30.48), of which the properties will be thoroughly discussed in Chapter 34. The spin sector remains gapless and we can write down expressions for the fields which have power law correlation functions. For reasons which will become clear later I will do this in the next chapter.

References

Brazovsky, S. A. and Kirova, N. N. (1984). *Soviet Scientific Reviews*, Vol. 5, p. 100. Harwood Academic, New York.

Frahm, H. and Korepin, V. V. (1990). *Phys. Rev. B*, **42**, 10553.

Kawakami, N. and Yang, S. K. (1991). *J. Phys. Condens. Matter*, **3**, 5983.

Lieb, E. H. and Wu, F. Y. (1968). *Phys. Rev. Lett.*, **20**, 1445.

Yang, C. N. (1989). *Phys. Rev. Lett.*, **63**, 2144.

Yang, C. N. and Zhang, S. C. (1990). *Mod. Phys. Lett. B*, **4**, 759.

31

Kac–Moody algebras:
Wess–Zumino–Novikov–Witten model

In this chapter I repeat and generalize the arguments given in Chapter 30. We have already seen that for fermions with spin $1/2$ at low energies, the Hamiltonian is represented as a sum of commuting parts. This representation is based on the so-called Sugawara construction where the Hamiltonian of free fermions is represented in terms of current operators.

In order to achieve a deeper understanding of matters related to the Sugawara construction, it is logical to study the algebraic properties of current operators first. Let us therefore consider currents of *free* fermions $\psi^+_{\alpha,n}, \psi_{\alpha,n}$, where $n = 1, 2, \ldots, k$ and the index α is transformed according to some Lie group G. The reason for introducing the second index n will become clear later. The corresponding current operators are defined as follows:

$$J^a(z) = \sum_{n=1}^{k} \psi^+_{R,\alpha,n}(z) \tau^a_{\alpha\beta} \psi_{R,\beta,n}(z)$$

$$J^a(\bar{z}) = \sum_{n=1}^{k} \psi^+_{L,\alpha,n}(\bar{z}) \tau^a_{\alpha\beta} \psi_{L,\beta,n}(\bar{z})$$

$$(31.1)$$

where τ^a are matrices, generators of the Lie algebra $\mathcal{G}$. They satisfy the following relations:

$$[\tau^a, \tau^b] = i f^{abc} \tau^c$$

$$\mathrm{Tr}\, \tau^a \tau^b = \frac{1}{2} \delta^{ab}$$

$$(31.2)$$

For the case of the SU(2) group $\tau^a = S^a$ are the matrices of spin $S = 1/2$ and $f^{abc} = \epsilon^{abc}$.

Since the fermions with different chiralities commute, so do the currents with different chiralities; currents with the same chirality satisfy the following Wilson operator expansion:

$$J^a(z)J^b(z') = \frac{k}{8\pi^2(z-z')^2} \delta^{ab} + i f^{abc} \frac{J^c(z')}{2\pi(z-z')} + \cdots \qquad (31.3)$$

To check the validity of this very important identity, one should consider two diagrams: one for the current-current correlation function and another for the correlation function of three currents. It is also very convenient to rewrite the operator expansion (31.3) in commutators:

$$[J^a(x), J^b(y)] = \frac{ik}{4\pi} \delta^{ab} \delta'(x-y) + i f^{abc} J^c(y)\delta(x-y) \qquad (31.4)$$

This is the most general form of *Kac–Moody* algebra $\mathcal{G}_k$ associated with algebra $\mathcal{G}$. We have already came across the $SU_1(2)$ case in Chapter 30.

Exercise

Check that the currents (31.1) satisfy the algebra (31.4). The second term in the right-hand side of (31.4) is an ordinary commutator and can be obtained by a straightforward calculation. The term with a derivative of the delta-function is anomalous; it appears only in infinite systems. In order to calculate it one has to recall the derivation given in Chapter 14 and calculate the pair correlation function of the currents.

Often the Kac–Moody algebra is written for the Fourier components of current operators. In this case we assume that the system is defined on the strip $0 < x < L$ with periodic boundary conditions and expand the current operators into the series

$$J^a(x) = \frac{1}{L} \sum_n e^{-2i\pi nx/L} J_n^a$$

Substituting this expansion into (31.4) we get the Kac–Moody algebra for the components:

$$\left[J_n^a, J_m^b \right] = \frac{nk}{2} \delta^{ab} \delta_{n+m,0} + i f^{abc} J_{n+m}^c \tag{31.5}$$

The Kac–Moody algebra includes the original algebra $\mathcal{G}$ as its subalgebra; this is the algebra of the zeroth components of the currents:

$$\left[J_0^a, J_0^b \right] = i f^{abc} J_0^c \tag{31.6}$$

Therefore one can think of these zeroth components as about matrices, generators of the group G.

Let us now construct a stress energy tensor for a conformal theory which has an additional symmetry group G. It is important to realize that we can use for this purpose the original fermions: their symmetry group is larger than G. Therefore we have to use the currents only. We have several clues: (i) we have seen that in the simplest case of $k = 1$ the Hamiltonian is the quadratic form of the currents; let us suppose that this holds for general k, and (ii) the Hamiltonian is related to the zeroth components of the stress energy tensor

$$\hat{H} = \frac{2\pi v}{L}(L_0 + \bar{L}_0) \tag{31.7}$$

and therefore the stress energy tensor is also quadratic in the currents. The only G-invariant quadratic form one can have is

$$T(z) = \frac{1}{k + c_v} : J^a(z) J^a(z) :$$

$$\bar{T}(\bar{z}) = \frac{1}{k + c_v} : \bar{J}^a(\bar{z}) \bar{J}^a(\bar{z}) : \tag{31.8}$$

The numerical value of the coefficient $1/(k + c_v)$, where c_v is the Kazimir operator in the adjoint representation, is defined by the identity

$$f_{abc} f_{\bar{a}bc} = c_v \delta_{a\bar{a}}$$

This numerical value is not dictated by these general properties and is derived from the fact that the stress energy tensors (31.8) satisfy the Virasoro algebra (26.19). The latter was proved by Knizhnik and Zamolodchikov (1984), who also found the expression for the conformal charge:

$$C = \frac{kD}{k + c_v} \tag{31.9}$$

where D is the number of generators of the group G. The values of D, c_v and the Kazimir operators C_{rep} for all simple Lie groups are presented in Table III, Appendix 9.C in the book by Itzykson and Drouffe.

Thus we have a new conformal field theory which I shall refer to as the Wess–Zumino–Novikov–Witten (WZNW) model. Its Hamiltonian is given by

$$\hat{H}(k, G) = \frac{2\pi v}{(k + c_v)L} \left[J_{0,R}^a J_{0,R}^a + 2 \sum_{n>0} J_{-n,R}^a J_{n,R}^a + (\text{R} \to \text{L}) \right] \tag{31.10}$$

where $J_{n,R}^a$, $J_{n,L}^b$ satisfy the Kac–Moody algebra (31.5). The SU(2)-symmetric conformal field theory discussed in the previous chapter is a particular case of the WZNW model with G = SU(2), $k = 1$.

The Hilbert space of the WZNW model (31.10) is a G-invariant subspace of the Hilbert space of the free fermions. It is instructive to construct the full basis of its eigenstates. One should do this using the current operators *only*, because only in this way can one be sure that all eigenstates are G-invariant. To simplify this task, we need to know how the current operators commute with the Hamiltonian. The corresponding commutation relations follow from (26.20)[1] and the fact that currents $J(z)$, $\bar{J}(\bar{z})$ are conformal fields with dimensions (1, 0) and (0, 1), respectively:

$$\left[L_n, J_m^a \right] = m J_{n+m}^a \tag{31.11}$$

Substituting this expression into (31.7) we get

$$\left[H, J_m^a \right] = \frac{2\pi v m}{L} J_m^a \tag{31.12}$$

Now we have everything we need to construct the eigenstates. Let us define the vacuum vectors $|h\rangle$ as states annihilated by positive Fourier components of the currents:

$$J_m^a |h\rangle = 0 \qquad \bar{J}_m^a |h\rangle = 0 \tag{31.13}$$

The Hamiltonian (31.10) acting on these states is reduced to

$$H_{\text{reduced}} = \frac{2\pi v}{(k + c_v)L} \left(J_0^a J_0^a + \bar{J}_0^a \bar{J}_0^a \right) \tag{31.14}$$

---

[1] Remember the shift of index in the definition of A_n in (26.18).

Therefore $|h\rangle$ are eigenstates of the Hamiltonian if they realize irreducible representations of the left and right G groups:

$$J_0^a J_0^a |h\rangle = C_{\text{rep}}|h\rangle \qquad \bar{J}_0^a \bar{J}_0^a |h\rangle = \bar{C}_{\text{rep}}|h\rangle \qquad (31.15)$$

Now one can use the positive components of the currents as creation operators. According to (31.12), the vectors

$$\bar{J}_{-m_1}^{a_1} \cdots \bar{J}_{-m_p}^{a_p} J_{-n_1}^{a_1} \cdots J_{-n_q}^{a_q} |h\rangle \qquad (31.16)$$

are eigenvectors of the Hamiltonian (31.10) with the energies

$$E_{pq}(h) = \frac{2\pi v}{L}\left(\frac{C_{\text{rep}}}{k+c_v} + \sum_{i=1}^{q} n_i + \frac{\bar{C}_{\text{rep}}}{k+c_v} + \sum_{i=1}^{p} m_i\right) \qquad (31.17)$$

As we have seen in Chapter 25, each eigenstate in conformal field theories is associated with some conformal field, whose conformal dimensions are related to the eigenvalues of energy and momentum via (25.9) and (25.10). In the given case we have

$$\Delta = \frac{C_{\text{rep}}}{k+c_v} + \sum_{i=1}^{q} n_i$$

$$\bar{\Delta} = \frac{\bar{C}_{\text{rep}}}{k+c_v} + \sum_{i=1}^{p} m_i \qquad (31.18)$$

The basis of states (31.16) is overcomplete. To make it complete one has to restrict the number of vacuum states $|h\rangle$ choosing a finite number of irreducible representations of G. For example, in the case G = SU(2) where the irreducible representations are representations of spin operators with $C_{\text{rep}} = S(S+1)$, $S = 1/2, 1, \ldots$, the basis is composed by $S = 1/2, 1, \ldots, k/2$ (Fateev and Zamolodchikov, 1986). Each vacuum vector $|h\rangle$ can be considered as a state created from the lowest vacuum $|0\rangle$ by a corresponding primary field $g_h(\tau, x)$. Indeed, according to (25.7) we have

$$|h\rangle = \lim_{\tau \to +\infty} e^{\tau(E_h - E_0)}(L/2\pi)^{(2\Delta_h + 2\bar{\Delta}_h)} g_h(\tau, x = 0)|0\rangle \qquad (31.19)$$

There is one more restriction on the vacuum states, namely a requirement that physical fields must have integer or half-integer conformal spins. If these fields are built exclusively from the WZNW fields, this means that

$$\frac{C_{\text{rep}}}{k+c_v} - \frac{\bar{C}_{\text{rep}}}{k+c_v} = n/2 \qquad (31.20)$$

where n is an integer. In general, this requirement is difficult to satisfy except for $n = 0$. In this latter case $C_{\text{rep}} = \bar{C}_{\text{rep}}$ and the primary fields are $\dim C_{\text{rep}} \times \dim C_{\text{rep}}$ matrices g_{ab}, tensors realizing irreducible representations of the group G. This is the case considered in the original paper by Knizhnik and Zamolodchikov (1984). There are other cases, however, where physical fields are products of fields of several WZNW models (we have seen similar examples in the previous chapter when we discussed spin-charge separation). Then the restriction on right and left representations is different. I shall return to this issue later.

Knizhnik–Zamolodchikov (KZ) equations

Now we shall derive differential equations for multi-point correlation functions of the WZNW primary fields. The derivation is based on the fact that the WZNW stress energy tensor is quadratic in currents (31.8). Writing this expression in components, we get

$$L_n = \frac{1}{c_v + k} \sum_m : J_m^a J_{n-m}^a : \tag{31.21}$$

where the normal ordering assumes that the operators J_m^a with a positive subscript (annihilation operators) stay on the right. Among the Virasoro generators there is one whose action on operators is particularly simple – it is $L_{-1} \equiv \partial_z$. For $n = -1$ we have

$$\partial_z \equiv L_{-1} = \frac{2}{c_v + k} \left[J_{-1}^a J_0^a + \sum_{m=1}^\infty J_{-m-1}^a J_m^a \right] \tag{31.22}$$

Since primary fields are vacuum states, they are annihilated by positive components of current operators (31.13). Therefore from (31.22) it follows that any primary field satisfies the identity

$$\left(\partial_z - \frac{2}{c_v + k} J_{-1}^a J_0^a \right) \phi_h(z) = 0 \tag{31.23}$$

We already know that J_0^a acts simply as the generator of the group (see (31.15)). To proceed further we need to know the action of J_{-1}. This can be extracted from the Ward identity

$$\langle J^a(z)\phi(1) \cdots \phi(N) \rangle = \sum_j \frac{\tau_j^a}{z - z_j} \langle \phi(1) \cdots \phi(N) \rangle \tag{31.24}$$

Thus we have

$$\langle J_{-1}^a \phi(1) \cdots \phi(N) \rangle = \frac{1}{2\pi i} \int_C dz(z_1 - z)^{-1} \langle J^a(z)\phi(z_1) \cdots \phi(N) \rangle$$

$$= \sum_{j \neq 1} \frac{\tau_j^a}{z_1 - z_j} \langle \phi(1) \cdots \phi(N) \rangle \tag{31.25}$$

where the contour C encircles z.

Combining these results we conclude that the N-point function of primary fields satisfies the following system of equations:

$$\left[\frac{1}{2}(c_v + k)\partial_{z_i} - \sum_{j \neq i} \frac{\tau_i^a \tau_j^a}{z_i - z_j} \right] \langle \phi(1) \cdots \phi(N) \rangle = 0 \tag{31.26}$$

where a matrix τ_i^a acts on the indices of the ith operator.

Let us consider this equation for the case $N = 2$. Then we have

$$\left[\frac{1}{2}(c_v + k)\partial_{z_1} - \frac{\tau_1^a \tau_2^a}{z_1 - z_2} \right] \langle \phi(1)\phi(2) \rangle = 0$$

$$\left[\frac{1}{2}(c_v + k)\partial_{z_2} + \frac{\tau_1^a \tau_2^a}{z_1 - z_2} \right] \langle \phi(1)\phi(2) \rangle = 0 \tag{31.27}$$

From these two equations it follows that $\langle \phi(1)\phi(2) \rangle = G(z_{12})$ and the function $G(z)$ satisfies the following equation:

$$\partial_z G^{\beta,\beta'}_{\alpha,\alpha'}(z) - \frac{2\tau^a_{\alpha,\gamma}\tau^a_{\beta,\delta}}{(c_v + k)z} G^{\delta,\beta'}_{\gamma,\alpha'}(z) = 0 \tag{31.28}$$

It is natural to suggest that a nonzero solution exists only if the second field is the Hermitian conjugate of the first one: $\phi(2) = \phi^+(1)$. It is essential that τ^a_2 acting on this operator gives a minus sign. Then $\tau^a_1 \tau^a_2$ becomes a Kazimir operator:

$$\tau^a_1 \tau^a_2 = -(\tau^a)^2 = -C_{\text{rep}}$$

and we get

$$\langle \phi(1)\phi^+(2) \rangle = z_{12}^{-2\Delta_h} \tag{31.29}$$

with the correct conformal dimension (31.18).

Solutions of KZ equations for four-point correlation functions of primary fields in the fundamental representation of the SU(N) group were found by Knizhnik and Zamolodchikov (1984). Solutions for four-point functions of all primary fields of the SU$_k$(2) WZNW model were found by Fateev and Zamolodchikov (1986). There is a regular procedure for finding multi-point correlation functions, called the Wakimoto construction, which is based on the representation of the WZWN operators in terms of free bosonic fields (Wakimoto, 1986). The details of this procedure can be found in Dotsenko (1990).

Conformal embedding

Let us now recall that the currents (31.1) were originally constructed as bilinears of free fermions. Later we managed to get rid of these fermions and to define the new Hamiltonian and its eigenstates exclusively in terms of the current operators. What about the original fermions, can we reintroduce them in our new formalism? In order to do this, we have to rewrite the Hamiltonian of free fermions in terms of currents belonging to three groups. I write down the corresponding expression only for the case G = SU(N):

$$\int dx[-i\psi^+_{R,\alpha,n}\partial_x\psi_{R,\alpha,n} + i\psi^+_{L,\alpha,n}\partial_x\psi_{L,\alpha,n}] = H[U(1)] + H[SU_k(N)] + H[SU_N(k)] \tag{31.30}$$

$$H[U(1)] = 2\pi \int dx[:J(x)J(x): + :\bar{J}(x)\bar{J}(x):] \tag{31.31}$$

$$H[SU_k(N)] = \frac{2\pi}{(N+k)} \sum_{a=1}^{D_N} \int dx[:J^a(x)J^a(x): + :\bar{J}^a(x)\bar{J}^a(x):] \tag{31.32}$$

$$H[SU_N(k)] = \frac{2\pi}{(N+k)} \sum_{\lambda=1}^{D_k} \int dx[:\mathcal{J}^\lambda(x)\mathcal{J}^\lambda(x): + :\bar{\mathcal{J}}^\lambda(x)\bar{\mathcal{J}}^\lambda(x):] \tag{31.33}$$

where J, J^a and $\mathcal{J}^\lambda$ are respectively U(1), $SU_k(N)$ and $SU_N(k)$ currents. This representation corresponds to the decomposition of the nonsimple symmetry group of the free fermions into the product of simple groups $U(1) \times SU(N) \times SU(k)$. I have taken into account the fact that for the SU(N) group $c_v = N$. The decomposition (31.30) is a generalization for a case of fermions with $U(k) \times SU(N)$ symmetry of decomposition (30.10), (30.11) considered in the previous chapter.

Let us make sure that the decomposition (31.30) really works. First of all, it has to reproduce the free energy of the original Hamiltonian. For conformal theories where the free energy is proportional to the central charge, this means that the sum of the central charges of the WZNW models is equal to Nk, the central charge of free fermions with Nk indices. According to (31.9) the central charge of the SU(N)-invariant WZNW model is given by

$$C = \frac{k(N^2 - 1)}{k + N} \tag{31.34}$$

where the number of generators is $D_N = N^2 - 1$. Thus we have

$$Nk = 1 + \frac{k(N^2 - 1)}{k + N} + \frac{N(k^2 - 1)}{k + N}$$

which is identically satisfied. Let us now find out what happens with the primary fields, for example, with fermionic bilinears. These objects are complex fields (the U(1) group!) and they belong to the fundamental representations (i.e. to representations with minimal possible dimension) of both SU(N) and SU(k) groups. Therefore we can write them as a product of primary fields transforming according to the fundamental representations of the groups U(1), SU(N) and SU(k) respectively:

$$\psi^+_{\alpha,n}(\bar{z})\psi_{\beta,m}(z) = \exp\left[i\sqrt{(4\pi/Nk)}\Phi(z,\bar{z})\right] \sum_{a,b=1}^{2} U^a_{\alpha\beta}(z,\bar{z})G^b_{nm}(z,\bar{z})C_{ab} \tag{31.35}$$

Here the fields U and G are primary fields of the $SU_k(N)$ and $SU_N(k)$ WZNW models and C_{ab} is a constant matrix. The reader may be confused by the fact that these fields carry an additional superscript. I will explain this after I demonstrate that the composite field (31.35) satisfies a necessary condition possessing the correct conformal dimension $1/2$. The eigenvalue of the Kasimir operator C_R in the fundamental representation of the SU(N) group is equal to

$$C^f_{rep} = \frac{1}{2}\left(N - \frac{1}{N}\right) \tag{31.36}$$

Substituting this in the expression for the conformal dimension (31.18) we find that

$$\Delta_U = \frac{1}{2(N + k)}\left(N - \frac{1}{N}\right)$$

$$\Delta_G = \frac{1}{2(N + k)}\left(k - \frac{1}{k}\right) \tag{31.37}$$

Thus we reproduce the conformal dimension of the free termion field:

$$\frac{1}{2} = \frac{1}{2Nk} + \Delta_U + \Delta_G$$

A more detailed discussion of decomposition (31.35) can be found in Affleck and Ludwig (1991) (see also a very interesting paper by Ludwig and Maldacena (1997)).

Now let us return to the meaning of the superscripts in (31.35). Recall that the arguments leading to the decomposition of Hamiltonian (31.30) are based on properties of the corresponding *algebras*. Therefore they do not allow us to distinguish between the group SU(N) and its complexification SL(N, C). Both these groups have the same algebras. Therefore we do not actually know a priori whether we get the WZNW models on SU(N) or on SL(N, C) groups. In order to understand whether this subtlety may lead to any difference, I have to discuss the concept of complex group. Since such groups are not often used in condensed matter physics, I think it is better to discuss a simple example of complexification of the SU(2) group. Matrices from its fundamental representation can be parametrized by a unit four-dimensional *real* vector:

$$\hat{g} = n_0 \hat{I} + i \vec{\sigma}\, \mathbf{n} \qquad n_0^2 + \mathbf{n}^2 = 1 \tag{31.38}$$

The fundamental representation of the complex group SL(2, C) has the same representation, but the vector components are complex. In order to fulfil the property $\hat{g}\hat{g}^+ = I$, we need

$$|n_0|^2 + \mathbf{n}\mathbf{n}^* = 1$$

More generally a matrix from a group G can be written in terms of generators of its algebra $\mathcal{G}$:

$$\hat{g} = \exp[i\alpha_a \hat{I}^a] \tag{31.39}$$

For real groups such as SU(N), parameters α_a are real. For complex groups they are complex. Complexification lifts certain restrictions on the group matrices. For instance, for any SU(2) matrix one has

$$\hat{g} + \hat{g}^+ \sim I \tag{31.40}$$

For SL(2, C) this property is not satisfied. In the final section of this chapter, I shall demonstrate explicitly that the adding of two SU$_1$(2) models yields not SU$_2$(2), but the SL$_2$(2, C) model. A similar demonstration can be given for general N and k. The result is that U and G operators are actually primary fields of SL$_k$(N, C) and SL$_N$(k, C) WZNW models. They can be represented as sums of 'real' and 'imaginary' parts labelled by the superscripts in (31.35). Operators $U^{(1)}$ and $U^{(2)}$ ($G^{(1)}$, $G^{(2)}$) are mutually nonlocal, belonging to different sectors of the SU(N) (SU(k)) WZNW models respectively. These operators are allowed to appear simultaneously because they enter in (31.35) as products and their nonlocality is mutually quenched. In other words, correlation functions of $U^{(1)}U^{(2)}$ and $G^{(1)}G^{(2)}$ are not uniquely defined, but correlation functions of their products are. In the final section of this chapter the role of such operators is played by the order and disorder parameter operators (σ, μ) of the Ising model.

U(k) × SU(N)-invariant model of interacting fermions

Let us now introduce an interaction. The simplest model we can solve is the one with current-current interaction:

$$H_{\text{int}} = 4c \int dx \sum_{\lambda=1}^{D_k} : J_R^\lambda(x) J_L^\lambda(x) : \qquad (31.41)$$

This interaction is relevant if $c > 0$ and since the interaction includes only currents from the SU(k) group, the SU(k) sector acquires a gap and the spectra of the $H[\text{U}(1)]$ and $H[\text{SU}_k(N)]$ Hamiltonians are not affected. Thus we have a peculiar situation where the conformal symmetry is destroyed in one sector, but remains intact in two others. Effectively the interaction removes the Hamiltonian $H[\text{SU}_N(k)]$ from the lower energy sector. What remains is $H[\text{U}(1)]$ and $H[\text{SU}_k(N)]$. That is the low energy degrees of freedom are described by the theory of the free bosonic field and the WZNW model. The model with interaction (31.41) was solved exactly by the author via the Bethe ansatz (Tsvelik, 1987). Here we shall discuss only those properties which can be understood without complicated calculations.

The most interesting problem is how the gap influences correlation functions of physical fields. Naturally, correlation functions of the SU(N) currents remain as they were. That is, response functions at small momenta are the same. As we shall see in a moment, the correlations at $q = 2k_F$ are strongly affected. Let us consider, for example, the field

$$Q_{\alpha\beta}(z, \bar{z}) = \sum_n \psi_{\alpha,n}^+(\bar{z}) \psi_{\beta,n}(z) \qquad (31.42)$$

When the SU(k) sector acquires a gap, two things happen: (i) some combinations of the fields G^a acquire finite averages, and (ii) fluctuations around average values become exponentially decaying with the correlation length inversely proportional to the spectral gap. One may compare this with the situation for the sine-Gordon model. In the massive phase generated by the $\cos \beta \Phi$ term, averages $\langle \cos \gamma \Phi \rangle$ are finite for all γ and $\langle \sin \gamma \Phi \rangle = 0$. The analogy with Z_2 symmetry breaking in the Ising model, when either σ or μ acquires an expectation value, would serve even better. This condensation breaks the degeneracy between 'real' and 'imaginary' sectors of the SL(k, C) group such that

$$\langle \text{Tr} G^a(z, \bar{z}) \rangle \sim \delta_{a,1} M^{1-d} \qquad (31.43)$$

where $M \sim \exp(-2\pi/kc)$ is the spectral gap and d is the scaling dimension of the remaining operators. The exponent $(1 - d)$ is chosen in such a way that the entire expression has the correct scaling dimension 1. Now one can rewrite (31.42), keeping only the terms which have power law correlations:

$$Q_{\alpha\beta}(z, \bar{z}) = \exp\left[i\sqrt{(4\pi/Nk)}\Phi(z, \bar{z})\right] U_{\alpha\beta}(z, \bar{z}) M^{1-d} \qquad (31.44)$$

Here $U_{\alpha\beta}(z, \bar{z})$ is the primary field of the SU$_k$(N) (!) WZNW model whose conformal dimension Δ_U is given by (31.37). Thus the corresponding pair correlation function at distances larger than M^{-1} is

$$\langle Q(1)Q^+(2)\rangle \sim \frac{M^{2-2d}}{|z_1 - z_2|^{2d}} + \mathcal{O}[\exp(-M|z_{12}|)]$$

$$d = \frac{1}{Nk} + \frac{N - 1/N}{k + N}$$

(31.45)

This correlation function is a dynamical response function on the wave vector $q \approx 2k_F$ (an analogue of staggered magnetic susceptibility). After the Fourier transformation we have at $\omega^2 + k^2 \ll M^2$:

$$\langle Q_{\alpha\beta}(\omega, k)Q^+_{\gamma\delta}(\omega, k)\rangle \sim \left(\frac{M^2}{\omega_n^2 + k^2}\right)^{1-d} \delta_{\alpha\gamma}\delta_{\beta\delta}$$

(31.46)

For noninteracting fermions the scaling dimension of Q is equal to 1, which means that the response function (31.46) has only a weak logarithmic singularity at $q = 2k_F$. We see that the opening of the gap leads to the enhancement of this singularity. I have already discussed this mechanism (see the discussion around (30.47)).

It is curious that the scaling dimension d remains finite even when $N \to 0$: $d(N = 0) = 1/k^2$. This means that the fermionic model on the $U(k) \times SU(N)$ group has a nontrivial limit at $N \to 0$. Such limits usually play an important role in the theory of disordered systems. We refer the reader to the paper by Bhaseen *et al.* (2001), where this limit is studied, and references therein.

$SU_1(2)$ WZNW model and spin $S = 1/2$ Heisenberg antiferromagnet

The discussion of the previous section remains valid for a very important particular case $k = 1, N = 2$. The $SU_1(2)$-invariant subsector describes the low energy dynamics of the isotropic spin $S = 1/2$ Heisenberg antiferromagnet (XXX model). Though the latter system can be efficiently treated by Abelian bosonization, sometimes it is more convenient to use the non-Abelian approach.

Let me recall the basic facts about the relationship between XXX and $SU_1(2)$ WZNW models scattered throughout the book.

The isotropic Heisenberg Hamiltonian can be obtained from the Hubbard model at half-filling by a canonical transformation in the limit $t \ll U$ (t is the hopping integral):

$$H_{\text{Hubbard}} = \frac{1}{2}\sum_{i,j} t_{ij}(C^+_{\sigma,i}C_{\sigma,j} + \text{H.c.}) + U N_{i,\uparrow}N_{i,\downarrow} \to$$

$$H_{\text{XXX}} = \frac{1}{2}\sum_{i,j} J_{ij}\mathbf{S}_i\mathbf{S}_j$$

(31.47)

This result is valid in any space dimension. The exchange integral is

$$J_{ij} = \frac{t_{ij}^2}{U}$$

(31.48)

The charge excitations do not contribute to the low energy sector since at half-filling
there is a spectral gap; at $U \gg t$ the size of the gap $\sim U$ far exceeds the value of the
exchange.

In one dimension this result is reinforced by spin-charge separation and therefore does
not require large U/t (though at smaller U/t expression (31.48) for J_{ij} is modified).
However, from Chapter 30 we know that the spin sector is always described by
the $SU_1(2)$ WZNW model perturbed by a marginally irrelevant current-current
interaction.

Bosonization of the $SU_1(2)$ Kac–Moody algebra explained in Section 3.1 establishes a
connection between Abelian and non-Abelian descriptions of the continuous limit
of XXX chains.

Now let us recall the general discussion of antiferromagnetism in Part III. In Chapter 16 I
derived an effective field theory for the low energy sector of a spin S antiferromagnet. This
theory has the form of the O(3) nonlinear sigma model with a topological term (16.24). The
coefficient at the topological term is proportional to S such that for an integer S this term
does not contribute to the partition function and can be omitted. In contrast, for half-integer
S the topological term radically changes the low energy behaviour of the system. This gives
rise to the famous distinction between integer and half-integer spin antiferromagnets in
one dimension discussed in Chapter 16. The low energy behaviour of a half-integer spin
antiferromagnet is universal and is described by the $SU_1(2)$ WZNW model.

The attentive reader might notice something unusual about this situation. Indeed, in the
ultraviolet we start from the O(3) nonlinear sigma model and in the infrared we finish with
the WZNW model. This makes something of a mockery of the concept of renormalization
because the corresponding action does not reproduce itself in the process.

Moreover, even the manifold on which the order parameter is defined changes: for the
sigma model the order parameter, a unit vector, lives on SU(2)/U(1) symmetric space and
for the WZNW model it is a 2×2 SU(2) matrix living on the group. The latter means
that the order parameter field of a spin-1/2 chain is not just a staggered magnetization.
Indeed, the entries of matrix $\hat{g}$ include not only vector components of the magnetization n^a
($a = x, y, z$), but also a scalar field ϵ:

$$\hat{g}(x) = \epsilon(x)I + i\sigma^a n^a(x) \tag{31.49}$$

This field is identified with the dimerization field

$$\epsilon = (-1)^n (\mathbf{S}_n \mathbf{S}_{n+1}) \tag{31.50}$$

Though one needs three parameters to parametrize an SU(2) matrix, for the case of the
$SU_1(2)$ model this can be done in a more economical way since the $SU_1(2)$ WZNW model
is equivalent to the Gaussian model with $K = 1$. As a consequence the field $\hat{g}$ can also be
represented in terms of chiral bosonic fields φ, $\bar{\varphi}$ governed by actions (22.27):

$$\hat{g}_{\sigma\sigma'} = \frac{1}{\sqrt{2}} C_{\sigma\sigma'} : \exp[-i\sqrt{2\pi}(\sigma\varphi + \sigma'\bar{\varphi})] := \frac{1}{\sqrt{2}} C_{\sigma\sigma'} z_\sigma \bar{z}_{\sigma'} \tag{31.51}$$

$$C_{\sigma\sigma'} = e^{i\pi(1-\sigma\sigma')/4}$$

Figure 31.1. The failure of perturbative RG to reach the critical point.

where

$$z_\sigma = \exp[i\sigma\sqrt{2\pi}\varphi] \qquad \bar{z}_\sigma = \exp[-i\sigma\sqrt{2\pi}\bar{\varphi}] \qquad (\sigma = \pm 1) \tag{31.52}$$

In particular,

$$\epsilon \sim \cos(\sqrt{2\pi}\,\Phi) \tag{31.53}$$

Expression (31.51) can be obtained from expressions (30.34), (30.35). For this we just need to set $2k_F = \pi$ (half-filling). Then the Umklapp processes generate a spectral gap in the charge sector. According to the final section of Chapter 30, the charge sector of the Hubbard model at half-filling is described by the sine-Gordon model (30.48). The effective potential of this model has a minimum at $\sqrt{8\pi}\,\Phi_c = \pi$. Therefore at temperatures much smaller than the charge gap we can just replace the charge field by its value at the minimum (this procedure is called *field freezing*). As a result fluctuations of certain operators (like the charge density) are suppressed, but others remain power law. For the case under consideration we have

$$K(x) = \psi_\sigma^+(x)\psi_\sigma(x+a) = (-1)^{x/a_0}\epsilon(x) = (-1)^{x/a_0}A\mathrm{Tr}(\hat{g} + \hat{g}^+) \qquad (31.54)$$

$$S(x) = \mathbf{J} + \bar{\mathbf{J}} + (-1)^{x/a_0}\mathbf{n}(x) = \mathbf{J} + \bar{\mathbf{J}} + i(-1)^{x/a_0}B\mathrm{Tr}\sigma(\hat{g} - \hat{g}^+) \qquad (31.55)$$

where A, B are nonuniversal amplitudes. This derivation is a particular case of the one described in the previous section.

From the qualitative point of view the most important feature of a spin $S = 1/2$ antiferromagnet is that the staggered energy density operator there is as singular as the staggered magnetization (they have equal scaling dimensions equal to $1/2$). Therefore the system of weakly coupled chains has equally strong tendencies toward antiferromagnetism and spin-Peierls ordering.

$SU_2(2)$ WZNW model and the Ising model

The case of $SU_2(2)$ Kac–Moody algebra is very special because, as we are going to show, it can be represented as a current algebra of three species of free Majorana fermions. This can be conjectured from the fact that, according to (31.34), the WZNW central charge in this case ($N = 2, k = 2$) is equal to $3/2$. Since a single Majorana mode carries central charge $C = 1/2$, we conclude that the $SU_2(2)$ WZNW model is equivalent to the theory of three species of massless Majorana fermions. It is straightforward to check that the currents being expressed in terms of a triplet of Majorana fermionic fields χ^a,

$$J_{R,L}^a = \frac{i}{2}\epsilon^{abc}\chi_{R,L}^b\chi_{R,L}^c \qquad (31.56)$$

satisfy the algebra (31.4) with $k = 2$ and $f_{abc} = \epsilon_{abc}$. Here we assume that the Majorana fields have the Hamiltonian (28.10) with $m = 0$, so that

$$\langle\langle\chi(z)\chi(0)\rangle\rangle = 1/2\pi z \qquad (31.57)$$

As we know from (31.1), however, the current algebra $SU_2(2)$ can be realized using four species of Dirac fermions with spin and flavour (or chain index) $R_{\sigma,n}$ ($n = 1, 2; \sigma = \pm 1/2$). Naturally, it is interesting to know how these 'normal' fermions are related to these Majorana modes.

Let us recall some basic facts about our problem. The central charge of the model of $U(2) \times SU(2)$-symmetric free massless fermions is equal to four. So the problem is equivalent to the Gaussian model of four bosonic fields. From the available fermions one can construct three sorts of currents: the $U(1)$ charge current and $SU_2(2)$-symmetric spin and flavour currents respectively. Both spin and flavour sectors can be described by Majorana fermions which we denote χ^a ($a = 1, 2, 3$) and η^j ($j = 1, 2, 3$) respectively.

Let us show that one can express all possible chiral fermion bilinears in terms of χ and η. Consider right-moving (or left-moving, it does not matter) Dirac fermions transforming according to the U(2) × SU(2) group. For each flavour we can write the spin current:

$$J_n^a = L_{\sigma,n}^+ S_{\sigma,\sigma'}^a L_{\sigma',n} \tag{31.58}$$

where S^a are matrices of spin $1/2$. The combination obeying the $SU_2(2)$ Kac–Moody algebra is $\mathbf{I} = \mathbf{J}_1 + \mathbf{J}_2$. The expression for $\mathbf{I}$ in terms of Majoranas is already available (31.56). The remaining problem is to express the 'wrong' combination $\mathbf{K} = \mathbf{J}_1 - \mathbf{J}_2$. Using the fact that $\mathbf{J}_1$, $\mathbf{J}_2$ mutually commute and separately satisfy the $SU_1(2)$ Kac–Moody algebra, we derive the following commutation relations:

$$[K^a(x), K^b(y)] = i\epsilon^{abc}\delta(x - y)J^c(y) + \frac{i\delta_{ab}}{2\pi}\delta'(x - y) \tag{31.59}$$

$$[I^a(x), K^b(y)] = i\epsilon^{abc}\delta(x - y)K^c(y) \tag{31.60}$$

Now taking into account (31.56), it is easy to check that these relations are satisfied if we assume

$$K^a = i\chi^a \eta^3 \tag{31.61}$$

The choice of index of the flavour Majorana fermion corresponds to the fact that $K^a = R^+ S^a \tau^3 R$ where τ^a are spin-$1/2$ matrices acting on flavour indices. The symmetry considerations dictate

$$R^+ S^a \tau^j R = i\chi^a \eta^j \tag{31.62}$$

Thus I have demonstrated that the $su_2(2)$ Kac–Moody algebra can be constructed out of currents of three Majorana fermions. The question now is whether we can express the primary fields of the $SU_2(2)$ WZNW models. The sector with zero conformal spin is represented by the two multiplets of primary fields: $\Phi_{\alpha\beta}^{(1/2)}$ and $\Phi_{ab}^{(1)}$ with the conformal dimensions $(3/16, 3/16)$ and $(1/2, 1/2)$ respectively. These fields are 2×2 and 3×3 matrices, tensors realizing the representations of the SU(2) group with isospins $S = 1/2$ and $S = 1$ respectively. The $S = 1/2$ field is nonlocal in terms of the fermions. To write it down we have to recall that the massless Majorana fermion is equivalent to the critical Ising model (see Chapter 28, second section). The calculation done by Fateev and Zamolodchikov (1986) establishes that the correlation functions of $\Phi^{(1/2)}$ are identical to correlation functions of the products of the Ising model order and disorder parameter fields. So we have the following identity:

$$\Phi^{(1/2)}(z, \bar{z}) = \hat{\tau}^0 \sigma_1 \sigma_2 \sigma_3 + i(\hat{\tau}^1 \mu_1 \sigma_2 \sigma_3 + \hat{\tau}^2 \sigma_1 \mu_2 \sigma_3 + \hat{\tau}^3 \sigma_1 \sigma_2 \mu_3) \tag{31.63}$$

where $\hat{\tau}_j$ ($j = 1, 2, 3$) are Pauli matrices, $\hat{\tau}_0$ is the identity matrix and σ_a, μ_a ($a = 1, 2, 3$) are the order and disorder parameter fields of the Ising models. The $\Phi^{(1)}$ operator is just a tensor composed of fermion bilinears:

$$\Phi_{ab}^{(1)} = \chi_a \bar{\chi}_b \tag{31.64}$$

Thus we have managed to express the $SU_2(2)$ WZNW model as a sum of three critical Ising models. Now let us address another problem: what model will emerge if one adds up two $SU_1(2)$ WZNW models. On the level of Hamiltonians we have the following identity:

$$\frac{2\pi}{3} \int dx \, (:J_1^2: + :J_2^2:) = \frac{\pi}{2} \int dx : \mathbf{I}^2 : -\frac{i}{2} \int dx \chi_0 \partial_x \chi_0 = -\frac{i}{2} \int dx \sum_{a=0}^{3} \chi_a \partial_x \chi_a$$

(31.65)

Thus we can recast the operators of two $SU_1(2)$ models in terms of operators of four Ising models or in terms of one Ising model and $SU_2(2)$ or $SL_2(2,C)$ WZNW model. From the discussion on conformal embedding, we know that there is an uncertainty of what group to choose. Since the $SU_1(2)$ model is so simple, the problem can be resolved by a straightforward calculation. This calculation uses the Abelian bosonization of the $SU_1(2)$ model described in the previous section, and is described further in the second section of Chapter 36. The staggered magnetizations of models 1 and 2 are expressed in terms of order and disorder parameter fields of the four Ising models:

$$\mathbf{n}_{1,2} = \sigma_0 Tr[\sigma(\hat{g} + \hat{g}^+)] \pm i\mu_0 Tr[\sigma(\hat{g} - \hat{g}^+)]$$

(31.66)

where

$$\hat{g} = \sum_{j=0}^{3} \hat{t}_j G_j(z, \bar{z})$$

$$G_0 = \sigma_1\sigma_2\sigma_3 + i\mu_1\mu_2\mu_3 \qquad G_1 = \sigma_1\mu_2\mu_3 + i\mu_1\sigma_2\sigma_3$$

(31.67)

$$G_2 = \mu_1\sigma_2\mu_3 + i\sigma_1\mu_2\sigma_3 \qquad G_3 = \mu_1\mu_2\sigma_3 + i\sigma_1\sigma_2\mu_3$$

This matrix field is the spin-1/2 primary field of the $SL_2(2,C)$ WZNW model. As was demonstrated by Fateev and Zamolodchikov (1986), this matrix field satisfies all the WZNW model fusion rules. The choice of the group (not discussed in the original publication) is dictated by the fact that $Tr[\sigma(\hat{g} + \hat{g}^+)] \neq 0$, as clearly follows from (31.67). Notice also that

$$(\hat{g} - \hat{g}^+) = \Phi^{(1/2)}$$

where the latter field is given by (31.63).

References

Affleck, I. and Ludwig, A. W. W. (1991). *Nucl. Phys. B*, **352**, 849; *Phys. Rev. Lett.*, **67**, 3160 (1991).
Bhaseen, M. J., Caux, J.-S., Kogan, I. I. and Tsvelik, A. M. (2001). *Nucl. Phys. B*, **618**, 465.
Dotsenko, Vl. S. (1990). *Nucl. Phys. B*, **338**, 747; **358**, 547 (1990).
Fateev, V. A. and Zamolodchikov, A. B. (1986). *Yad. Fiz. (Sov. Nucl. Phys.)*, **43**, 657.
Knizhnik, V. G. and Zamolodchikov, A. B. (1984). *Nucl. Phys. B*, **247**, 83.
Ludwig, A. W. W. and Maldacena, H. (1997). *Nucl. Phys. B*, **506**, 565.
Tsvelik, A. M. (1987). *Zh. Eksp. Teor. Fiz.*, **93**, 1329 (*JETP*, **66**, 221 (1987)).
Wakimoto, M. (1986). *Commun. Math. Phys.*, **104**, 605.

Wess–Zumino–Novikov–Witten model in the Lagrangian form: non-Abelian bosonization

In many cases it is necessary to have the Wess–Zumino–Novikov–Witten model in its Lagrangian form. In this chapter I reproduce the derivation of the WZNW Lagrangian given in the original paper by Polyakov and Wiegmann (1983).

Let us consider relativistic fermions described by creation and annihilation operators $\psi_{n\alpha}^{+}$, $\psi_{n\alpha}$ ($n = 1, \ldots, k$; $\alpha = 1, \ldots, N$) interacting with a non-Abelian gauge field

$$A_{\mu}^{\alpha\beta} = A_{\mu}^{a} \tau_{a}^{\alpha\beta}$$

where τ_a are matrix-generators of some Lie group G. These matrices act only on the Greek indices; the interaction is diagonal with respect to the English indices. The fermionic part of the action has the standard form:

$$S_{\mathrm{f}} = \int \mathrm{d}^2 x \sum_{n=1}^{k} \bar{\psi}_{n\alpha} \gamma_{\mu} \left(\partial_{\mu} \delta_{\alpha\beta} + \mathrm{i} A_{\mu}^{\alpha\beta} \right) \psi_{n\beta} \tag{32.1}$$

Let us integrate over the fermions and consider the effective action for the gauge field:

$$\exp\{-S_{\mathrm{eff}}[A]\} = Z^{-1} \int \mathrm{D}\bar{\psi} \mathrm{D}\psi \mathrm{e}^{-S_{\mathrm{f}}[\bar{\psi},\psi,A]} \equiv \mathrm{e}^{kW[A]} \tag{32.2}$$

$$W[A] = \ln \det \left[\gamma_{\mu} \left(\partial_{\mu} \delta_{\alpha\beta} + \mathrm{i} A_{\mu}^{\alpha\beta} \right) \right] - \ln \det(\gamma_{\mu} \partial_{\mu}) \tag{32.3}$$

The field interacts only with the SU(N) currents; using the results of the previous chapter on the non-Abelian bosonization, we can separate the relevant degrees of freedom and rewrite the above effective action as an average over the Wess–Zumino–Novikov–Witten action:

$$\mathrm{e}^{-S_{\mathrm{eff}}[A]} = \left\langle \exp \left(-\mathrm{i} \int \mathrm{d}^2 x A_{\mu}^{a} J^{a,\mu} \right) \right\rangle_{\mathrm{w}} \tag{32.4}$$

At the present stage we are not yet in a position to write down an explicit expression for this average in path integral form. The reason is that we do not know the expression for the WZNW Lagrangian. In order to find it, we need to do some preparatory work.

Our first step is to calculate $W[A]$. A similar problem for the Abelian case was considered in Chapter 14 where we discussed $(1 + 1)$-dimensional quantum electrodynamics. Let us define the current:

$$j_{\mu}(x) = \mathrm{i} k \frac{\delta W[A]}{\delta A_{\mu}(x)}$$

The functional $W[A]$ is completely determined by the two equations satisfied by its derivatives $j_\mu(x)$:

$$\partial_\mu j_\mu + i[A_\mu, j_\mu] = 0 \tag{32.5}$$

$$\epsilon_{\mu\nu}(\partial_\mu j_\nu + i[A_\mu, j_\nu]) = -\frac{k}{2\pi}\epsilon_{\mu\nu}F_{\mu\nu} \tag{32.6}$$

where

$$F_{\mu\nu} = \partial_\mu A_\nu - \partial_\nu A_\mu + i[A_\mu, A_\nu]$$

In the operator language these equations should be understood as equations of motion for the current operators. In the path integral approach we understand them as identities for the correlation functions. Equations (32.5) and (32.6) were derived by Johnson (1963) and solved by Polyakov and Wiegmann (1984). The first one, (32.5), follows from the gauge invariance and expresses the fact that the non-Abelian charge

$$Q^a = \int dx_1 J_0^a(x)$$

is a conserved quantity. The total charge is the sum of charges created by left- and right-moving particles. Naively one can expect that for a massless theory, where there are no transitions between fermionic states with different chirality, right- and left-moving charges are conserved independently. We have already seen in Chapter 14 that this expectation is false because the measure is invariant only with respect to simultaneous gauge transformations of right and left particles. The phenomenon of nonconservation of chiral components of the charge is called the *chiral anomaly*. As was shown by Johnson, the difference between right and left charges is proportional to the total flux of the gauge field, and (32.6) expresses this fact. Thus we have two equations (32.5) and (32.6) for two unknown functions j_μ and can attempt to solve this system. In order to do this we rewrite the equations in a more convenient form, namely, in the light cone coordinates. The corresponding notation is:

$$\bar{j} = j_0 + ij_x \qquad j = j_0 - ij_x$$
$$\bar{A} = A_0 + iA_x \qquad A = A_0 - iA_x$$
$$\partial_\tau = \partial + \bar{\partial} \qquad \partial_x = i(\partial - \bar{\partial})$$

The equations (32.5) and (32.6) are gauge invariant; we shall solve them in a particular gauge:

$$\bar{A} = 0 \qquad A = -2ig^{-1}\partial g \tag{32.7}$$

where g is a matrix from the G group. In this notation (32.5) and (32.6) acquire the following form:

$$\partial\bar{j} + \frac{i}{2}[A, \bar{j}] + \bar{\partial}j = 0 \tag{32.8}$$

$$\partial\bar{j} + \frac{i}{2}[A, \bar{j}] - \bar{\partial}j = -\frac{ik}{\pi}\bar{\partial}A \tag{32.9}$$

Substituting $\partial \bar{j} + \frac{i}{2}[A, \bar{j}]$ from the first equation into the second one we find

$$j = \frac{ik}{2\pi} A = \frac{ik}{\pi} g^{-1} \partial g \qquad (32.10)$$

Then the first equation transforms into

$$\partial \bar{j} + [g^{-1}\partial g, \bar{j}] + \frac{ik}{\pi} \bar{\partial}(g^{-1}\partial g) = 0 \qquad (32.11)$$

whose solution is given by

$$\bar{j} = -\frac{ik}{\pi} g^{-1}\bar{\partial} g \qquad (32.12)$$

In order to find the effective action $W[A]$ one has to integrate the equation

$$k\delta W = \frac{i}{2} \int d^2 x \, \mathrm{Tr}(\bar{j}\delta A) = -\frac{k}{\pi} \int d^2 x \, \mathrm{Tr}[\delta g g^{-1} \partial(g^{-1}\bar{\partial} g)] \qquad (32.13)$$

(in the derivation of this expression I have used the invariance of trace under cyclic permutations and the identity $\delta g^{-1} = -g^{-1}\delta g g^{-1}$).

Exercise

Check that the solution of this problem is given by the following functional described by Novikov (1982) and Witten (1984):[1]

$$S_{\mathrm{eff}} = -kW(g)$$
$$W(g) = \frac{1}{16\pi} \int d^2 x \, \mathrm{Tr}(\partial_\mu g^{-1} \partial_\mu g) + \Gamma[g] \qquad (32.14)$$

$$\Gamma[g] = -\frac{i}{24\pi} \int_0^\infty d\xi \int d^2 x \epsilon^{\alpha\beta\gamma} \mathrm{Tr}(g^{-1}\partial_\alpha g g^{-1}\partial_\beta g g^{-1}\partial_\gamma g) \qquad (32.15)$$

The appearance of the third coordinate in the second term Γ requires explanation (this term is usually called the Wess–Zumino term after the people who introduced it first). It is supposed that the field $g(\xi, x)$ is defined on a three-dimensional hemisphere whose boundary coincides with the two-dimensional plane where the original theory is defined, so that $g(\xi = 0, x) = g(x)$. The three-dimensional integral does not really depend on $g(\xi)$, being a total derivative.

To illustrate the latter statement, let us consider an example of the SU(2) group. In this case it is relatively easy to find a suitable parametrization for the g matrix and rewrite the effective action in terms of the group coordinates instead of matrices. The current example is especially important because the SU(2) group appears frequently in applications. Let the matrix g belong to the fundamental representation of the SU(2) group. I choose the Euler parametrization (4.36).

We shall use the following useful identities which I leave to be proved as an excercise.

---

[1] Do not forget that in our notation $4\partial\bar{\partial} = \partial_\mu^2$ and $\mathrm{Tr}\tau^a\tau^b = 1/2\delta_{ab}$.

Exercise

$$\Omega^3_\mu \equiv -i\mathrm{Tr}[\sigma^z g^{-1}\partial_\mu g] = \partial_\mu \phi + \cos\theta \partial_\mu \psi$$

$$\Omega^\pm_\mu \equiv -i\mathrm{Tr}[\sigma^\pm g^{-1}\partial_\mu g] = e^{\pm i\phi}[\partial_\mu \theta \pm i\sin\theta \partial_\mu \psi] \tag{32.16}$$

With these identities we can express $ig^{-1}\partial_\mu g$ as follows:

$$-ig^{-1}\partial_\mu g = \frac{1}{2}\sigma_a \Omega^a_\mu \tag{32.17}$$

Substituting the latter expression into (32.15) we find that the integrand is equal to the Jacobian of the transformation from ψ, θ, ϕ to ξ, x:

$$\Gamma_{SU(2)} = -\frac{i}{96\pi}\int_0^\infty d\xi \int d^2 x \, \epsilon^{\alpha\beta\gamma} \epsilon_{abc}\Omega^a_\alpha \Omega^b_\beta \Omega^c_\gamma$$

$$= \frac{i}{4\pi}\int d\xi d^2 x \frac{\partial(\phi,\theta,\psi)}{\partial(\xi,x_1,x_2)} = \frac{i}{4\pi}\int d^2 x \epsilon_{\mu\nu}\phi \sin\theta \, \partial_\mu\theta\partial_\nu\psi \tag{32.18}$$

Thus the three-dimensional integral depends on the boundary values only. However, the result of integration cannot be expressed locally in terms of $g(x)$; g is periodic in ϕ, but $\Gamma_{SU(2)}$ is not: when ϕ changes on $2\pi q$ it changes on

$$\delta\Gamma_{SU(2)} = \frac{iq}{2}\int d^2 x \epsilon_{\mu\nu} \sin\theta \, \partial_\mu\theta\partial_\nu\psi \tag{32.19}$$

The latter integral is equal to an integer number, the number of times the vector field $\mathbf{n} = (\cos\theta, \sin\theta\cos\psi, \sin\theta\sin\psi)$ covers the sphere.

Now we are in a position to determine the action for the WZNW model. The following theorem holds.

Theorem. The Euclidean action for the Sugawara Hamiltonian

$$\hat{H} = \frac{2\pi}{k+c_v}\sum_{a=1}^D \int dx \left[:J^a_R(x)J^a_R(x): + :J^a_L(x)J^a_L(x):\right]$$

where the currents J^a satisfy the Kac–Moody algebra for the group G with the central extension k defined by (31.4), is given by $S = kW(U)$, where U is a matrix from the fundamental representation of the group G and $W(U)$ is defined by (32.15). Correlation functions of the current operators J^a coincide with correlation functions of the fields

$$J_L = \frac{k}{2\pi}U\bar{\partial}U^{-1} \qquad J_R = -\frac{k}{2\pi}U\partial U^{-1} \tag{32.20}$$

To prove this theorem we use the following remarkable property of the WZNW action (32.15):

$$W(gU) = W(U) + W(g) + \frac{1}{2\pi}\int d^2 x \mathrm{Tr}(g^{-1}\partial g U\bar{\partial}U^{-1}) \tag{32.21}$$

Let us show that the integral (32.4) can be written as the path integral with the WZNW action:

$$e^{-S_{\text{eff}}[A]} = \left\langle \exp\left(-i \int d^2x\, A_\mu^a\, J^{a,\mu}\right)\right\rangle_{\text{w}}$$

$$= \frac{\int [U^{-1}DU]\exp[-kW(U) - (k/2\pi)\int d^2x\,\text{Tr}(g^{-1}\partial g U \bar\partial U^{-1})]}{\int [U^{-1}DU]\exp[-kW(U)]} \quad (32.22)$$

where $[U^{-1}DU]$ means the measure of integration on the group G. Now using the identity (32.21) and invariance of the measure with respect to the transformations

$$U \to Ug$$

we can rewrite the integral as follows:

$$e^{-S_{\text{eff}}[A]} = \frac{\int [U^{-1}DU]\exp[kW(g) - kW(Ug)]}{\int [U^{-1}DU]\exp[-kW(U)]} = \exp[kW(g)] \quad (32.23)$$

and get the answer obtained before. The theorem is proven.

It is also possible to identify the following fields.

(i) U-field, a primary field transforming according to the fundamental representation of the group G. If G = SU(N) ($c_v = N$) its scaling dimension is equal to (see (31.18))

$$\Delta_U = \bar\Delta_U = \frac{N^2 - 1}{2N(N + k)} \quad (32.24)$$

(ii) $\phi_1^{ab} = \text{Tr}(U^{-1}t^a U t^b)$ where t^a are generators of G. This field is also a primary one and belongs to the adjoint representation; its scaling dimensions are equal to

$$\Delta_1 = \bar\Delta_1 = \frac{c_v}{c_v + k} \quad (32.25)$$

(iii) The 'wrong currents' $K^a \sim \text{Tr}(t^a U^{-1}\partial U)$, $\bar K^a \sim \text{Tr}(t^a \bar\partial U U^{-1})$ whose scaling dimensions are equal to $(\Delta_1 + 1, \Delta_1)$ and $(\Delta_1, \Delta_1 + 1)$, respectively. These fields are not primary.

(iv) The Lagrangian density $\text{Tr}(\partial_\mu U^{-1}\partial_\mu U)$ is also not a primary field and has scaling dimensions equal to $(\Delta_1 + 1, \Delta_1 + 1)$.

As we see, the Lagrangian density is an irrelevant operator. Therefore the general theory with the action

$$W(c; g) = \frac{1}{2c} \int d^2x\, \text{Tr}(\partial_\mu g^{-1}\partial_\mu g) + k\Gamma[g] \quad (32.26)$$

scales to the critical point $c = 8\pi/k$. The crossover was described exactly by Polyakov and Wiegmann who solved this model by the Bethe ansatz (1984). In fact, it is correct to call this model the Wess–Zumino–Novikov–Witten model; then the theory described by the action (32.15) should be called 'the WZNW model at critical point' or *critical* WZNW model. There are at least two remarkable facts about the WZNW model. The first is that if we

Figure 32.1. Paul Wiegmann.

remove the topological term from the action (32.26), the model becomes an asymptotically free theory. The model with $k = 0$ is one of the nonlinear sigma models (see Chapter 9) and its excitations are massive particles. The second remarkable fact is that the topological term, responsible for the critical point, does not contribute to the perturbation expansion in powers of c. Therefore at small c both models have the same beta-function and the critical point is essentially nonperturbative.

Non-Abelian bosonization: nontrivial determinant

As we have seen from the previous discussion certain combinations of fermionic fields governed by the Dirac Hamiltonian can be written in terms of fields of the WZNW model. In particular, we have the explicit equivalencies for currents. At the end of the previous chapter we have also shown that the slow components of the $\psi_R^+ \psi_L$-field in the U(N)$\times$SU(k) model with the current-current interaction are proportional to the Wess–Zumino matrix U. The latter property also holds for noninteracting Dirac fermions with the symmetry U(N) (see Knizhnik and Zamolodchikov, 1984) and for noninteracting Majorana fermions with the symmetry O(N) (Witten, 1984). Thus we can compose the bosonization table for non-Abelian theories.

Using Table 32.1 and the properties of the WZNW action we can calculate the following fermionic determinant (Tsvelik, 1994):

$$D[U] = \operatorname{Tr} \ln[\gamma_\mu \partial_\mu + (1 + \gamma_5)mU/2 + (1 - \gamma_5)mU^+/2] \qquad (32.27)$$

where U is an external matrix field from the SU(N) group and m is a parameter of the dimension of mass. We assume that the field $U(\tau, x)$ varies slowly on the scale m^{-1} and will be interested in the expansion of $D[U]$ in powers of m^{-1}. Such problems appear frequently in applications (see, for example, Chapter 33).

Table 32.1. Non-Abelian bosonization

Critical $U(N)$ WZNW model with $k = 1$	Massless Dirac fermions
Action	Action
$\dfrac{1}{2} \int d^2x (\partial_\mu \Phi)^2 + W[g]; g \in \mathrm{SU}(N)$	$2 \int d^2x (\psi_{R,\alpha}^+ \partial_{\bar{z}} \psi_{R,\alpha} + \psi_{L,\alpha}^+ \partial_z \psi_{L,\alpha})$
Operators	Operators
$1/(2\pi a)\exp(i\sqrt{4\pi/N}\Phi)g_{\alpha\beta}$	$\psi_{R,\alpha}^+ \psi_{L,\beta}$
$i\sqrt{N/4\pi}\,\partial\Phi$	$\dfrac{1}{2}:\psi_{R,\alpha}^+ \psi_{R,\alpha}:$
$-i\sqrt{N/4\pi}\,\bar{\partial}\Phi$	$\dfrac{1}{2}:\psi_{L,\alpha}^+ \psi_{L,\alpha}:$
$1/2\pi \, \mathrm{Tr}(\tau^a g \partial g^{-1})$	$:\psi_{R,\alpha}^+ \tau_{\alpha\beta}^a \psi_{R,\beta}:$
$-1/2\pi \, \mathrm{Tr}[\tau^a g \bar{\partial} g^{-1}]$	$:\psi_{L,\alpha}^+ \tau_{\alpha\beta}^a \psi_{L,\beta}:$
Critical $O(N)$ WZNW model with $k = 1$	Massless real fermions
Action	Action
$W[g]; g \in O(N)$	$2 \int d^2x \sum_{\alpha=1}^{N} (\chi_{R,\alpha} \partial_{\bar{z}} \chi_{R,\alpha} + \chi_{L,\alpha} \partial_z \chi_{L,\alpha})$
Operators	Operators
$1/(2\pi a)g_{\alpha\beta}$	$\chi_{R,\alpha} \chi_{L,\beta}$
$1/2\pi (g\partial g^{-1})_{\alpha\beta}$	$:\chi_{R,\alpha} \chi_{R,\beta}:$
$-1/2\pi (g\bar{\partial} g^{-1})_{\alpha\beta}$	$:\chi_{L,\alpha} \chi_{L,\beta}:$

Using the table we can rewrite the determinant as a path integral over the matrix field g with the WZNW action:

$$e^{D[U]} = \int Dg D\Phi \exp\left\{-\int d^2x \left[\frac{m}{2\pi a}\mathrm{Tr}(Ug^+e^{-i\gamma\Phi} + gU^+e^{i\gamma\Phi})\right]\right.$$
$$\left. - W(g) - \frac{1}{2}\int d^2x(\partial_\mu\Phi)^2\right\} \tag{32.28}$$

where $\gamma = \sqrt{(4\pi/N)}$. Now we make a shift of variables in the path integral introducing a new variable G:

$$g = UG$$

This shift leaves the measure of integration unchanged. Using the identity (32.21) we get the following result:

$$e^{D[U]} = \int DG D\Phi \exp\{-S[G, \Phi] - S_{\mathrm{int}}[U, G] - W[U]\} \tag{32.29}$$

$$S[G, \Phi] = \int d^2x \left[\frac{m}{2\pi a}\mathrm{Tr}(e^{i\gamma\Phi}G + G^+e^{-i\gamma\Phi})\right] + W(G) + \frac{1}{2}\int d^2x(\partial_\mu\Phi)^2 \tag{32.30}$$

$$S_{\mathrm{int}} = \frac{1}{2\pi}\int d^2x \mathrm{Tr}(U^{-1}\partial U G\bar{\partial}G^{-1}) \tag{32.31}$$

The only term in the action connecting G and U is the interaction term S_{int}. Let us show that at small energies much less than m this term is irrelevant. In order to make it completely obvious we fermionize the new action using Table 32.1. Then we have

$$e^{D[U]} = e^{-W(U)} \int D\bar{\psi} D\psi \exp\left\{-\int d^2x [\bar{\psi}\gamma_\mu \partial_\mu \psi + m\bar{\psi}\psi + \text{Tr}(U^{-1}\partial U J_{\text{L}})]\right\} \quad (32.32)$$

where J_{L} is the left current of free *massive* fermions. Calculating the first diagram in the expansion in $U^{-1}\partial U$ we get the term of order of

$$\frac{1}{m^2}\text{Tr}[\bar{\partial}(U^{-1}\partial U)]^2$$

which is irrelevant at low energies. Therefore from (32.29) we find (Tsvelik, 1994)

$$D[U]/D[I] = -W(U) + \mathcal{O}(m^{-2}) \quad (32.33)$$

The latter result is nontrivial since it includes the topological term which is difficult to get by more conventional methods.

References

Johnson, K. (1963). *Phys. Lett.*, **5**, 253.
Knizhnik, V. G. and Zamolodchikov, A. B. (1984). *Nucl. Phys. B*, **247**, 83.
Novikov, S. (1982). *Usp. Mat. Nauk*, **37**, 3.
Polyakov, A. M. and Wiegmann, P. B. (1983). *Phys. Lett. B*, **131**, 121.
Polyakov, A. M. and Wiegmann, P. B. (1984). *Phys. Lett. B*, **141**, 223.
Tsvelik, A. M. (1994). *Phys. Rev. Lett.*, **72**, 1048.
Witten, E. (1984). *Commun. Math. Phys.*, **92**, 455.

33

Semiclassical approach to
Wess–Zumino–Novikov–Witten models

The concept of Kac–Moody algebra represents an extension of ordinary algebra; in the same way the WZNW model Hamiltonian (31.10) is a generalization of the Hamiltonian of a quantum top (31.14). From elementary courses on quantum mechanics we know that the problem of the quantum top can be treated algebraically, using the commutation relations of the angular momentum operators, and by solving the Schrödinger equation in a coordinate representation. Both approaches have advantages. The algebraic approach to the WZNW model was developed in Chapter 32; in this chapter I will explain the coordinate approach.

The WZNW action in its most general form can be written as

$$S = \int d^2x [\sqrt{g}g^{\mu\nu}G_{ab}[X]\partial_\mu X^a \partial_\nu X^b + \epsilon^{\mu\nu}B_{ab}[X]\partial_\mu X^a \partial_\nu X^b] \tag{33.1}$$

where X^a are fields representing the coordinates on some group (or coset) manifold, G_{ab} is the metric tensor on this manifold, and B_{ab} is an antisymmetric tensor. We define the action on a curved (world sheet) surface with a metric $g_{\mu\nu}$ ($\mu, \nu = 1, 2$).

A very important feature of the action (33.1) is that the second term does not contain the world sheet metric. Consequently, the classical stress energy tensor, $T_{\mu\nu} = \delta S/\delta g^{\mu\nu}$, is determined solely by the first term. In particular, the most important components for the critical model are given by

$$T_{zz} = G_{ab}[X]\partial_z X^a \partial_z X^b \qquad T_{\bar{z}\bar{z}} = G_{ab}[X]\partial_{\bar{z}} X^a \partial_{\bar{z}} X^b \tag{33.2}$$

where $z = x_0 + ix_1, \bar{z} = x_0 - ix_1$. Here, the reader should not get the false impression that the Wess–Zumino term is not important. This cannot be the case because it contributes to the equations of motion. Since the model (33.1) is supposed to be *critical*, these equations (to be understood as identities for correlation functions in the quantum theory) are

$$T_{z\bar{z}} = 0 \qquad \partial_{\bar{z}}T_{zz} = 0 \qquad \partial_z T_{\bar{z}\bar{z}} = 0 \tag{33.3}$$

Though the Wess–Zumino term does not enter into the form of $T, \bar{T}$, the fulfilment of these equations depends on it through the dynamics of the underlying fields X^a.

In the infinite plane, parametrized by coordinates $(z, \bar{z})$, conformal invariance restricts the two-point function of primary fields of conformal dimension $(h, \bar{h})$ to be of the form[1]

$$\langle \phi(z_1, \bar{z}_1)\phi(z_2, \bar{z}_2)\rangle = (z_1 - z_2)^{-2h}(\bar{z}_1 - \bar{z}_2)^{-2\bar{h}} \tag{33.4}$$

Under a conformal transformation of the plane, $w = w(z)$, this correlation function transforms like a tensor of rank $(h, \bar{h})$:

$$\langle \phi(w_1, \bar{w}_1)\phi(w_2, \bar{w}_2)\rangle = \prod_{i=1}^{2} \left[\frac{dz(w_i)}{dw}\right]^{h} \left[\frac{d\bar{z}(\bar{w}_i)}{d\bar{w}}\right]^{\bar{h}} \langle \phi[z(w_1), \bar{z}(w_1)]\phi[z(w_2), \bar{z}(w_2)]\rangle$$

$$\tag{33.5}$$

One may pass from the infinite plane to a strip of width $2\pi R$ by means of the conformal transformation $w = R \ln z$. Combining this transformation with (33.4) and (33.5), one obtains the two-point function in the strip geometry:

$$\langle \phi(w_1, \bar{w}_1)\phi(w_2, \bar{w}_2)\rangle = [2R \sinh(w_{12}/2R)]^{-2h} [2R \sinh(\bar{w}_{12}/2R)]^{-2\bar{h}} \tag{33.6}$$

Introducing coordinates (τ, σ) along and across the strip respectively ($w = \tau + i\sigma$, $\bar{w} = \tau - i\sigma$; $-\infty < \tau < \infty$, $0 < \sigma < 2\pi R$), one may expand (33.6) in the following manner:

$$\langle \phi(\tau_1, \sigma_1)\phi(\tau_2, \sigma_2)\rangle = \sum_{n=0}^{\infty} \sum_{m=-\infty}^{\infty} C_{nm} \, e^{-(h+\bar{h}+n)|\tau_{12}|/R} \, e^{-i(h-\bar{h}+m)|\sigma_{12}|/R} \tag{33.7}$$

One may also obtain the two-point function in the operator formalism, leading to the Lehmann expansion:

$$\langle \phi(\tau_1, \sigma_1)\phi(\tau_2, \sigma_2)\rangle = \sum_{\alpha} |\langle 0|\hat{\phi}|\alpha\rangle|^2 \, e^{-E_\alpha|\tau_{12}|-iP_\alpha|\sigma_{12}|} \tag{33.8}$$

where E_α and P_α are the eigenvalues of the Hamiltonian and the momentum operator respectively, in the state $|\alpha\rangle$. Comparing (33.7) and (33.8) one obtains a relationship between the eigenvalues of the Hamiltonian and the momentum operator, and the scaling dimensions in the corresponding conformal field theory:

$$E_\alpha = \frac{h + \bar{h} + n}{R} \qquad P_\alpha = \frac{h - \bar{h} + m}{R} \tag{33.9}$$

Restricting our attention to fields with $h = \bar{h} = d/2$, one may rewrite (33.6) in the form

$$\langle \phi(\tau_1, \sigma_1)\phi(\tau_2, \sigma_2)\rangle = R^{-2d} [2\cosh(\tau_{12}/R) - 2\cos(\sigma_{12}/R)]^{-d} \tag{33.10}$$

One observes that for $\tau_{12} \gg R$ the asymptotic form of the correlation function is *independent* of σ,

$$\langle \phi(\tau_1, \sigma_1)\phi(\tau_2, \sigma_2)\rangle \sim R^{-2d} \exp(-d\tau_{12}/R) \qquad (\tau_{12} \gg R) \tag{33.11}$$

and in this limit one should set $n = m = 0$ in (33.7) and (33.9).

---

[1] Here I repeat some of the discussion of Chapter 25.

Let us now place the model (33.1) on a thin strip of width $2\pi R$, and neglect any σ-dependence of the fields X_a. As we may see from (33.10) and (33.11), such a procedure preserves the (large τ) asymptotics of the correlation functions. The caveat is that from now on we exclude from consideration all conformal descendants of the primary fields. The action (33.1) becomes

$$S = 2\pi R \int d\tau \, G_{ab}[X]\partial_\tau X^a \partial_\tau X^b \qquad (33.12)$$

which may be recognized as the action for a free, nonrelativistic particle, of mass $m = 4\pi R$. The corresponding Hamiltonian is the Laplace–Beltrami operator (multiplied by $-1/2m$),

$$\hat{H} = \frac{-1}{8\pi R\sqrt{G}} \frac{\partial}{\partial X^a} \left(\sqrt{G} G^{ab} \frac{\partial}{\partial X^b} \right) \qquad (33.13)$$

For a WZNW model this Hamiltonian is equivalent to the truncated Hamiltonian (31.14). As we have already established, the eigenvalues of this Hamiltonian are related to the spectrum of scaling dimensions in our conformal field theory by (33.9). By considering σ-independent fields we have truncated the Hilbert space of the conformal field theory down to the space of primary fields. Solution of this Schrödinger equation allows one to obtain explicit expressions for the eigenstates, $|\alpha\rangle$, in a coordinate representation.

Exercise

A metric on a group space can be introduced as

$$(ds)^2 = \text{Tr}(g^{-1}dg)^2 \qquad G_{ab} = \text{Tr}[(g^{-1}\partial_a g)(g^{-1}\partial_b g)]$$

where $g(X)$ is a matrix from the fundamental representation. Show that $g(X)$ satisfies the Schrödinger equation (33.13). In case of any difficulty consider a particular case, for example considering the SU(2) group space and using Euler parametrization (4.36) for the matrix g.

The semiclassical approach can be useful when one has to deal with complicated groups or coset spaces. One can find useful illustrations of this point in Bhaseen et al. (1999).

Reference

Bhaseen, M. J., Kogan, I. I., Solovyev, O., Taniguchi, N. and Tsvelik, A. M. (1999). *Nucl. Phys. B*, **580**, 688.

34

Integrable models: dynamical mass generation

In our discussion of one-dimensional physics, more than once we have encountered situations when some relevant interaction scales to strong coupling. In such a situation the original description becomes inapplicable at low energies and must be replaced by some other description. In all previous cases I have restricted the discussion by some qualitative analysis, promising to provide more details later. Now the time is ripe to fulfil the promise.

In fact, almost all our understanding of strong coupling physics comes from exact solutions of just a few models. The most ubiquitous among them is the sine-Gordon model,[1] which we have already encountered in this book many times. Anyone who thoroughly understands this model and possibly a few others (such as the O(3) nonlinear sigma model and the off-critical Ising model) may consider themself an expert in the area of strongly correlated systems.

The sine-Gordon model belongs to a category of integrable field theories. This means that it has an infinite number of constants of motion. On the one hand this fact makes it possible to solve this model exactly and describe its thermodynamics and correlation functions; on the other hand it makes the results less general than one might wish. First of all, not all interesting models are integrable, and one may wonder how and in what respect their behaviour may differ from the behaviour of integrable models. Integrable models possess some exceptional physical properties (such as ballistic transport) which are destroyed when the integrability is violated.

There are several important review articles which one can read to familiarize oneself with the subject. Thus the review article by the Zamolodchikovs (1979) gives an excellent introduction to the scattering theory and the bootstrap approach (see below). Those who are interested in microscopic derivation of Bethe ansatz equations and the thermodynamics are advised to read the review article written by the author and Wiegmann (1983). As far as the correlation functions are concerned, the reader may read papers by Lukyanov and co-authors (see references at the end of the chapter). An excellent and rather pedagogical review is also given by Babujian *et al.* (1999, 2002).

---

[1] There are several important models such as the SU(2) invariant Thirring model and the $SU_1(2)$ WZNW model perturbed by the current-current interaction which are equivalent to the sine-Gordon model.

Figure 34.1. Alexander (Sasha) Zamolodchikov.

The subject of exactly solvable models is very rich and complicated. I cannot possibly give it the space it deserves – a serious discussion would take another volume or two. At the same time I cannot resist the temptation to discuss the problem on some level. Below I will give a review of the main results about exactly solvable problems in general and the sine-Gordon model in particular.

General properties of integrable models

All exactly solvable $(1 + 1)$-dimensional models share certain common features. In what follows I shall discuss only the models whose low energy spectrum displays Lorentz invariance. I shall restrict the present consideration to the models with spectral gaps. In particular, the sine-Gordon model has a spectral gap in the entire interval $\beta^2 < 8\pi$. The spectrum of relativistic massive theories can be conveniently parametrized by the parameter called 'rapidity' θ:

$$\epsilon(\theta) = m \cosh \theta \qquad p(\theta) = m \sinh \theta \qquad (34.1)$$

As a consequence of Lorentz invariance, the two-particle scattering matrix must be a function of $\theta_1 - \theta_2$, where $\theta_{1,2}$ are the rapidities of colliding particles.

As a consequence of integrability, colliding particles conserve their rapidities (momenta). This means that ⟨in| and |out⟩ states are characterized by the same set of rapidities. In addition, multi-particle scattering is factorizable, that is an N-particle scattering matrix is represented as a product of two-particle ones. Thus complete information about the interaction is contained in the two-particle S matrix and *can be included in the commutation relations*. Such a description of integrable theories was introduced by the Zamolodchikovs (see their review article published in 1979).

Let $Z_a^+(\theta)$ and $Z_a(\theta)$ be creation and annihilation operators for a particle state characterized by the isotopic index a and rapidity θ. These operators satisfy what is called the

Zamolodchikov–Faddeev algebra:

$$Z_a^+(\theta_1)Z_b^+(\theta_2) = S_{a,b}^{\bar{a},\bar{b}}(\theta_1 - \theta_2)Z_{\bar{b}}^+(\theta_2)Z_{\bar{a}}^+(\theta_1)$$

$$Z^a(\theta_1)Z^b(\theta_2) = S_{\bar{a},\bar{b}}^{a,b}(\theta_1 - \theta_2)Z^{\bar{b}}(\theta_2)Z^{\bar{a}}(\theta_1) \tag{34.2}$$

$$Z^a(\theta_1)Z_b^+(\theta_2) = S_{b,\bar{a}}^{\bar{b},a}(\theta_2 - \theta_1)Z_{\bar{b}}^+(\theta_2)Z^{\bar{a}}(\theta_1) + \delta(\theta_1 - \theta_2)\delta_a^b$$

The commutation relations are self-consistent, that is the algebra has a property of associativity if the S matrix satisfies the Yang–Baxter relations:

$$S_{\bar{a}_1,\bar{a}_2}^{a_1,a_2}(\theta_{12})S_{b_1,\bar{a}_3}^{\bar{a}_1,a_3}(\theta_{13})S_{b_2,b_3}^{\bar{a}_2,\bar{a}_3}(\theta_{23}) = S_{\bar{a}_2,\bar{a}_3}^{a_2,a_3}(\theta_{23})S_{\bar{a}_1,b_3}^{a_1,\bar{a}_3}(\theta_{13})S_{b_1,b_2}^{a_1,\bar{a}_2}(\theta_{12}) \tag{34.3}$$

Looking at (34.2) one can see that excitations in integrable models are neither bosons nor fermions.

If the Yang–Baxter relations are fulfiled, the N-body S matrix is automatically factorizable. Therefore the Yang–Baxter relations are necessary and sufficient conditions of integrability. As such they can be used in two ways. The direct way would be to take a given model, to calculate its two-body S matrix and to check the model for integrability by checking whether the S matrix satisfies the Yang–Baxter equation. The other way would be to do the reverse: to find all possible solutions of the Yang–Baxter equations with a given symmetry and use your skill and luck to find an integrable model they may correspond to. The latter method is called *bootstrap*; it has been perfected by the Zamolodchikov brothers and Fateev.

The Zamolodchikov–Faddeev description (34.2) is universal for most integrable models (there are those where a simple particle description is not possible and (34.2) should be modified, but I will not dwell on this at the moment). The only thing which changes from theory to theory is the spectrum content and the form of S matrix. However, the S matrix also has certain universal features. In particular

$$S_{\bar{a},\bar{b}}^{a,b}(0) = -\delta_{\bar{a}}^b\delta_{\bar{b}}^a \tag{34.4}$$

Therefore taking the limit $\theta_1 - \theta_2 = 0$ in (34.2) we arrive at the statement

$$Z^a(\theta)Z^b(\theta) = -Z^a(\theta)Z^b(\theta) = 0$$

which means that states with the same rapidity cannot be occupied more than once. This looks somewhat like the Pauli principle, though it is not, because the double occupancy is forbidden *irrespective* of the isotopic indices. So it is not a Pauli principle, but the statement that the repulsion between particles with equal momenta becomes *infinitely strong*. The other universal property is related to the asymptotic behaviour of the S matrix at large momentum transfer:

$$\lim_{\theta \to \infty} S_{\bar{a},\bar{b}}^{a,b}(\theta) = \delta_{\bar{a}}^a\delta_{\bar{b}}^b \exp(-i\varphi_{ab}\mathrm{sign}\theta) \tag{34.5}$$

where φ_{ab} does not depend on θ. That is, at large momenta the scattering becomes diagonal and the commutation relations acquire a more conventional form:

$$Z^a(\theta_1)Z^b(\theta_2) = e^{i\varphi_{ab}\mathrm{sign}\theta_{12}} Z^b(\theta_2)Z^a(\theta_1) \tag{34.6}$$

Even the asymptotic form of the commutation relations does not coincide with the standard bosonic or fermionic ones, unless $\varphi = 0, \pi$.

Example

Let us consider the spin sector of the SU(2) symmetric Thirring model given by the $SU_1(2)$ WZNW model with the relevant current-current interaction (30.19). This model also constitutes the limiting case of the sine-Gordon model with $\beta^2 \to 8\pi - 0$. The spectrum consists of massive particles with spin 1/2 (solitons and antisolitons); their S matrix is

$$S_{\bar{a},\bar{b}}^{a,b}(\theta) = -S_0(\theta)\frac{\theta\delta_{\bar{a}}^a\delta_{\bar{b}}^b + i\pi\delta_{\bar{a}}^b\delta_{\bar{b}}^a}{\theta + i\pi} \tag{34.7}$$

$$S_0(\theta) = \frac{\Gamma(1/2 - i\theta/2\pi)\Gamma(1 + i\theta/2\pi)}{\Gamma(1/2 + i\theta/2\pi)\Gamma(1 - i\theta/2\pi)} \tag{34.8}$$

In this particular case $\varphi = \pi/2$.

There is a standard procedure which establishes a connection between the two-body S matrix and the thermodynamics. As the first step one solves the problem of scattering of an arbitrary number of particles. Suppose we have an integrable system where the excitation consists of a single massive particle with mass m (the fundamental particle) and (possibly) its bound states. The energy of the N-particle state is

$$E = m\sum_{j=1}^{N}\cosh\theta_j \tag{34.9}$$

where bound states correspond to complex θ_j. Let us put the system into a box of size L with periodic boundary conditions. This will impose restrictions on possible values of θ_j. If the S matrix is nondiagonal in isotopic indices, the problem of diagonalization of the N-particle S matrix becomes nontrivial. To understand this point better, let us consider the case where this problem has the easiest solution, namely the case of two particles $N = 2$. When a particle with rapidity θ_j ($j = 1, 2$) goes around the system, the wave function acquires a factor $\exp[iLp(\theta_j)]$ which is compensated by the S matrix such that

$$\exp[-iLp(\theta_1)]\xi(\sigma_1, \sigma_2) = S_{\bar{s}_1,\bar{s}_2}^{\sigma_1,\sigma_2}(\theta_1 - \theta_2)\xi(\bar{s}_1, \bar{s}_2)$$
$$\exp[-iLp(\theta_2)]\xi(\sigma_1, \sigma_2) = S_{\bar{s}_2,\bar{s}_1}^{\sigma_2,\sigma_1}(\theta_2 - \theta_1)\xi(\bar{s}_1, \bar{s}_2) \tag{34.10}$$

where $\xi(\sigma_1, \sigma_2)$ is the wave function in isotopic space. For two particles with spin 1/2 this wave function can be either a singlet or a triplet one. For a singlet the eigenvalue of the permutation operator

$$P_{\bar{a},\bar{b}}^{a,b} = \delta_{\bar{a}}^b\delta_{\bar{b}}^a$$

is equal to -1, for a triplet it is $+1$. Therefore depending on the value of the total spin we get either

$$\exp(-imL\sinh\theta_1) = -S_0(\theta_1 - \theta_2) \qquad \exp(-imL\sinh\theta_2) = -S_0(\theta_2 - \theta_1) \tag{34.11}$$

for a *triplet* or

$$\exp(-imL\sinh\theta_1) = S_0(\theta_{12})\frac{-\theta_{12} + i\pi}{\theta_{12} + i\pi}$$

$$\exp(-imL\sinh\theta_2) = S_0(\theta_{21})\frac{-\theta_{12} + i\pi}{\theta_{12} + i\pi}$$
(34.12)

for a *singlet*.

For a general N the spin-wave function must satisfy the following set of equations:

$$e^{-imL\sinh\theta_j}\xi(a_1, \cdots a_N) = S_{\sigma_1,\bar{a}_1}^{a_j,a_1}(\theta_j - \theta_1)S_{\sigma_2,\bar{a}_2}^{\sigma_1,a_1}(\theta_j - \theta_2)\cdots S_{\bar{a}_j,\bar{a}_N}^{\sigma_{N-1},a_N}(\theta_j - \theta_N)\xi(\bar{a}_1, \cdots \bar{a}_N)$$
(34.13)

The fact that the S matrix satisfies the Yang–Baxter equations guarantees that this system has a solution. This solution has a rather characteristic form with auxilary rapidities; in particular for the model described by the S matrix (34.7), (34.8) we have the *Bethe ansatz equations*

$$e^{-imL\sinh\theta_j} = \prod_{k=1,k\neq j}^{N} S_0(\theta_j - \theta_k)\prod_{\alpha=1}^{M}\frac{\theta_j - \lambda_\alpha - \pi i/2}{\theta_j - \lambda_\alpha + \pi i/2}$$
(34.14)

$$\prod_{j=1}^{N}\frac{\lambda_\alpha - \theta_j - \pi i/2}{\lambda_\alpha - \theta_j + \pi i/2} = \prod_{\beta=1,\beta\neq\alpha}^{M}\frac{\lambda_\alpha - \lambda_\beta - \pi i/2}{\lambda_\alpha - \lambda_\beta + \pi i/2}$$
(34.15)

where M is smaller than or equal to $N/2$ and is related to the projection of the total spin:

$$S^z = N/2 - M$$
(34.16)

Solutions of these equations determine eigenvalues of energy (34.9).

The auxiliary rapidities λ_α parametrize the spin-wave function. Their appearance is completely natural since for $M > 1$ the spin-wave function cannot be reduced to just a singlet or a triplet, but becomes nontrivial. These rapidities are also constants of motion, as θ_j.

Exercise

Check that for the $N = 2$ system (34.14), (34.15) reproduces either (34.11) ($M = 0$) or (34.12) ($M = 1$).

To study the thermodynamics one needs to consider the case $L \to \infty$, N/L, $M/L =$ const. There is a well trodden path leading from discrete Bethe ansatz equations to the so-called thermodynamic Bethe ansatz (TBA) equations which determine the free energy. The procedure was introduced by Yang and Yang (1965); I shall illustrate it using as an example one of the simplest integrable models, the so-called *sinh-Gordon* model:

$$S = \int d\tau dx\left[\frac{1}{2}(\partial_\mu\Phi)^2 + \frac{m_0^2}{\beta^2}\cosh(\beta\Phi)\right]$$
(34.17)

The excitation spectrum of this model consists of a single scalar massive particle. There are no bound states. In that respect this model is very similar to the off-critical Ising model with which it shares many properties (see Ahn *et al.*, 1993). In the limit $\beta \to 0$ one can expand the cosh term and the theory becomes a theory of a massive noninteracting scalar bosonic field. The S matrix was derived by Fateev (1990) from the S matrix of breathers in the sine-Gordon one by the analytic continuation $\beta \to i\beta$ (see the discussion in the next section):

$$S(\theta) = \frac{\sinh\theta - i\sin\gamma}{\sinh\theta + i\sin\gamma} \qquad \gamma = \frac{\pi\beta^2}{8\pi + \beta^2} \tag{34.18}$$

This S matrix is unitary and crossing-symmetric:

$$S(\theta)S(-\theta) = 1 \qquad S(i\pi - \theta) = S(\theta) \tag{34.19}$$

The discrete Bethe ansatz equations are

$$\exp(imL\sinh\theta_j) = \prod_{k=1,k\neq j}^{N} \frac{\sinh(\theta_j - \theta_k) + i\sin\gamma}{\sinh(\theta_j - \theta_k) - i\sin\gamma} \tag{34.20}$$

and the energy is given by (34.9). Here the mass of physical particles m is a complicated function of the bare parameters m_0, β.[2] These equations have only real solutions, which is related to the absence of bound states. Taking the logarithm of both sides of (34.20) we arrive at the following equations:

$$mL\sinh\theta_j + \sum_{k=1,k\neq j}^{N} K(\theta_j - \theta_k) = 2\pi J_j \tag{34.21}$$

where

$$K(\theta) = 2\tan^{-1}[\sinh\theta/\sin\gamma]$$

and J_j are integer numbers, eigennumbers characterizing a state with a given energy. In general these numbers do not occupy the axis of integers densely; between occupied positions ('particles') there are unoccupied positions ('holes'). It is convenient to extend (34.21) and define a rapidity u_l for *every* integer l:

$$mL\sinh u_l + \sum_{k=1,k\neq j}^{N} K(u_l - \theta_k) = 2\pi l \tag{34.22}$$

Notice that the summation is over particles only. Therefore (34.22) does not deny (34.21), but serves merely as a definition of an extended set of rapidities. This set includes the old one as a subset: if l is equal to one of the eigennumbers J_j, $u_{J_j} = \theta_j$.

In the thermodynamic limit it is convenient to introduce the distribution function of rapidities of particles:

$$\rho(\theta) = \frac{1}{L}\sum_j \delta(\theta - \theta_j) \tag{34.23}$$

---

[2] Only in the limit $\beta \to 0$ does it become simple: $m = m_0$.

such that summation over θ_j is replaced by integration:

$$\sum_j f(\theta_j) \to L \int d\theta \rho(\theta) f(\theta) \tag{34.24}$$

In addition to that we can introduce the density of unoccupied places or 'holes' such that summation over all integer numbers is replaced by

$$\sum_l f(u_l) \to L \int du f(u)[\rho(u) + \tilde{\rho}(u)] \tag{34.25}$$

Substituting in this formula $f(u) = \Theta(\theta - u)$ we obtain the expression which establishes a correspondence between l and θ:

$$l = L \int^\theta du[\rho(u) + \tilde{\rho}(u)] \tag{34.26}$$

Differentiating (34.22) with respect to θ one obtains the following integral equation:

$$\tilde{\rho}(\theta) + \rho(\theta) = \frac{m}{2\pi} \cosh \theta + \int d\theta' A(\theta - \theta') \rho(\theta') \tag{34.27}$$

where

$$A(\theta) = \frac{1}{2\pi} \frac{dK(\theta)}{d\theta} = \frac{1}{2\pi i} \frac{d\ln S(\theta)}{d\theta} \tag{34.28}$$

The number of states in the interval $(\theta, \theta + d\theta)$ is equal to $L\rho(\theta)d\theta$. The total number of integers in this interval is $L[\rho(\theta) + \tilde{\rho}(\theta)]d\theta$. Therefore the degeneracy of this state is

$$G = \frac{\{L[\rho(\theta) + \tilde{\rho}(\theta)]d\theta\}!}{[L\rho(\theta)d\theta]![L\tilde{\rho}(\theta)d\theta]!} \tag{34.29}$$

and its entropy (in the thermodynamic limit) is

$$S = \ln G \approx L d\theta[(\rho + \tilde{\rho}) \ln(\rho + \tilde{\rho}) - \rho \ln \rho - \tilde{\rho} \ln \tilde{\rho}] \tag{34.30}$$

Now we can write down the expression for the free energy:

$$F/L = \int d\theta \{m \cosh \theta \, \rho(\theta) - T\rho(\theta) \ln[1 + \tilde{\rho}(\theta)/\rho(\theta)] - T\tilde{\rho}(\theta) \ln[1 + \rho(\theta)/\tilde{\rho}(\theta)]\} \tag{34.31}$$

This expression is quite general and is valid for any integrable model with a relativistic spectrum (if the spectrum is nonrelativistic, one just needs to substitute $m \cosh \theta$ with the corresponding dispersion law). All information about the model is contained in (34.27).

In order to find the state of thermodynamic equilibrium, we have to minimize the free energy with respect to $\rho(\theta)$ taking into account that variations of ρ and $\tilde{\rho}$ are related through (34.27):

$$\delta\tilde{\rho}(\theta) + \delta\rho(\theta) = \int d\theta' A(\theta - \theta') \delta\rho(\theta') \tag{34.32}$$

Exercise

Using (34.31), (34.32) show that the equation

$$\frac{\delta F}{\delta \rho(\theta)} = 0$$

is equivalent to the following equation for the function $\epsilon \equiv T \ln[\tilde{\rho}/\rho]$: *thermodymanic Bethe ansatz (TBA) equations*

$$m \cosh \theta = \epsilon(\theta) + T \int_{-\infty}^{\infty} d\theta' A(\theta - \theta') \ln\left[1 + e^{-\epsilon(\theta')/T}\right] \tag{34.33}$$

$$F/L = -\frac{Tm}{2\pi} \int d\theta \ln\left[1 + e^{-\epsilon(\theta)/T}\right] \tag{34.34}$$

Exercises

(1) Show that in the limit $\beta \to 0$ when the kernel in (34.33) becomes a δ-function, (34.34) does become the expression for the free energy of massive bosonic particles.
(2) Find the first terms in the expansion of F/L in $\exp(-m/T)$.

TBA equations for other integrable models have a structure very similar to (34.33), (34.34). They are nonlinear equations whose kernels are related to the S matrix. The number of equations depends on the model; thus for the sinh-Gordon model the free energy depends on just a single dispersion function $\epsilon(\theta)$, at the same time TBA equations corresponding to (34.14), (34.15) include an infinite number of dispersions: *TBA equations for the* SU(2) *invariant model (30.19)*

$$\epsilon(\theta) = m \cosh \theta - T A_0 * \ln\left[1 + e^{-\epsilon(\theta)/T}\right] + T a_n * \ln\left[1 + e^{-\epsilon_n(\theta)/T}\right] \tag{34.35}$$

$$\ln\left[1 + e^{\epsilon_n(\theta)/T}\right] - A_{nm} * \ln\left[1 + e^{-\epsilon_m(\theta)/T}\right] = a_n * \ln\left[1 + e^{-\epsilon(\theta)/T}\right] \tag{34.36}$$

where the star stands for conjugation:

$$f * g(\theta) \equiv \int d\theta' f(\theta - \theta') g(\theta')$$

The Fourier images of the kernels defined as

$$A(\theta) = \int \frac{d\omega}{2\pi} e^{-i\omega\theta/\pi} \tilde{A}(\omega)$$

are as follows:

$$\tilde{A}_{nm}(\omega) = \coth |\omega|/2\{\exp[-|n - m||\omega|/2] - \exp[-(n + m)|\omega|/2]\}$$
$$\tilde{a}_n(\omega) = \exp[-n|\omega|/2] \qquad \tilde{A}_0(\omega) = [\exp(|\omega|) + 1]^{-1} \tag{34.37}$$

My purpose here is just to outline a general picture of the field. Therefore I am not going to discuss any details related to TBA equations. The details can be found in various papers on exactly solvable models including the review articles I mentioned at the beginning of the chapter.

Correlation functions: the sine-Gordon model

Thermodynamics of integrable models is a relatively easy subject which has been around since the beginning of the 1960s. A more complicated and at the same time more interesting topic is correlation functions of integrable models. Naturally, for integrable models with a gapless spectrum this problem can be tackled using the methods of conformal field theory and bosonization. I have spent enough time discussing these methods and here will concentrate exclusively on models with a spectral gap. At present the general procedure adopted to calculate the correlation functions is based on the Lehmann expansion familiar to us from Part I. The foundations of the method where laid by Karowski *et al.* (1977, 1978), and in the 1980s and 1990s the method was perfected (almost single-handedly) by Smirnov with the results being summarized in his book. The important recent developments can be found in the paper by Lukyanov and Zamolodchikov (1997). An excellent overview of the method and the results can be found in the articles by Babujian *et al.* (1999, 2002).

Generalities

Let me describe general aspects of this procedure. As the first step we have to describe the Hilbert space of multi-particle states. An arbitrary multi-particle state in terms of the Zamolodchikov–Fadeev creation operators acting on the vacuum state $|0\rangle$:

$$|(\theta_1, a_1) \cdots (\theta_N, a_N)\rangle = Z_{a_1}^+(\theta_1) \cdots Z_{a_N}^+(\theta_N)|0\rangle \tag{34.38}$$

In order to prevent overcounting, we assume that rapidities of particles of the same kind are ordered: $\theta_1 > \theta_2 > \cdots$

It is assumed that all physical operators can be expanded in terms of Z and Z^+. Let us consider some operator $A(\tau, x)$ and its matrix element between the vacuum and an N-particle eigenstate:

$$\langle 0|A(\tau, x)|(\theta_1, a_1) \cdots (\theta_N, a_N)\rangle = e^{-\tau E - ixP}\langle 0|A(0, 0)|(\theta_1, a_1) \cdots (\theta_N, a_N)\rangle$$
$$\equiv e^{-\tau E - ixP} F_{a_1, \dots, a_N}(A; \theta_1, \dots, \theta_N) \tag{34.39}$$

The quantity $F_{a_1, \dots, a_N}(A; \theta_1, \dots, \theta_N)$ is called the 'formfactor'.

The task of calculation of formfactors turns out to be not so daunting under close inspection. This is because formfactors satisfy certain general requirements greatly restricting their functional form. Below we shall confine our discussion to the one- and two-particle formfactors. The expressions for multi-particle formfactors with a detailed discussion of their derivation can be found in the book by Smirnov cited in the select bibliography, and also in Babujian *et al.* (1999, 2000) (see also the paper by Lukyanov (1995)).

Here I recall the basic requirements for formfactors.

Formfactors of a given operator must transform under Lorentz transformations in accordance with the representation this operator belongs to.[3] Thus if the operator is a Lorentz scalar, its formfactors are invariant under Lorentz transformations. This is possible only if they are functions of differences of rapidities $\theta_i - \theta_j$.

---

[3] Recall that Lorentz transformation is equivalent to a simultaneous shift of the rapidities of all particles.

The second requirement is for the asymptotics at $\theta_i \to \infty$. If at the critical point $m \to 0$ the operator A has a conformal dimension Δ_A, then its formfactor satisfies the following inequality:

$$F(A; \theta_1, \ldots, \theta_N)_{\theta_i \to \infty} \leq \exp(\Delta_A \theta_i) \tag{34.40}$$

It is clear from the definition of the formfactor and commutation relations (34.2) that

$$F_{a_1,a_2}(A; \theta_1, \theta_2) S_{\bar{a}_1, \bar{a}_2}^{a_1, a_2}(\theta_{12}) = F_{\bar{a}_1, \bar{a}_2}(A; \theta_2, \theta_1) \tag{34.41}$$

$$F_{a_1,a_2}(A; \theta_1, \theta_2 + 2i\pi) = F_{a_2, a_1}(A; \theta_2, \theta_1) \tag{34.42}$$

Combining the last two equations we obtain the Riemann problem for the two-particle formfactor:

$$F_{a_1,a_2}(A; \theta_1, \theta_2 + 2i\pi) = F_{\bar{a}_1, \bar{a}_2}(A; \theta_1, \theta_2) S_{a_1, a_2}^{\bar{a}_1, \bar{a}_2}(\theta_{12}) \tag{34.43}$$

If particles create bound states their formfactors have poles in the complex plane of θ. One can use this property to derive formfactors of breathers in the sine-Gordon model from soliton formfactors. The following rule holds:

$$\mathrm{Res} F_{a_1,a_2}(A; \theta_1, \theta_2) = \pm [\mathrm{Res} S(\theta_{12})]^{1/2} F_{\mathrm{br}}(A) \tag{34.44}$$

The sine-Gordon model

The sine-Gordon model is one of the simplest and best studied exactly solvable noncritical models. The spectrum of the sine-Gordon model was obtained semiclassically (Dashen et al., 1975) and from the exact solution (Takhtadjan and Faddeev, 1975). The S matrix was calculated by Zamolodchikov (1977). A regularized integrable version of the sine-Gordon model was proposed and studied in detail by Japaridze et al. (1984). The formactors which define the correlation functions were derived by Smirnov in the series of papers published between 1985 and 1992 (Smirnov, 1984, 1985, 1990; Reshetikhin and Smirnov, 1990). A review of these results is given by Babujian et al. (1999, 2002). I shall use this well understood model to illustrate the problems one encounters outside the critical point.

To fix the notation I write down the explicit expression for the sine-Gordon model Hamiltonian density:

$$\mathcal{H} = \frac{1}{2}[(\hat{\Pi})^2 + (\partial_x \Phi)^2] - \frac{m_0^2}{\beta^2} \cos(\beta \Phi) \tag{34.45}$$

where $[\Pi(x), \Phi(y)] = -i\delta(x - y)$. The notation corresponds to the Minkovsky metric. The spectrum and the properties of the sine-Gordon model change rather dramatically throughout the interval $8\pi > \beta^2 > 0$ where the cosine term is relevant. The most essential difference appears between the intervals $8\pi > \beta^2 > 4\pi$ and $4\pi > \beta^2$.

Before going into detail, however, we shall first consider the simplest point, namely the point $\beta^2 = 4\pi$ where the sine-Gordon model is equivalent to free massive Dirac fermions:

$$e^{i\sqrt{4\pi}\phi} \sim R^+ L \qquad e^{-i\sqrt{4\pi}\phi} \sim L^+ R$$

Thus the correlation functions of the bosonic exponents with $\beta = \sqrt{4\pi}$ can be easily calculated. One can go further however and calculate two-point correlation functions of all
bosonic exponents. This problem was solved by Bernard and LeClair (1994) who applied
to this problem the methods developed earlier for the Ising model (see Wu *et al.*, 1976).
The corresponding expressions for the exponents with $\beta < \sqrt{4\pi}$ are

$$\langle \exp[i\sqrt{4\pi}\alpha\phi(\tau, x)] \exp[-i\sqrt{4\pi}\alpha'\phi(\tau, x)]\rangle \equiv \exp[\Sigma(\tau, x)] \tag{34.46}$$

$$\nabla^2\Sigma = -m^2 \sinh^2 \rho \qquad 2\nabla^2\rho = m^2 \sinh 2\rho + 2(\alpha + \alpha')^2 \sinh \rho / r^2 \cosh^3 \rho \tag{34.47}$$

Equations (34.47) allow us to add to Σ any analytic or antianalytic function. The solution
depends on parameters α, α' via the boundary conditions which can be extracted from the
long- or short-distance behaviour of the correlation function. In particular, for $\alpha = \alpha'$ we
must have

$$\Sigma(r \to 0) = -2\alpha^2[\ln(mz) + \ln(m\bar{z})] + \mathcal{O}(1)$$

Alternatively, the correlation function can be expressed as an infinite series:

$$G_{\alpha,\alpha'}(\tau, x) = m^{2\Delta + 2\Delta_2} C_\alpha C_{\alpha'} \sum_{n=0}^{\infty} \frac{m^{2n\Delta_1 + 2n\Delta_2} C_\alpha^n C_{\alpha'}^n \sin^n(\pi\alpha) \sin^n(\pi\alpha')}{\pi^{2n}(n!)^2} \int d\theta_1 \cdots \theta_{2n}$$

$$\times \exp\left[-m\tau \sum_j \cosh\theta_j - imx \sum_j \sinh\theta_j\right] F_\alpha(\theta_1, \ldots \theta_{2n}) F_{\alpha'}^*(\theta_1, \ldots \theta_{2n}) \tag{34.48}$$

$$F_\alpha(\theta_1, \ldots \theta_{2n}) = \left\{\prod_{j=1}^{n} \exp[(1/2 - \alpha)\theta_j + (1/2 + \alpha)\theta_{j+n}]\right\}$$

$$\times \frac{\prod_{i<j\leq n}(e^{\theta_i} - e^{\theta_j}) \prod_{n+1\leq i<j}(e^{\theta_i} - e^{\theta_j})}{\prod_{r=1}^{n} \prod_{s=n+1}^{2n}(e^{\theta_r} + e^{\theta_s})} \tag{34.49}$$

where the coefficients C_α are determined from the one-point function

$$\langle \exp[i\alpha\sqrt{4\pi}\phi(\tau, x)]\rangle = C_\alpha m^{2\Delta} \tag{34.50}$$

Now we return to the problem of calculating the correlation functions with general β.
In order to get an idea about the excitations, one may solve the corresponding classical
equation, which is the Lagrange equation for the sine-Gordon action:

$$-(\partial_t^2 - \partial_x^2)\Phi + \beta^{-1}m^2 \sin\beta\Phi = 0 \tag{34.51}$$

Since β and m can be removed by rescaling of Φ and the coordinates respectively, the
classical equation is insensitive to these parameters. In the following discussion of the
classical solution we shall put $\beta = 1, m = 1$.
   One of the simplest solutions of (34.51) is the so-called 'kink' ('antikink')

$$\Phi_{k,a}(x, t) = 4\tan^{-1}\left[\exp\left(\pm\frac{x - x_0 - vt}{\sqrt{1 - v^2}}\right)\right] \tag{34.52}$$

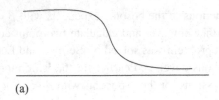

(a)

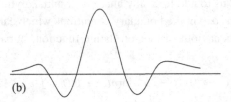

(b)

Figure 34.2. Classical solutions of the sine-Gordon equations corresponding to (a) a kink, (b) a breather.

(see Fig. 34.2). This solution has a nonzero 'topological charge'

$$Q = \frac{1}{2\pi} \int_{-\infty}^{\infty} \partial_x \Phi dx \qquad (34.53)$$

($Q = 1$ for a kink and -1 for an antikink).

In fact, kinks and antikinks look like moving domain walls. For this reason they are also called solitary waves, or 'solitons'. One can observe such waves on shallow water (and thus they were first observed and described) where they sometimes reach colossal size; in Japan such giant solitary waves are called 'tsunami'.

Another elementary solution of the sine-Gordon equation is called a 'breather'. Its analytical form is given by

$$\Phi = 4 \tan^{-1} \left\{ \frac{\Im m\lambda}{|\Re e\lambda|} \frac{\sin\left[\frac{\Re e\lambda}{|\lambda|}\left(\frac{t - t_0 - x/v}{\sqrt{1-v^2}}\right)\right]}{\cosh\left[\frac{\Im m\lambda}{|\lambda|}\left(\frac{x - x_0 - vt}{\sqrt{1-v^2}}\right)\right]} \right\} \qquad (34.54)$$

where

$$v = \frac{4|\lambda|^2 - 1}{4|\lambda|^2 + 1}$$

and x_0, t_0 are parameters. Breathers appear as soliton–antisoliton bound states and look more similar to ordinary waves, especially if $\Im m\lambda \ll |\lambda|$.

As soon as we return to the quantum world, the actual magnitude of β becomes important. As we have mentioned, the spectrum is different at $8\pi > \beta^2 > 4\pi$ and at $4\pi > \beta^2$. In the first sector breathers do not form and the spectrum consists solely of kinks and antikinks. In the latter sector breathers emerge and the number of types depends on the magnitude of the coupling constant β. In general the coupling constant enters into S matrices and the spectrum in a combination

$$\gamma = \frac{\pi \Delta}{1 - \Delta} = \frac{\pi \beta^2}{8\pi - \beta^2} \tag{34.55}$$

We shall consider first the special points in the second sector, namely

$$\gamma = \pi/\nu \qquad \nu = 1, 2, \ldots$$

at which both the spectrum and the S matrices are especially simple.

The simplicity stems from the fact that all scattering matrices are diagonal, that is upon scattering colliding particles acquire only phase shifts, but do not change their isotopic numbers:

$$S_{a,b}^{\bar{a},\bar{b}}(\theta) = \delta_a^{\bar{a}} \delta_b^{\bar{b}} S_{a,b}(\theta) \tag{34.56}$$

so that Yang–Baxter equations (34.3) are trivially satisfied. The spectrum consists of kinks and antikinks with the mass M_s and their bound states. There are $\nu - 1$ breathers with different masses:

$$M_n = 2M_s \sin(\pi n/2\nu) \qquad n = 1, 2, \ldots, (\nu - 1) \tag{34.57}$$

The particle masses are functions of β and g (Zamolodchikov, 1995):

$$M_s = \frac{2\Gamma(\gamma/2)}{\sqrt{\pi}\Gamma(1/2 + \gamma/2)} \left[\frac{m_0^2 \Gamma(1 - \Delta)}{16\Gamma(1 + \Delta)} \right]^{1/(2-2\Delta)} \tag{34.58}$$

where γ, Δ are related to β^2 through (34.55). This formula constitutes a considerable improvement of the renormalization group estimate $M_s \sim m_0^{1/(1-\Delta)}$. The operators have the following transformation properties under parity conjugation:

$$C\Phi C^{-1} = -\Phi$$
$$C Z_s(\theta) C^{-1} = Z_{\bar{s}}(\theta) \tag{34.59}$$
$$C Z_n(\theta) C^{-1} = (-1)^n Z_n(\theta) \qquad n = 1, \ldots, \nu - 1$$

In this case the soliton–soliton, soliton–antisoliton and antisoliton–antisoliton S matrices are all equal to each other:

$$S_{ss}(\theta) = S_{s\bar{s}}(\theta) = S_{\bar{s}\bar{s}}(\theta) \equiv S_0(\theta)$$

$$S_0(\theta) = -\exp\left\{ -i \int_0^\infty dx \frac{\sin(x\theta)\sinh[(\pi - \gamma)x/2]}{x\cosh(\pi x/2)\sinh(\gamma x/2)} \right\}$$

$$= \prod_{m=0}^\infty \frac{\Gamma[\nu(2m - i\theta/\pi)]\Gamma[\nu(2m + 1 + i\theta/\pi) - 1]\Gamma[\nu(2m + 2 - i\theta/\pi) - 1]}{\Gamma[\nu(2m + i\theta/\pi)]\Gamma[\nu(2m + 1 - i\theta/\pi) - 1]\Gamma[\nu(2m + 2 + i\theta/\pi) - 1]}$$

$$\times \frac{\Gamma[\nu(2m + 1 + i\theta/\pi)]}{\Gamma[\nu(2m + 1 - i\theta/\pi)]} \tag{34.60}$$

This S matrix has a pole on the physical sheet $\theta = i(\pi - \gamma)$ which corresponds to the first bound state, the first breather with the mass $2M \sin(\gamma/2)$. The S matrix of first breathers is

$$S_{11}(\theta) = \frac{\sinh\theta + i\sin\gamma}{\sinh\theta - i\sin\gamma} \tag{34.61}$$

This S matrix also has a pole at $\theta = i\gamma$ which means that two first breathers may create a bound state, a second breather etc.

Now recall that these S matrices determine commutation relations of the creation and annihilation operators. We see that all S matrices have the same asymptotic behaviour: $S(|\theta| \to \infty) = 1$ and $S(0) = -1$. This means that the excitations with rapidities far apart in rapidity space view each other as Bose particles, while those which are close become fermions! This explains why states with the same rapidity cannot be occupied twice: in a remarkable way the Pauli principle is generated by the interactions.

For the sine-Gordon model the poles are formed at $\theta_{12} = i(\pi - m\gamma)$.

Let us find the simplest formfactors of the operators $A = \cos\beta\Phi$ and $\sin\beta\Phi$. The first operator is a Lorentz scalar and the second one is a pseudoscalar, i.e. changes sign under charge conjugation $\Phi \to -\Phi$. Thus their formfactors are functions of θ_{ij}. At the critical point both of them have conformal dimensions (Δ, Δ). The fact that the operators behave differently under parity conjugation introduces a further restriction for their formfactors. According to (34.59), the expansion of $\cos\beta\Phi$ may contain only even breathers and the expansion of $\sin\beta\Phi$ only odd ones. Since breathers are bound states of solitons and anti-solitons, these requirements also give us information about analytic properties of the soliton formfactors.

The two-soliton sector is most representative. For two particles there are two invariant subspaces: C-even and C-odd. The S matrices are the same in both cases, but the formfactors have different singularities in the θ plane. From (34.59) we conclude that the odd state wave function changes its sign under parity conjugation and the even one does not. Since the wave function of the first breather has negative parity, it can emerge only as a bound state in the odd channel. Therefore the odd formfactor $F_o(\theta)$ has a pole at $\theta = i(\pi - \gamma)$ and the even formfactor $F_e(\theta)$ does not.

In the invariant subspaces the matrix Riemann problem (34.43) becomes a scalar one:

$$F_{e,o}(A; \theta - 2i\pi) = F_{e,o}(A; \theta)S_{e,o}(\theta) \tag{34.62}$$

A general solution of this problem is given by

$$F_a(\theta) = \mathcal{R}_{A,a}(e^\theta)\exp\left\{\int_{-\infty}^{\infty}\frac{dx}{x}\frac{1 - e^{ix\theta}}{e^{2\pi x} - 1}K_a(x)\right\}$$

$$K_a(x) = \frac{1}{2\pi i}\int d\theta e^{i\theta x}\frac{d}{d\theta}\ln S_a(\theta) \tag{34.63}$$

where $a = $ o,e and $\mathcal{R}_{A,a}(y)$ is a rational function which is determined by the requirements specific for a given operator. For instance, as we have mentioned above, the singlet formfactor of $\sin \beta \Phi$ must have a pole at $\theta = i(\pi - \gamma)$. Therefore it is convenient to extract from $F_a(\theta)$ a function which does not have poles on the physical strip:

$$F_0(\theta) = -i \sinh(\theta/2) \exp \left\{ \int_0^\infty dx \frac{\sin^2[x(\theta + i\pi)/2] \sinh[(\pi - \gamma)x/2]}{x \sinh \pi x \cosh(\pi x/2) \sinh(\gamma x/2)} \right\} \quad (34.64)$$

At large θ we have $F_0(\theta \gg 1) \sim \exp(\theta)$.

Then the soliton–antisoliton formfactor of the $\cos \beta \Phi$ operator is

$$F_{s\bar{s}}^{\cos}(\theta) = Z^{1/2} \frac{i \cosh(\theta/2)}{\sinh[(\pi/2\gamma)(\theta - i\pi)]} F_0(\theta) \quad (34.65)$$

where the normalization factor $Z \sim M$. This solution is 'minimal', that is contains the necessary pole and has the mildest asymptotic behaviour at infinity (Karowski *et al.*, 1977, 1978; Delfino and Mussardo, 1996).

For $\sin \beta \Phi$ we have

$$F_{s\bar{s}}^{\sin}(\theta) = Z^{1/2} \frac{\cosh(\theta/2)}{\cosh[(\pi/2\gamma)(\theta - i\pi)]} F_0(\theta) \quad (34.66)$$

Using these formfactors and the rule (34.44) we can obtain formfactors for breathers. For the first breather we have

$$F_{b1}^{\sin} = \{\text{Res} S[\theta = i(\pi - \gamma)]\}^{-1/2} \text{Res} F_{s\bar{s}}^{\sin}[\theta = i(\pi - \gamma)] \equiv Z^{1/2} Z_1^{1/2}$$

$$Z_1^{1/2} = \frac{\gamma \sin \gamma}{\pi} \exp \left(-\int_0^\infty \frac{dx \sinh \gamma x/2 \sinh(\pi - \gamma)x/2}{x \sinh \pi x \cosh \pi x/2} \right) \quad (34.67)$$

The formfactor (34.65) has a pole at the position of the second breather and so we have:

$$F_{b2}^{\cos} = \{\text{Res} S[\theta = i(\pi - 2\gamma)]\}^{-1/2} \text{Res} F_{s\bar{s}}^{\cos}[\theta = i(\pi - 2\gamma)] \equiv Z^{1/2} Z_2^{1/2}$$

$$Z_2^{1/2} = \frac{\gamma \sin 2\gamma}{\pi} \exp \left(-2 \int_0^\infty \frac{dx \sinh^2 \gamma x \sinh(\pi - \gamma)x/2}{x \sinh \pi x \cosh \pi x/2 \sinh \gamma x/2} \right) \quad (34.68)$$

We are now in a position to describe single breather and kink–antikink contributions to the correlation functions of $\sin \beta \Phi$ and $\cos \beta \Phi$. For the sake of compactness we write down only the imaginary part of the corresponding retarded correlation functions:

$$\Im m D^{\sin}(\omega, q) = Z \left\{ 2\pi \sum_{j=0} Z_{2j+1} \delta \left(s^2 - M_{2j+1}^2 \right) \right.$$

$$\left. + \Re e \frac{1}{s\sqrt{s^2 - 4M^2}} |F^{\sin}[\theta(s)]|^2 + \cdots \right\} \quad (34.69)$$

$$s^2 = \omega^2 - q^2 \qquad \theta(s) = 2\ln(s/2M + \sqrt{s^2/4M^2 - 1})$$

and similarly

$$\Im m D^{\cos}(\omega, q) = Z \left\{ 2\pi \sum_{j=1} Z_{2j} \delta \left(s^2 - M_{2j}^2 \right) + \Re e \frac{1}{s\sqrt{s^2 - 4M^2}} |F^{\cos}[\theta(s)]|^2 + \cdots \right\}$$

$$(34.70)$$

The dots here stand for multi-breather contributions which we have omitted.

In the sector $\gamma > \pi/2$ there are only kinks and antikinks, and the correlation functions have two-particle thresholds. We shall not discuss this sector in detail, referring the reader to Smirnov's book.

It is interesting to discuss how the shape of the spectral functions depends on β. According to (34.69), (34.70), at low enough energies these spectral functions consist of series of delta-functions originating from emissions of breathers, and two-particle thresholds. Let us imagine that the lowest threshold corresponds to emission of a kink–antikink pair (this imposes a certain constraint on β which we leave to the reader to derive). Let us discuss the shape of the spectral functions at the threshold where at $s - 2M \ll M$ we have

$$\theta(s) \approx \sqrt{8(s/2M - 1)} \qquad\qquad (34.71)$$

At small θ we always have $F_0(\theta) \sim \theta$, but the trigonometric prefactors in (34.65), (34.66) behave differently for different values of γ. Namely, the sine formfactor is finite at $\theta = 0$ when $\gamma = \pi/(1 + 2k)$ and the cosine formfactor is finite when $\gamma = \pi/2k$. Substituting (34.71) into (34.65), (34.66) and then into (34.69), (34.70) we obtain the following expressions for the spectral functions at $s - 2M \ll M$:

$$\Im m D^{\cos}(\omega, q) \sim \frac{\sqrt{s^2 - 4M^2}}{s^2/4M^2 - 1 + [(\gamma/\pi)\cos(\pi/2\gamma)]^2} \qquad\qquad (34.72)$$

$$\Im m D^{\sin}(\omega, q) \sim \frac{\sqrt{s^2 - 4M^2}}{s^2/4M^2 - 1 + [(\gamma/\pi)\sin(\pi/2\gamma)]^2} \qquad\qquad (34.73)$$

These expressions are applicable when $(\gamma/\pi)\cos(\pi/2\gamma)$ or $(\gamma/\pi)\sin(\pi/2\gamma)$ are small numbers.

Let us follow closely the change of the two-particle emission contribution upon changing the coupling constant γ. Without loss of generality we can choose the spectral function of sines. Let us start from $\gamma = \pi/(1 + 2k_1)$ where k_1 is some integer number. The spectral function has an $x^{-1/2}$ singularity at the threshold (see Fig. 34.3(a)). At this point the spectrum includes $2k_1$ breathers and the last one, being even by number, does not appear in the sin–sin correlation function. Instead its presence in the spectrum is signified by the square root singularity.

According to (34.73) this singularity is smeared out when γ decreases. However, as soon as $\gamma < \pi/(1 + 2k_1)$, a new breather emerges with mass $M_{(2k_1+1)} = 2M \sin[(2k_1 + 1)\gamma/2]$ and since its number is odd, it contributes to the sin–sin spectral function. Therefore the disappearance of the singularity is compensated by the appearance of a new peak near the two-particle threshold (see Fig. 34.3(b)). When γ moves further towards $\pi/(2k_1 + 2)$ the splitting between this peak and the two-particle threshold grows and the spectral function at the threshould becomes featureless (Fig. 34.3(c)).

I conclude this section by citing a remarkable result obtained by Lukyanov and Zamolodchikov (1997) who calculated the exact expectation values of bosonic exponents in the

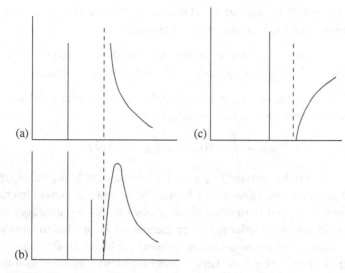

Figure 34.3. The shape of the spectral function of sines at different values of γ.

sine-Gordon model:

$$\langle e^{i\alpha\sqrt{8\pi}\,\Phi}\rangle = \left[\frac{m_b a\Gamma(1/2+\gamma/2)\Gamma(1-\gamma/2)}{4\sqrt{\pi}}\right]^{2\alpha^2}$$

$$\times \exp\left\{\int_0^\infty \frac{dx}{x}\left[\frac{\sinh^2(2\alpha\tilde{\beta}x)}{2\sinh(\tilde{\beta}^2 x)\sinh x\cosh((1-\tilde{\beta}^2)x)} - 2\alpha^2 e^{-2x}\right]\right\} \quad (34.74)$$

valid for $2\beta|\Re e\alpha| < 1$. Here $8\pi\tilde{\beta}^2 = \beta^2$. It is assumed that the correlation function of the two exponents at small distances behaves as

$$\langle\langle e^{i\alpha\sqrt{8\pi}\,\Phi(x)}e^{-i\alpha\sqrt{8\pi}\,\Phi(y)}\rangle\rangle = \left(\frac{a}{|x-y|}\right)^{4\alpha^2}$$

Perturbations of spin $S = 1/2$ Heisenberg chain: confinement

The problem I am going to discuss in this section is whether a system without gapless excitations may remain sensitive to small external perturbations. The naive answer is no; indeed, how can it be if all correlation functions decay exponentially? This naive answer is incorrect; the examples I am going to give demonstrate that sometimes even small perturbations can cause a complete restructuring of the excitation spectrum. I have already touched on this matter in the final section of Chapter 28; here I will elaborate further.

An excellent example fully suited for my purpose is the spin $S = 1/2$ Heisenberg model with relevant perturbations. An isotropic spin $S = 1/2$ Heisenberg antiferromagnet in $2(1+1)$ dimensions provides a textbook example of a quantum critical system and we have discussed it at length in this book. As for any critical point in two dimensions, conformal field theory supplies us with a complete list of relevant perturbations. In the given case when the continuous limit of the model is described by the $SU_1(2)$ WZNW model (see Chapter 31, penultimate section), the relevant fields include

components of the tensor fields g, g^+ with conformal dimensions $(1/4, 1/4)$,
bilinear products of the current operators $J^a \bar{J}^b$ with conformal dimensions $(1, 1)$.[4]

If we require that the perturbation leaves the global $SU(2)$ symmetry unbroken, it leaves us with the following general form for the perturbation:

$$V_{su(2)} = \int d^2x [\lambda_1 \text{Tr}(g + g^+) + \lambda_2 \mathbf{J}\bar{\mathbf{J}}] \qquad (34.75)$$

According to (31.50) the operator $\text{Tr}(g + g^+)$ is a continuous limit, the staggered energy density $(-1)^j \mathbf{S}_j \bar{\mathbf{S}}_{j+1}$. As we know from Chapter 29, the current-current interaction is already present in the original Heisenberg chain, though with a negative coupling constant which renders it irrelevant. In order to change the sign of λ_2 one has to introduce an antiferromagnetic exchange with next-nearest neighbours (Haldane, 1982; Affleck *et al.*, 1989; White and Affleck, 1996). Hence the lattice model which with continuous limit yields the $SU_1(2)$ WZNW model perturbed by (34.75) is the spin $S = 1/2$ chain with a dimerization and a next-nearest-neighbour exchange:

$$\hat{H} = \sum_j \{[1 + \delta(-1)^j]\mathbf{S}_j\mathbf{S}_{j+1} + \gamma\mathbf{S}_j\mathbf{S}_{j+2}\} \qquad (34.76)$$

where δ, γ are related to the coupling constants in (34.75). According to the numerical calculations (White and Affleck, 1996) λ_2 becomes positive when γ exceeds the value ≈ 0.2411. Model (34.76) has applications to real magnets. In real magnets the explicit dimerization δ may either originate from the chemical composition of the material or appear as a result of the spin-Peierls phase transition due to the interaction with three-dimensional phonons.[5]

At $\lambda_1 \neq 0$, $\lambda_2 = 0$ and $\lambda_2 \neq 0$, $\lambda_1 = 0$ the continuous model is integrable. Let us compare these two limiting cases; they show very different excitation spectra. The first case corresponds to the magnet with explicit dimerization. The action in the continuous limit is

$$S = W[SU_1(2)] + \lambda \int d^2x \text{Tr}(g + g^+) = \int d^2x \left[\frac{1}{2}(\partial_\mu \Phi)^2 + \tilde{\lambda} \cos(\sqrt{2\pi}\Phi) \right] \qquad (34.77)$$

where $\tilde{\lambda} \sim \lambda$ and the equivalence to the sine-Gordon model follows from (31.53). As we know from the previous section, the sine-Gordon model at $\beta^2 = 2\pi$ has in its spectrum the soliton, antisoliton and their bound state (the first breather) with equal masses $M_{\text{tr}} \sim \delta^{2/3}$ and the second breather with the mass $M_s = \sqrt{3}M_{\text{tr}}$. The particles with equal masses constitute a triplet corresponding to a massive excitation with spin 1. The physically interesting correlation functions of staggered magnetization and energy density are described in detail in Chapter 22 of the book by Gogolin *et al.* cited in the select bibliography.

---

[4] This perturbation is marginal; its relevance depends on the sign of the coupling constant.
[5] I would like to emphasize that one-dimensional phonons cannot produce dimerization; this issue was discussed by Delfino *et al.* (1997).

(a)

(b)

(c)

Figure 34.4.

At $\lambda_1 = 0$ the continuous model coincides with model (30.19) which is also equivalent to the sine-Gordon model, though at $\beta^2 = 8\pi - 0$. We have already discussed this model in Chapter 30, penultimate section. The excitations are massive solitons and antisolitons constituting the SU(2) doublet (in other words, they carry spin $1/2$). Their scattering matrix is given by (34.8), and the thermodynamics is described by TBA equations (34.36). The remarkable fact is that dimerization here is not triggered by an external potential, as in model (34.77), but appears spontaneously at $T = 0$. At zero temperature the system chooses one of the minima of the potential $-\cos(\sqrt{8\pi}\,\Phi)$ and 'gets stuck' in it. The dimerization operator $\epsilon \sim \cos(\sqrt{2\pi}\,\Phi)$ changes sign under the shift $\Phi \to \Phi + \sqrt{\pi/2}$ and therefore distinguishes between the two sets of vacua. Its sign depends on whether the system chooses the minimum with $\sqrt{8\pi}\,\Phi = 4n\pi$ or $\sqrt{8\pi}\,\Phi = 2\pi + 4n\pi$.

When both coupling constants are nonzero the model is no longer integrable. It coincides with the double sine-Gordon model described in Chapter 28, final section. The role of a weak explicit dimerization is that it confines spinons. Any non-zero λ_1 leads to *confinement* of spinons: excitations with quantum numbers of spinons disappear from the spectrum.

The following simple arguments describe the qualitative side of the problem. At $\gamma = 1$, $\delta = 0$ the lattice model has a particularly simple ground state. In the thermodynamic limit it is twice degenerate: one ground state consists of products of singlets on sites $(2n, 2n + 1)$, the other ground state consists of singlets on sites $(2n - 1, 2n)$ (see Fig. 34.4(a,b)). Excitations are domain walls between these ground states and, as follows from Fig. 34.4(c), a flip of a single spin ($\delta S^z = 1$) produces two domain walls. Hence one wall carries spin $1/2$. Any finite dimerization lifts the ground state degeneracy and confines spinons. Now excitations correspond to breaking of the weakest singlets. These particles are obviously triplets.

Let us now replace the dimerization with a staggered magnetic field $\mathbf{h}(-1)^j \mathbf{S}_j$. In the continuous limit this corresponds to the following perturbation:

$$V = i\mathbf{h}\text{Tr}[\sigma(g - g^+)] \tag{34.78}$$

This perturbation breaks the SU(2) symmetry down to U(1). Choosing the $\hat{z}$-axis along the direction of **h** and using (31.53) we obtain the following double sine-Gordon model:

$$S = \int d^2x \left[\frac{1}{2}(\partial_\mu \Phi)^2 + h \sin(\sqrt{2\pi K}\,\Phi) - \gamma \cos(\sqrt{8\pi K}\,\Phi) \right] \tag{34.79}$$

where $K = (1 + \gamma/2\pi v)^{-1}$.

References

Affleck, I., Gepner, D., Schulz, H. J. and Ziman, T. (1989). *J. Phys. A*, **22**, 511.

Ahn, C., Delfino, G. and Mussardo, G. (1993). *Phys. Lett. B*, **317**, 593.

Babujian, H. and Karowski, M. (2002). *Nucl. Phys. B*, **620**, 407.

Babujian, H., Fring, A., Karowski, M. and Zapletal, A. (1999). *Nucl. Phys. B*, **538**, 535.

Bernard, D. and LeClair, A. (1994). *Nucl. Phys. B*, **426**, 534.

Dashen, R. F., Hasslacher, B. and Neveu, A. (1975). *Phys. Rev. D*, **11**, 3424.

Delfino, G. and Mussardo, G. (1996). *Nucl. Phys. B*, **455**, 724.

Delfino, G., Essler, F. H. L. and Tsvelik, A. M. (1997). *Phys. Rev. B*, **56**, 11001.

Fateev, V. A. (1990). *Int. J. Mod. Phys. A*, **5**, 1025.

Haldane, F. D. M. (1982). *Phys. Rev. B*, **25**, 4925.

Japaridze, G. E., Nersesyan, A. A. and Wiegmann, P. B. (1984). *Nucl. Phys. B*, **230**, 10.

Karowski, M., Thun, H.-J., Truong, T. T. and Weisz, P. H. (1977). *Phys. Lett. B*, **67**, 321; Karowski, M. and Weisz, P., *Nucl. Phys. B*, **139**, 455 (1978).

Lukyanov, S. (1995). *Commun. Math. Phys.*, **167**, 183.

Lukyanov, S. and Zamolodchikov, A. (1997). *Nucl. Phys. B*, **493**, 571.

Reshetikhin, N. Yu. and Smirnov, F. A. (1990). *Commun. Math. Phys.*, **131**, 157.

Smirnov, F. A. (1984). *J. Phys. A*, **17**, L873; **19**, L575 (1985); *Nucl. Phys. B*, **337**, 156 (1990).

Takhtadjan, L. A. and Faddeev, L. D. (1975). *Sov. Teor. Math. Phys.*, **25**, 147.

Tsvelik, A. M. and Wiegmann, P. B. (1983). *Adv. Phys.*, **32**, 453.

White, S. R. and Affleck, I. (1996). *Phys. Rev. B*, **54**, 9862.

Wu, T. T., McCoy, B., Tracy, C. A. and Barouch, E. (1976). *Phys. Rev. B*, **13**, 316.

Yang, C. N. and Yang, C. P. (1965). *Phys. Rev. B*.

Zamolodchikov, A. B. (1977). *Pisma Zh. Eksp. Teor. Fiz.*, **25**, 499.

Zamolodchikov, Al. B. (1995). *Int. J. Mod. Phys. A*, **10**, 1125.

Zamolodchikov, A. B. and Zamolodchikov, Al. B. (1979). *Ann. Phys. (NY)*, **120**, 253.

35

A comparative study of dynamical mass generation
in one and three dimensions

As I have pointed out in the previous chapter, under certain circumstances a gap is generated in the spectrum of an interacting one-dimensional electron liquid. For instance, this happens when the interaction matrix element at $2k_F$ is negative. In dimensions higher than one, the gap opening would be accompanied by a symmetry breaking and a phase transition. In one dimension a continuous symmetry cannot be broken even at $T = 0$; however the gap still opens. It is very instructive to compare the corresponding scenario in three and one dimensions.

The phase transition I am going to discuss is called the Peierls transition. It is associated with formation of a charge density wave with a finite wave vector $\mathbf{Q}$ incommensurate with vectors of the Bravé lattice. At weak coupling the instability occurs only when the Fermi surface possesses a special property called 'nesting'. This means that there is a portion of the Fermi surface in the vicinity of which the single-particle dispersion satisfies (see Fig. 35.1)

$$\epsilon(\mathbf{k}) = -\epsilon(\mathbf{Q} + \mathbf{k}) \tag{35.1}$$

To simplify the discussion, I will consider an open Fermi surface consisting of two sheets which can be superimposed on each other by a shift on vector $\mathbf{Q}$. Such an open Fermi surface is generated in a system of weakly coupled chains. In the one-dimensional limit when the two sheets become flat, $\mathbf{Q}$ identifies with $2k_F$.

Electron creation and annihilation operators at the right and left sheets will be denoted as R^+, R and L^+, L respectively. The noninteracting part of the Hamiltonian can be written in a form similar to the one used in the analysis of one-dimensional models:

$$H_0 = \sum_{\mathbf{k}} \epsilon(\mathbf{k})[R_\sigma^+(\mathbf{k})R_\sigma(\mathbf{k}) - L_\sigma^+(\mathbf{k})L_\sigma(\mathbf{k})] \tag{35.2}$$

The difference is that $\epsilon(\mathbf{k})$ depends now on all momenta projections. When the Fermi surface is nested, the bare charge susceptibility at wave vector $\mathbf{Q}$ (see Fig. 35.2) logarithmically diverges at low temperature and frequency:

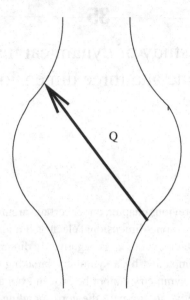

Figure 35.1. An example of a nested Fermi surface.

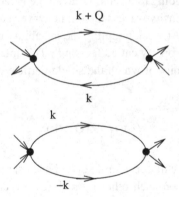

Figure 35.2. The logarithmically divergent diagrams in the Peierls and the Cooper channels.

$$\chi_P(\Omega, \mathbf{Q} + \mathbf{q}) = 2T \sum_n \int \frac{d^3 k}{(2\pi)^3} \frac{1}{[i(\Omega + \omega_n) + \epsilon(\mathbf{k} + \mathbf{q})][i\omega_n - \epsilon(\mathbf{k})]}$$

$$\approx -2\rho(\epsilon_F) \int \frac{d\epsilon \, d\hat{\mathbf{n}}}{4\pi} \frac{1 - n(\epsilon) - n(\epsilon + v\mathbf{nq})}{i\Omega + v\mathbf{nq} + 2\epsilon} \sim -2\rho(\epsilon_F) \ln[\Lambda/\max(\Omega, v|q|, T)]$$

$$(35.3)$$

where the factor 2 comes from the summation over the spin indices and $d\hat{\mathbf{n}}$ is the infinitesimal solid angle. This divergence signals a possible instability at this wave vector (the so-called Peierls instability). There is another logarithmically divergent diagram (the so-called Cooper

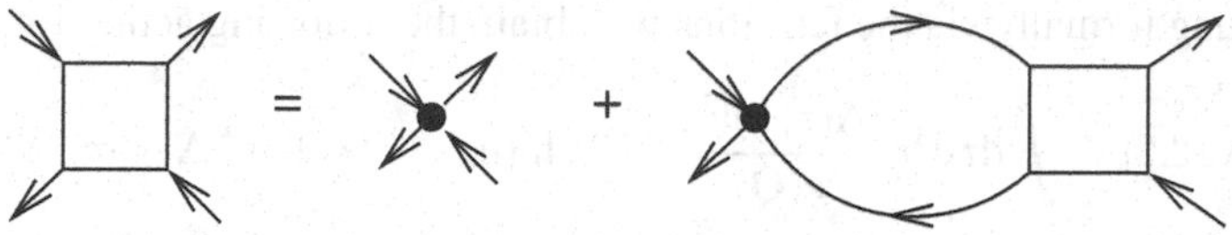

Figure 35.3. The diagram series contributing to the divergence of the charge susceptibility.

channel diagram) which may lead to the superconducting instability:

$$\chi_{sc}(\Omega, \mathbf{q}) = T \sum_n \int \frac{d^3k}{(2\pi)^3} \frac{1}{[i(\Omega - \omega_n) - \epsilon(-\mathbf{k} + \mathbf{q})][i\omega_n - \epsilon(\mathbf{k})]}$$

$$\approx -2\rho(\epsilon_F) \int \frac{d\epsilon \, d\hat{n}}{4\pi} \frac{1 - n(\epsilon) - n(\epsilon - v\mathbf{n}\mathbf{q})}{i\Omega - v\mathbf{n}\mathbf{q} + 2\epsilon} \sim -2\rho(\epsilon_F) \ln[\Lambda/\max(\Omega, v|\mathbf{q}|, T)]$$

$$(35.4)$$

In general these instabilities may compete, which somewhat complicates the analysis (not in one dimension, recall the discussion in the previous chapter). To avoid this complication I shall assume that the bare coupling constant in the Cooper channel is repulsive; hence the superconducting instability is suppressed. The part of the electron–electron interaction responsible for the Peierls instability is

$$V = -\frac{1}{2}g(\mathbf{Q}) \sum_{\mathbf{p},\mathbf{k},\mathbf{q}} R_\sigma^+(\mathbf{k} + \mathbf{q})L_\sigma(\mathbf{k})L_{\sigma'}^+(\mathbf{p} - \mathbf{q})R_{\sigma'}(\mathbf{p}) \qquad (35.5)$$

where the summation includes only momenta close to the Fermi surface and is supposed to be restricted by the cut-off Λ. The bare coupling constant is attractive: $g(\mathbf{Q}) > 0$. Summation of the most divergent diagrams yields

$$\langle\langle (R_\sigma L_\sigma^+)_{\Omega,\mathbf{q}} (L_{\sigma'}^+ R_{\sigma'})_{\Omega,\mathbf{q}} \rangle\rangle \equiv \chi_Q(\Omega, \mathbf{q}) = \left[\chi_P^{-1}(\Omega, \mathbf{q}) - g(\mathbf{Q})\right]^{-1} \qquad (35.6)$$

At $\Omega = 0, q = 0$ we have

$$\chi_P(0, 0) = 2\rho(\epsilon_F) \ln(\Lambda/T)$$

and the susceptibility diverges at the temperature

$$T_c = \Lambda \exp[-1/2\rho(\epsilon_F)g(\mathbf{Q})] \qquad (35.7)$$

At this temperature the three-dimensional metal undergoes a phase transition into a charge density wave state characterized by the complex order parameter

$$\Delta = \langle R_\sigma^+(x)L_\sigma(x) \rangle \qquad \Delta^* = \langle L_\sigma^+(x)R_\sigma(x) \rangle \qquad (35.8)$$

The low temperature state can be conveniently described using the path integral representation. Namely, one should decouple the interaction term (35.5) in the path integral by the Hubbard–Stratonovich transformation which gives rise to the following action density:

$$\mathcal{S} = \frac{|\Delta(\tau, \mathbf{r})|^2}{2g(\mathbf{Q})} + (R_\sigma^+, L_\sigma^+) \begin{pmatrix} \partial_\tau - \hat{\epsilon} & \Delta \\ \Delta^* & \partial_\tau + \hat{\epsilon} \end{pmatrix} \begin{pmatrix} R_\sigma \\ L_\sigma \end{pmatrix} \qquad (35.9)$$

After integrating formally via the fermions we obtain the following action for the field Δ:

$$S[\Delta, \Delta^*] = \int d\tau d^3r \frac{|\Delta(\tau, \mathbf{r})|^2}{2g(\mathbf{Q})} - 2\text{Tr}\ln[\partial_\tau - \sigma^z\hat{\epsilon} + \sigma^+\Delta + \sigma^-\Delta^*] \quad (35.10)$$

where the Pauli matrices act in the space of chiral indices (right and left). At $T < T_c$ this action has a saddle point corresponding to a spatially uniform Δ. Its value is given by the self-consistency condition

$$1 = 2\rho(\epsilon_F)g(\mathbf{Q}) \int_0^\Lambda \frac{\tanh[E(\epsilon)/2T]}{E(\epsilon)} \qquad E(\epsilon) = \sqrt{\epsilon^2 + \Delta^2} \quad (35.11)$$

where $E(\mathbf{k})$ represents the quasi-particle spectrum. The symmetry broken low temperature state has *anomalous* Green's functions, that is ones which vanish above T_c:

$$\langle\langle RL^+\rangle\rangle_{\omega,\mathbf{k}} = \frac{\Delta^*}{\omega^2 + \epsilon^2(\mathbf{k})} \qquad \langle\langle LR^+\rangle\rangle_{\omega,\mathbf{k}} = \frac{\Delta}{\omega^2 + \epsilon^2(\mathbf{k})} \quad (35.12)$$

To establish whether the saddle point solution is stable with respect to fluctuations, one has to study the effective action (35.10). An explicit form of this action (up to the factors) can be written using the symmetry considerations. Indeed, the low temperature state has a quasi-particle gap which introduces in the problem a natural scale $\xi \sim |\Delta|^{-1}$. Since we are looking for the effective action describing fluctuations of Δ, Δ^* with wavelengths greater than ξ, such an action can be represented as an expansion in powers of $\xi\partial_a\Delta$, where ∂_a represent space and time gradients. The zeroth term in this expansion represents action (35.10) calculated on uniform configurations of Δ. These considerations together with the reality of the effective action dictate the following form for its density:

$$\mathcal{S}_{\text{eff}} = F(|\Delta|^2) + \frac{\rho}{2|\Delta|^2}(\partial_\tau\Delta^*\partial_\tau\Delta + D_{\mu\nu}\partial_\mu\Delta^*\partial_\nu\Delta) + \cdots \quad (35.13)$$

where the dots stand for higher powers of gradients and ρ, $D_{\mu\nu}$ are constants. Writing down the order parameter as

$$\Delta = |\Delta|e^{i\phi}$$

I obtain the following form of the effective action:

$$S_{\text{eff}} = S[|\Delta|] + \frac{\rho}{2}[(\partial_\tau\phi)^2 + D_{\mu\nu}\partial_\mu\phi\partial_\nu\phi] + \cdots \quad (35.14)$$

The amplitude fluctuations always have a spectral gap and will not be considered.
  We see that

  in the long-wavelength limit the phase fluctuations are decoupled from fluctuations of
      the modulus,
  the phase fluctuations are gapless with the linear spectrum

$$\omega^2 = D_{\mu\nu}q^\mu q^\nu \quad (35.15)$$

which corresponds to the fact that they represent a Goldstone mode, the ϕ-dependent part of (35.14) is just a three-dimensional version of the Gaussian action describing the gapless excitations in the charge sector of the one-dimensional model of interacting fermions (see the discussion in the previous chapter).

A fundamental difference between $D \geq 3$ and $D < 3$ comes from the fact that in the one-dimensional case (and in two dimensions at finite temperatures) the phase fluctuations destroy the order parameter (35.8). Indeed, in D dimensions we have

$$
\langle \Delta \rangle \approx |\Delta| \langle e^{i\phi} \rangle = |\Delta| \exp\left[-\frac{1}{2} \langle\langle \phi(0,0)\phi(0,0) \rangle\rangle \right]
$$

$$
= \Delta| \exp\left[-\frac{T}{2\rho} \sum_n \int \frac{d^D q}{(2\pi)^D} \frac{1}{\omega_n^2 + D_{\mu\nu}q^\mu q^\nu} \right]
$$

$$
= \Delta| \exp\left\{ -\frac{1}{4\rho} \int \frac{d^D q}{(2\pi)^D} \frac{\coth[E(q)/2T]}{E(q)} \right\} \qquad E^2 = D_{\mu\nu}q^\mu q^\nu
$$

$$
(35.16)
$$

Therefore in one and two dimensions the phase transition is shifted to zero temperature. In two dimensions at $T = 0$ the order parameter is finite; in one dimension where the integral in (35.16) diverges even at $T = 0$ the anomalous Green's functions $\langle\langle RL^+ \rangle\rangle$, $\langle\langle LR^+ \rangle\rangle$ are never formed.

Nevertheless we see that some features of the three-dimensional charge density wave, such as the existence of a gapless collective (phase) mode and the quasi-particle gap, survive even in the one-dimensional case. The spectrum of the phase mode is robust for all D, but at $D = 1$ its formation is not accompanied by formation of the global order parameter and anomalous Green's functions.

Single-electron Green's function in a one-dimensional charge density wave state

The most dramatic differences between $D = 1$ and higher dimensionalities occur in the area of the single-electron Green's function. Indeed, we know that for $D > 1$ single-electron excitations are good quasi-particles (as Bogolyubov quasi-particles in a superconductor). Due to the spin-charge separation this cannot be the case in $D = 1$ where the single-electron Green's function has branch cuts instead of poles.

Since the spin sector acquires a spectral gap, the Fermi surface is absent even in $D = 1$. Since the described phenomenon may occur at arbitrary band filling, one may wonder whether the Luttinger theorem is fulfiled. According to this theorem the particle density n is related to the volume in momentum space where $G(\omega = 0, p) > 0$ (see, for example, the book by Abrikosov et al.):

$$
n = 2 \int_{G(\omega=0,p)>0} \frac{d^D k}{(2\pi)^D}
$$

$$
(35.17)
$$

The electron Green's function changes sign at the boundary of this volume. It can do this either going through a divergence (this would correspond to the standard situation of a

Fermi surface) or it may vanish at the boundary. Very often people forget about the second possibility and give a *wrong* formulation of the Luttinger theorem: 'The particle density is proportional to the volume inside the Fermi surface.' The true formulation (35.17), however, relates the density to the surface restricted by singularities of $\ln G(\omega = 0, p)$ which do not distinguish between infinities and zeros of $G(\omega = 0, p)$). Therefore in the given case of a one-dimensional charge density wave state when $G(\omega = 0, p)$ does not have infinities the Luttinger theorem may still be fulfiled if $G(0, p)$ has zeros on the noninteracting Fermi surface. As I am going to demonstrate, the existence of such zeros in a one-dimensional state with dynamically generated mass is a universal fact.

The proof becomes easier if we assume that the spin and the charge velocities are equal. Then at low energies the system is Lorentz invariant. The following three facts are important for the proof.

The anomalous Green's functions in one dimension are equal to zero, as we have already established as a general fact.

The electron creation and annihilation operators belong to the spinor representation of the Lorentz group. This suggests that in Euclidean space the single-particle Green's function has the following form:

$$\langle R_\sigma(\tau, x) R_\sigma^+(0, 0) \rangle = e^{i\phi} \mathcal{F}(\rho)$$
$$\langle L_\sigma(\tau, x) L_\sigma^+(0, 0) \rangle = e^{-i\phi} \mathcal{F}(\rho)$$

(35.18)

where $v\tau = \rho \cos\phi$, $x = \rho \sin\phi$ and $v = v_c = v_s$.

Since the excitations have a gap at least in one sector, the Green's function falls off exponentially at large distances:

$$\mathcal{F}(\rho) \sim \exp(-m\rho)$$

At small distances where interactions vanish, the Green's function behaves as the function for free fermions and we have $\mathcal{F}(\rho) \sim \rho^{-1}$. Therefore the integral

$$\int_0^\infty d\rho \rho \mathcal{R}(\rho)$$

(35.19)

is finite.

From these facts it follows that

$$G_{R,L}(0, 0) = \int_{-\pi}^{\pi} d\phi e^{\pm i\phi} \int_0^\infty d\rho \rho \mathcal{R}(\rho) = 0$$

(35.20)

For a metallic state integral (35.19) would diverge and the Green's function has singularity rather than zero.

I emphasize that the above derivation is quite general. It is interesting however to calculate the single-particle Green's function for some particular model with dynamical mass generation. I do this for the model of electrons with attractive interaction away from half-filling. The charge sector is gapless, the spin sector has a gap and is described by the Hamiltonian

(30.19). The bosonized expressions for the fermion creation and annihilation operators are given by (30.41). These expressions together with (30.42), (30.43) remain valid even if the gap is dynamically generated.

In order to calculate the Green's function, I use the formfactor approach and exploit the factorization (30.41):

$$G_{RR}^{(\sigma)}(\tau, x) = {}_c\langle 0|\mathcal{O}_c^\dagger(\tau, x)\, \mathcal{O}_c(0)|0\rangle_c\, {}_s\langle 0|e^{\mp\frac{i}{4}(\Phi_s+\Theta_s)(\tau,x)}\, e^{\pm\frac{i}{4}(\Phi_s+\Theta_s)(0)}|0\rangle_s \quad (35.21)$$

The spin part of (35.21) is easily evaluated as the charge sector is a free bosonic theory

$$G_{RR}^{(\sigma)}(\tau, x) = {}_c\langle 0|\mathcal{O}_c^\dagger(\tau, x)\, \mathcal{O}_c(0)|0\rangle_c\, (v_s\tau + ix)^{-\frac{1}{2}} \quad (35.22)$$

The large-distance asymptotics of the correlation function in the spin sector can be analysed using the formfactor approach. The spin sector is described by the SU(2) invariant Thirring model (30.19). The only single-particle excitations of this model are the soliton and antisoliton.

The first formfactor for the fermion operator is fixed by the Lorentz invariance and factorization (see Essler and Tsvelik, 2002a):

$$\langle 0|R|\theta_c, \theta_s\rangle = \exp[(\theta_c + \theta_s)/4] f(\theta_c - \theta_s) \quad (35.23)$$

The function $f(\theta)$ is periodic with period $2i\pi$ and does not contain poles. This leaves us with the only possibility $f = $ const. The numerical value of f was obtained by Lukyanov and Zamolodchikov (2002):

$$f = (Z_0/2\pi)^{1/2} \qquad Z_0 = 0.9218 \quad (35.24)$$

So, the charge part of the first term in the expansion is equal (up to a numerical factor) to

$$\int_{-\infty}^{\infty} d\theta\, e^{\theta/2} \exp[-mv_s\tau \cosh\theta - imx \sinh\theta]$$
$$= \left(\frac{v_s\tau - ix}{v_s\tau + ix}\right)^{1/4} K_{1/2}\left(m\sqrt{\tau^2 + x^2 v_s^{-2}}\right) = \frac{\exp\left[-m\sqrt{\tau^2 + x^2 v_s^{-2}}\right]}{\sqrt{v_s\tau + ix}} \quad (35.25)$$

Now I use the fact that the single-electron Green's function factorizes into the charge and the spin parts. The formfactor (35.23) corresponds to the emission of one charge soliton. Since charge solitons are massless, we have to sum up all multiple emission processes of these particles. It is obvious that (35.22), being the exact formula, represents the result of such summation. As far as the spin part is concerned, it turns out to be enough to take into account only single-soliton emission. The contribution of multi-soliton emissions gives a numerically small contribution to the Green's function. As a result we get

$$\langle R_\sigma(x, \tau)R_\sigma^\dagger(0, 0)\rangle = \frac{Z_0}{2\pi}\frac{\exp\left[-m\sqrt{\tau^2 + x^2 v_s^{-2}}\right]}{\sqrt{(v_s\tau + ix)(v_c\tau + ix)}} \quad (35.26)$$

and

$$\langle L_\sigma(x, \tau)L_\sigma^\dagger(0, 0)\rangle = \frac{Z_0}{2\pi}\frac{\exp\left[-m\sqrt{\tau^2 + x^2 v_s^{-2}}\right]}{\sqrt{(v_s\tau - ix)(v_c\tau - ix)}} \quad (35.27)$$

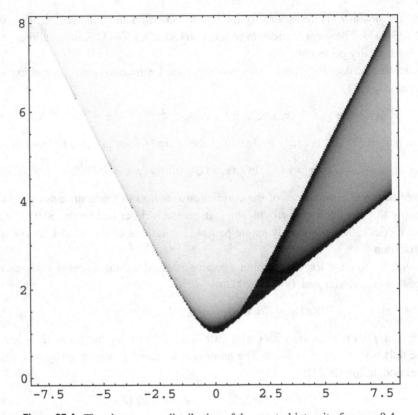

Figure 35.4. The phase space distribution of the spectral intensity for $\alpha = 0.4$.

One may compare this result with a similar expression for the spin-1/2 Tomonaga–Luttinger liquid (30.44). The power law function in the latter expression is replaced by the simple exponential and the square root term reamins unchanged.

The Fourier transform of these functions was calculated by Essler and Tsvelik (2002b):

$$G_{RR}^{(R)}(\omega, q) = -Z_0 \sqrt{\frac{2}{1+\alpha}} \frac{\omega + v_c q}{\sqrt{m^2 + v_c^2 q^2 - \omega^2}}$$

$$\times \left[\left(m + \sqrt{m^2 + v_c^2 q^2 - \omega^2} \right)^2 - \frac{1-\alpha}{1+\alpha}(\omega + v_c q)^2 \right]^{-\frac{1}{2}} \quad (35.28)$$

where $\alpha = v_c/v_s$. The imaginary part of this function is presented in Fig. 35.4.

At $\alpha = 1$ the above expression simplifies:

$$G_{RR}^{(R)}(\omega, q) = -\frac{Z_0}{\omega - vq} \left[\frac{m}{\sqrt{m^2 + v^2 q^2 - \omega^2}} - 1 \right] \quad (35.29)$$

As one may easily check, $G(\omega = 0, q \to 0) \sim q$ and the Green's function does vanish at the noninteracting Fermi surface.

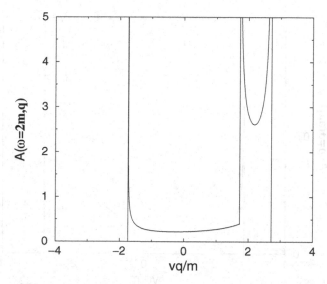

Figure 35.5. The spectral function as a function of momentum at $\alpha = 0.4$.

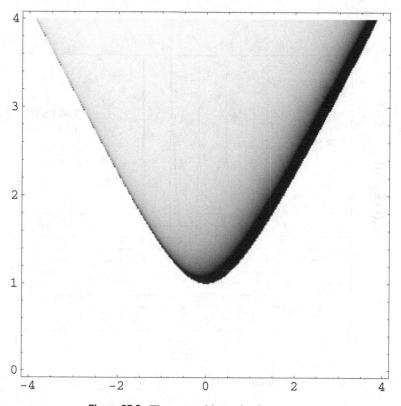

Figure 35.6. The spectral intensity for $v_c = v_s$.

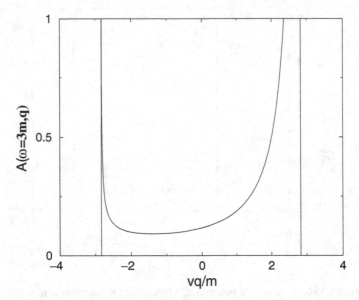

Figure 35.7. The spectral function as a function of momentum at $v_c = v_s$.

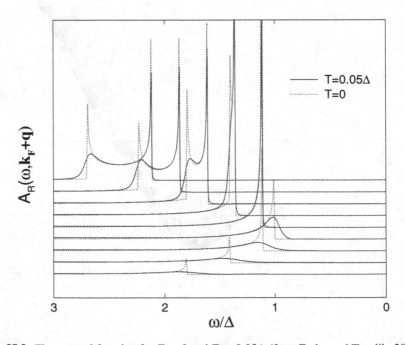

Figure 35.8. The spectral function for $T = 0$ and $T = 0.05\Delta$ (from Essler and Tsvelik, 2002b).

The results available for finite temperatures cover the area $T \gg m$ (the perturbation theory regime) and $T \ll m$ (Essler and Tsvelik, 2002b). Here I present a picture from this paper (see Fig. 35.8).

References

Essler, F. H. L. and Tsvelik, A. M. (2002a). *Phys. Rev. B*, **65**, 115117.
Essler, F. H. L. and Tsvelik, A. M. (2002b). *Phys. Rev. Lett.*, **88**, 096403.
Lukyanov, S. A. and Zamolodchikov, A. B. (2002). *Nucl. Phys. B*, **607**, 437.

36

One-dimensional spin liquids: spin ladder and spin $S = 1$ Heisenberg chain

The spin $S = 1$ Heisenberg chain is an interesting object. First of all, as we have already discussed in Part III, spin chains with integer and half-integer spins have very different low energy properties. The results of Chapter 16 suggest that chains with integer spin have a spectral gap[1] which is generated dynamically. The fact that there are many experimental realizations of quasi-one-dimensional antiferromagnets with $S = 1$ makes the situation even more interesting. The most well studied compound is $Ni(C_2H_8N_2)_2NO_2(ClO_4)$, abbreviated as NENP. Localized spins belong to magnetic Ni ions and all the other ingredients are necessary just to arrange them into well separated chains. The experiments show that there is indeed a spectral gap (Aijro *et al.*, 1989; Renard *et al.*, 1987; see Fig. 36.1). This gap is not very small in comparison with the bandwidth, but one can still hope that the continuum description is good enough. The Monte Carlo simulations give for the spin $S = 1$ Heisenberg chain the ratio of the correlation length to the lattice spacing $\xi/a = 6.2$ (Nomura, 1989) which can still be treated as a large number.

There is another intriguing fact about the $S = 1$ Heisenberg chain which makes it attractive for a theorist. It is the presence of a hidden, so-called *topological* order. This feature puts this theory in a much broader context of *spin liquids*, magnetically disordered systems with a hidden order.

Spin ladder

The spin $S = 1$ Heisenberg chain can be described as a limiting case of two ferromagnetically coupled $S = 1/2$ chains. The problem of coupled chains (so-called *spin ladders*) is interesting in its own right.

Let us consider the problem of two weakly coupled isotropic $S = 1/2$ Heisenberg chains. As I am going to demonstrate, the phase diagram of this model includes several different spin liquid phases. Thermodynamically all these phases are indistinguishable: they are all disordered with spectral gaps. In some of them even the excitation spectra coincide. The difference is in the correlation functions; that is, to distinguish one type of disorder from another, one has to study the dynamics.

---

[1] Though this result is essentially semiclassical and is rigorous only for $S \gg 1$, it is generally believed that the distinction between integer and half-integer spins is a general fact independent of any approximation.

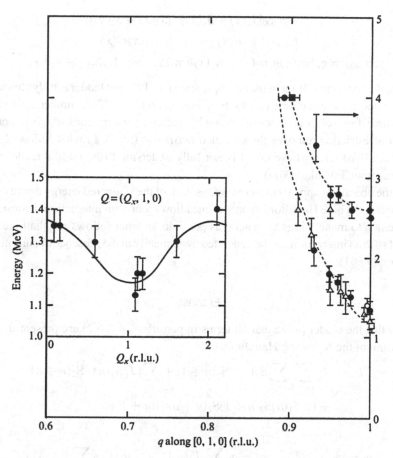

Figure 36.1. Dispersion curves of excitations in NENP below 4.2 K (from Renard *et al.*, 1987). The solid line in the insert is a guide to the eye. The dashed lines are fits to the relativistic dispersion relation.

In order to establish the phase diagram we have to classify all relevant perturbations which drive the system of two chains to strong coupling. In the low energy limit each spin $S = 1/2$ chain is described by the $SU_1(2)$ WZNW model or, alternatively, by the Gaussian model. Thus the bare Hamiltonian is given by

$$H_0 = \frac{v_s}{2} \sum_{j=1,2} \int dx \left[\Pi_j^2(x) + (\partial_x \phi_j(x))^2 \right] \tag{36.1}$$

where the velocity $v_s \sim J_\parallel a_0$ and Π_j are the momenta conjugate to ϕ_j. As we know from the previous chapters, for each chain j there are four primary fields, staggered components of energy density and magnetization ϵ_j, $\mathbf{n}_j$ all having the same scaling dimension 1/2. From bilinear combinations of these fields and from the corresponding current operators one can compose the following relevant perturbations:

$$H_\perp = \int dx (\mathcal{H}_J + \mathcal{H}_{\text{twist}} + \mathcal{H}_{n,\epsilon}) \tag{36.2}$$

$$\mathcal{H}_J = g_0[\mathbf{J}_1(x) + \bar{\mathbf{J}}_1(x)] \cdot [\mathbf{J}_2(x) + \bar{\mathbf{J}}_2(x)] \tag{36.3}$$

$$\mathcal{H}_{n,\epsilon} = g_s \mathbf{n}_1(x) \cdot \mathbf{n}_2(x) + u\epsilon_1(x)\epsilon_2(x) \tag{36.4}$$

$$\mathcal{H}_{\text{twist}} = g_1 [\mathbf{n}_1(x)\partial_x \mathbf{n}_2(x) - \mathbf{n}_2(x)\partial_x \mathbf{n}_1(x)] + g_2 [\epsilon_1 \partial_x \epsilon_2 - \epsilon_2 \partial_x \epsilon_1] \tag{36.5}$$

Therefore the continuous limit of the *isotropic* spin $S = 1/2$ spin ladder is fully characterized by five parameters, the coupling constants of interaction (36.2). This number can be reduced down to three if we restrict our consideration by interchain interactions which do not violate parity. This restriction removes the so-called *twist* term (36.5). In what follows I will not consider the twist term whose effect is not fully understood (though the reader may get some ideas from Tsvelik (2001)).

Since the operator ϵ appears as a continuous limit of the staggered energy density (31.50), it is present in the bare Hamiltonian only if one allows four-spin interaction. I consider such an interaction to make things as general as possible. In what follows we shall see that the effects of such an interaction can be generated dynamically in the presence of current-current interaction (36.61).

Exercise

I suggest that the reader prove that all terms in perturbation (36.2) are present in the continuous limit of the following Hamiltonian:

$$H = J_\parallel \sum_{j=1,2} \sum_n \mathbf{S}_j(n) \cdot \mathbf{S}_j(n+1) + \sum_n [J_\perp^a \mathbf{S}_1(n) \cdot \mathbf{S}_2(n+a)$$

$$+ U^a \mathbf{S}_1(n)\mathbf{S}_2(n+1)\mathbf{S}_2(n+a)\mathbf{S}_2(n+1+a)] \tag{36.6}$$

where

$$g_0 = a_0 \sum_a J_\perp^a \qquad g_s = a_0 \sum_a J_\perp^a(-1)^a \qquad u = a_0 \sum_a U^a(-1)^a$$

The exchange integral along the chains is antiferromagnetic ($J_\parallel > 0$); the interchain couplings ($|J_\perp|, |u| \ll J_\parallel$) are of arbitrary sign.

Let us start our analysis from the most relevant interaction (36.4). Using the bosonization formulas (30.37) for $\mathbf{n}_j(x)$ and (31.53) for $\epsilon_j(x)$, we get

$$H_\perp = \frac{g_s \lambda^2}{\pi^2 a_0} \int dx \left[\frac{1}{2}(u/g_s - 1)\cos\sqrt{2\pi}(\phi_1 + \phi_2) \right.$$

$$\left. + \frac{1}{2}(u/g_s + 1)\cos\sqrt{2\pi}(\phi_1 - \phi_2) + \cos\sqrt{2\pi}(\theta_1 - \theta_2) \right] \tag{36.7}$$

where $\theta_j(x)$ is the field dual to $\phi_j(x)$. Denote

$$m = \frac{g_s \lambda^2}{2\pi}$$

and introduce linear combinations of the fields ϕ_1 and ϕ_2:

$$\phi_\pm = \frac{\phi_1 \pm \phi_2}{\sqrt{2}} \tag{36.8}$$

The total (ϕ_+) and relative (ϕ_-) degrees of freedom decouple, and the Hamiltonian density of two identical Heisenberg chains transforms to a sum of two independent contributions (Schulz, 1986):

$$\mathcal{H} = \mathcal{H}_+ + \mathcal{H}_- \tag{36.9}$$

$$\mathcal{H}_+(x) = \frac{v_s}{2}(\Pi_+^2 + (\partial_x\phi_+)^2) + \frac{m}{\pi a_0}(u/g_s - 1)\cos\sqrt{4\pi}\,\phi_+ \tag{36.10}$$

$$\mathcal{H}_-(x) = \frac{v_s}{2}(\Pi_-^2 + (\partial_x\phi_-)^2) + \frac{m}{\pi a_0}(u/g_s + 1)\cos\sqrt{4\pi}\,\phi_- + \frac{2m}{\pi a_0}\cos\sqrt{4\pi}\theta_- \tag{36.11}$$

Let us turn back to (36.10) and (36.11). One immediately realizes that the critical dimension of all the cosine terms in (36.10), (36.11) is 1; therefore the model (36.9) is a theory of free massive fermions. The Hamiltonian H_+ describes the sine-Gordon model at $\beta^2 = 4\pi$; so it is equivalent to a free massive Thirring model. Let us introduce a spinless Dirac fermion related to the scalar field ϕ_+ via identification

$$\psi_{R,L}(x) \simeq (2\pi a_0)^{-1/2}\exp(\pm i\sqrt{4\pi}\,\phi_{+;R,L}(x)) \tag{36.12}$$

Using

$$\frac{1}{\pi a_0}\cos\sqrt{4\pi}\,\phi_+(x) = i\,[\psi_R^\dagger(x)\psi_L(x) - \text{H.c.}]$$

we get

$$H_+(x) = -iv_s(\psi_R^\dagger\partial_x\psi_R - \psi_L^\dagger\partial_x\psi_L) + im(u/g_s - 1)(\psi_R^\dagger\psi_L - \psi_L^\dagger\psi_R) \tag{36.13}$$

For future purposes, we introduce two real (Majorana) fermion fields

$$\xi_\nu^1 = \frac{\psi_\nu + \psi_\nu^\dagger}{\sqrt{2}} \qquad \xi_\nu^2 = \frac{\psi_\nu - \psi_\nu^\dagger}{\sqrt{2}i} \qquad \nu = \text{R, L} \tag{36.14}$$

to represent H_+ as a model of two degenerate massive Majorana fermions

$$H_+ = H_{m_t}[\xi^1] + H_{m_t}[\xi^2] \tag{36.15}$$

where

$$H_m[\xi] = -\frac{iv_s}{2}(\xi_R\,\partial_x\xi_R - \xi_L\,\partial_x\xi_L) - im\,\xi_R\xi_L \tag{36.16}$$

and

$$m_t = m(1 - u/g_s) \tag{36.17}$$

Now we shall demonstrate that the Hamiltonian H_- in (36.11) reduces to the Hamiltonian of two *different* Majorana fields. As before, we first introduce a spinless Dirac fermion

$$\chi_{R,L}(x) \simeq (2\pi a_0)^{-1/2}\exp(\pm i\sqrt{4\pi}\,\phi_{-;R,L}(x))$$

$$\frac{1}{\pi a_0}\cos\sqrt{4\pi}\,\phi_-(x) = i\,[\chi_R^\dagger(x)\chi_L(x) - \text{H.c.}] \tag{36.18}$$

$$\frac{1}{\pi a_0}\cos\sqrt{4\pi}\theta_-(x) = -i\,[\chi_R^\dagger(x)\chi_L^\dagger(x) - \text{H.c.}]$$

Apart from the usual mass bilinear term ('charge density wave' pairing), the Hamiltonian H_- also contains a 'Cooper pairing' term originating from the cosine of the dual field:

$$H_-(x) = -v_s(\chi_R^\dagger \partial_x \chi_R - \chi_L^\dagger \partial_x \chi_L) + im(1 + u/J_\perp)(\chi_R^\dagger \chi_L - \chi_L^\dagger \chi_R) + 2im(\chi_R^\dagger \chi_L^\dagger - \chi_L \chi_R) \tag{36.19}$$

We introduce two Majorana fields

$$\xi_v^3 = \frac{\chi_v + \chi_v^\dagger}{\sqrt{2}} \qquad \rho_v = \frac{\chi_v - \chi_v^\dagger}{\sqrt{2}i} \qquad v = R, L \tag{36.20}$$

The Hamiltonian H_- then describes two massive Majorana fermions, $\xi_{R,L}^3$ and $\rho_{R,L}$, with masses m_t and m_s, respectively:

$$H_- = H_{m_t}[\xi^3] + H_{m_s}[\rho] \tag{36.21}$$

where

$$m_s = -m(3 + u/g_s) \tag{36.22}$$

Now we observe that ξ^a, $a = 1, 2, 3$, form a triplet of Majorana fields with the same mass m_t. There is one more field ρ with a different mass, m_s. So, the total Hamiltonian is

$$H = H_{m_t}[\xi] + H_{m_s}[\rho] \tag{36.23}$$

with

$$H_{m_t}[\xi] = \sum_{a=1,2,3} \left\{ -\frac{iv_s}{2} \left(\xi_R^a \partial_x \xi_R^a - \xi_L^a \partial_x \xi_L^a \right) - im_t\, \xi_R^a \xi_L^a \right\} \tag{36.24}$$

The O(3)-invariant model $H_{m_t}[\xi]$ was suggested as a description of the $S = 1$ Heisenberg chain by Tsvelik (1990). This equivalence follows from the fact that, in the continuum limit, the integrable $S = 1$ chain with the Hamiltonian (Takhtajan, 1982; Babujan, 1983)

$$H = \sum_n [(S_n S_{n+1}) - (S_n S_{n+1})^2] \tag{36.25}$$

is described by the critical WZNW model on the SU(2) group at the level $k = 2$, and the latter is in turn equivalent to the model of three massless Majorana fermions, as follows from the comparison of conformal charges of the corresponding theories:

$$C_{\mathrm{SU}(2),k=2}^{\mathrm{WZNW}} = \frac{3}{2} = 3C_{\mathrm{Major.fermion}}$$

The $k = 2$ level, SU(2) currents expressed in terms of the fields ξ^a are given by

$$I_{R,L}^a = -\frac{i}{2}\epsilon^{abc}\, \xi_{R,L}^b \xi_{R,L}^c \tag{36.26}$$

When small deviations from criticality are considered, no single-ion anisotropy ($\sim D(S^z)^2$, $S = 1$) is allowed to appear due to the original SU(2) symmetry of the problem. So, the mass term in (36.24) turns out to be the only allowed relevant perturbation to the critical SU(2), $k = 2$ WZNW model.

Thus, the fields ξ^a describe triplet excitations related to the effective spin-1 chain. Remarkably, completely decoupled from them are singlet excitations described in terms of

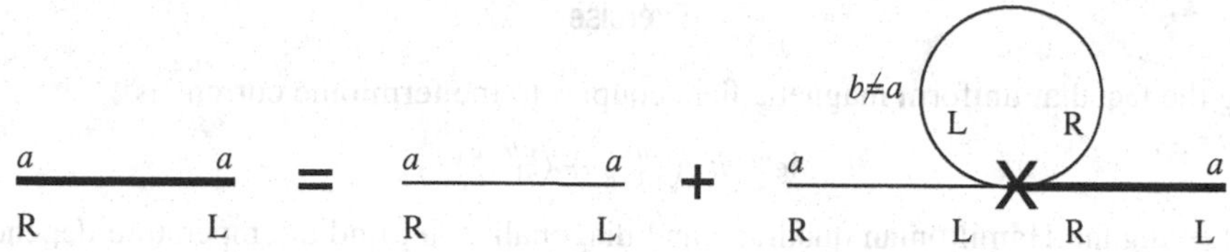

Figure 36.2. The Dyson equation for the Green's function of the Majorana fermions with Hamiltonian (36.27).

the field ρ. Another feature is that the form of the Hamiltonian in the continuous limit is independent of the signs of $J_\perp$ and u. This is because the corresponding parturbation (36.4) is strongly relevant.

Since the spectrum of the system is massive, the role of the so far neglected (marginal) part of the interchain coupling (36.61) is restricted to renormalization of the masses and velocities.[2] With logarithmic accuracy, neglecting the velocity renormalization coming from the interaction of currents with the same chirality, I obtain the following invariant form:

$$
\begin{aligned}
H_{\text{marg}} &= \frac{1}{2} g_0 \int dx \left[(I_R^a I_L^a) - (\xi_R^a \xi_L^a)(\rho_R \rho_L) \right] \\
&= \frac{1}{2} J_\perp a_0 \int dx \left[(\xi_R^1 \xi_L^1)(\xi_R^2 \xi_L^2) + (\xi_R^2 \xi_L^2)(\xi_R^3 \xi_L^3) + (\xi_R^3 \xi_L^3)(\xi_R^1 \xi_L^1) \right. \\
&\quad \left. - (\xi_R^1 \xi_L^1 + \xi_R^2 \xi_L^2 + \xi_R^3 \xi_L^3)(\rho_R \rho_L) \right]
\end{aligned}
\tag{36.27}
$$

In a theory of N massive Majorana fermions, with masses m_a ($a = 1, 2, \ldots, N$) and a weak four-fermion interaction

$$
H_{\text{int}} = \frac{1}{2} \sum_{a \neq b} g_{ab} \int dx \, (\xi_R^a \xi_L^a)(\xi_R^b \xi_L^b) \qquad g_{ab} = g_{ba}
$$

renormalized masses $\tilde{m}_a$ estimated in the first order in g with logarithmic accuracy are given by

$$
\tilde{m}_a = m_a + \sum_{b(\neq a)} \frac{g_{ab}}{2\pi v} m_b \ln \frac{\Lambda}{|m_b|}
\tag{36.28}
$$

Using (36.27) and (36.28), we find renormalized values of the masses of the triplet and singlet excitations:

$$
m_t = m_t^{(0)} + \frac{g_0}{4\pi v} \ln \frac{\Lambda}{|m|} (2m_t^{(0)} - m_s^{(0)})
\tag{36.29}
$$

$$
m_s = m_s^{(0)} + \frac{3 g_0}{4\pi v} \ln \frac{\Lambda}{|m|} m_t^{(0)}
\tag{36.30}
$$

This renormalization may lead to a *change of sign* of one of the masses. As we shall see, the sign of the product $m_t m_s$ determines the phase of the spin liquid. Therefore by changing the parameter $g_0 \sim \sum_a J_\perp^a$ one can reach the region of the phase diagram where even the purely Heisenberg spin ladder is in a non-Haldane phase (see the further discussion).

---

[2] It can also generate bound states of fundamental particles.

Exercise

Using the fact that uniform magnetic field couples to the fermionic current as

$$i\epsilon^{abc} h^a \left(\xi_R^b \xi_R^c + (\xi_L^b \xi_L^c) \right)$$

thus leaving the Hamiltonian quadratic and diagonalizable, find a temperature dependent magnetic susceptibility. Show that at $T \ll m_t$

$$\chi(T) \sim T^{-1/2} \exp(-m_t/T)$$

Correlation functions

Since the singlet excitation with mass m_s does not carry spin, its operators do not contribute to the slow components of the total magnetization. The latter is expressed in terms of the $k = 2$ SU(2) currents (36.26):

$$m^a \sim I_R^a + I_L^a \tag{36.31}$$

Therefore the two-point correlation function of spin densities at small wave vectors ($|q| \ll \pi/a_0$) is given by the simple fermionic loop (see Fig. 36.3).

A simple calculation gives the following expression for its imaginary part:

$$\Im m \chi^{(R)}(\omega, q) = \frac{2q^2 m^2 v_s^2}{s^3 \sqrt{s^2 - 4m_t^2}} \tag{36.32}$$

where $s^2 = \omega^2 - v_s^2 q^2$. Thus the dynamical magnetic susceptibility at small wave vectors has a threshold at $2m_t$.

It turns out that it is possible to calculate exactly the two-point correlation functions of the staggered magnetization. This is due to the fact that the corresponding operators of the Heisenberg chains are related (in the continuum limit) to the order and disorder parameter fields of two-dimensional Ising models; the correlation functions of the latter operators are known exactly even out of criticality (see Chapter 28).

Using formulas (30.37) the components of the total ($\mathbf{n}^{(+)} = \mathbf{n}_1 + \mathbf{n}_2$) and relative ($\mathbf{n}^{(-)} = \mathbf{n}_1 - \mathbf{n}_2$) staggered magnetization can be represented as

$$
\begin{aligned}
&n_x^{(+)} \sim \cos\sqrt{\pi}\theta_+ \cos\sqrt{\pi}\theta_- && n_x^{(-)} \sim \sin\sqrt{\pi}\theta_+ \sin\sqrt{\pi}\theta_- \\
&n_y^{(+)} \sim \sin\sqrt{\pi}\theta_+ \cos\sqrt{\pi}\theta_- && n_y^{(-)} \sim \cos\sqrt{\pi}\theta_+ \sin\sqrt{\pi}\theta_- \\
&n_z^{(+)} \sim \sin\sqrt{\pi}\phi_+ \cos\sqrt{\pi}\phi_- && n_z^{(-)} \sim \cos\sqrt{\pi}\phi_+ \sin\sqrt{\pi}\phi_-
\end{aligned}
\tag{36.33}
$$

The fields ϕ_+, θ_+ and ϕ_-, θ_- are governed by the Hamiltonians (36.10) and (36.11), respectively. Let us first consider exponentials $\exp(\pm i\sqrt{\pi}\phi_+)$, $\exp(\pm i\sqrt{\pi}\theta_+)$. Their correlation functions have been extensively studied in the context of the *noncritical* Ising model (see Chapter 28). It has been shown there that these bosonic exponents with scaling dimension $1/8$ are expressed in terms of the order (σ) and disorder (μ) parameters of two Ising models as follows:

$$
\begin{aligned}
\cos(\sqrt{\pi}\phi_+) &= \mu_1\mu_2 & \sin(\sqrt{\pi}\phi_+) &= \sigma_1\sigma_2 \\
\cos(\sqrt{\pi}\theta_+) &= \sigma_1\mu_2 & \sin(\sqrt{\pi}\theta_+) &= \mu_1\sigma_2
\end{aligned}
\tag{36.34}
$$

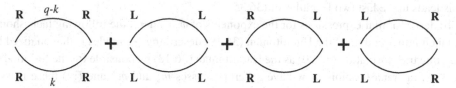

Figure 36.3. The fermionic loops.

Figure 36.4. A. A. Nersesyan.

Below we shall give an alternative derivation of (36.35).

As already discussed, the $\beta^2 = 4\pi$ sine-Gordon model H_+, (36.10), is equivalent to a model of two degenerate massive Majorana fermions, (36.15), (36.16). As is explained in Chapter 28, the theory of a massive Majorana fermion field describes long-distance properties of the two-dimensional Ising model, the fermionic mass being proportional to $m \sim t = (T - T_c)/T_c$. So, H_+ is equivalent to two decoupled two-dimensional Ising models. Let σ_j and μ_j ($j = 1, 2$) be the corresponding order and disorder parameters. At criticality (zero fermionic mass), four products $\sigma_1\sigma_2$, $\mu_1\mu_2$, $\sigma_1\mu_2$ and $\mu_1\sigma_2$ have the same critical dimension 1/8 as that of the bosonic exponentials $\exp(\pm i\sqrt{\pi}\phi_+)$, $\exp(\pm i\sqrt{\pi}\theta_+)$. Therefore there must be some correspondence between the two groups of four operators which should also hold at small deviations from criticality. To find this correspondence, notice that, as follows from (36.10), at $m > 0$ $\langle\cos\sqrt{\pi}\phi_+\rangle \neq 0$, while $\langle\sin\sqrt{\pi}\phi_+\rangle = 0$. Since the case $m > 0$ corresponds to the disordered phase of the Ising systems ($t > 0$), $\langle\sigma_1\rangle = \langle\sigma_2\rangle = 0$, while $\langle\mu_1\rangle = \langle\mu_2\rangle \neq 0$. At $m < 0$ (ordered Ising systems, $t < 0$) the situation is inverted: $\langle\cos\sqrt{\pi_+}\phi\rangle = 0$, $\langle\sin\sqrt{\pi}\phi_+\rangle \neq 0$, $\langle\sigma_1\rangle = \langle\sigma_2\rangle \neq 0$, $\langle\mu_1\rangle = \langle\mu_2\rangle = 0$. This explains the first two formulas of (36.35).

Clearly, the exponentials of the dual field θ_+ must be expressed in terms of $\sigma_1\mu_2$ and $\mu_1\sigma_2$. To find the correct correspondence, one has to take into account the fact that a local product of the order and disorder operators of a single Ising model results in the Majorana fermion operator, i.e.

$$\xi^1 \sim \cos\sqrt{\pi}(\phi_+ + \theta_+) \sim \sigma_1\mu_1 \qquad \xi^2 \sim \sin\sqrt{\pi}(\phi_+ + \theta_+) \sim \sigma_2\mu_2$$

This leads to the last two formulas of (36.35).

To derive similar expressions for the exponents of ϕ_- and θ_-, the following facts should be taken into account: (i) the Hamiltonian (36.19) describing '$-$'modes is diagonalized by the same transformation (36.20) as the Hamiltonian (36.13) responsible for the '$+$' modes; (ii) the Majorana fermions now have different masses m_s and m_t; and (iii) these masses may have different signs. In order to take proper account of these facts one should recall the following.

(a) A negative mass means that we are in the ordered phase of Ising model where $\langle \sigma \rangle \neq 0$.
(b) It follows from (ii) that '$-$' bosonic exponents are also expressed in terms of order and disorder parameters of two Ising models, the latter, however, being characterized by different t. We denote these operators σ_3, μ_3 (mass m_t) and σ, μ (mass m_s).
(c) Operators corresponding to a negative mass can be rewritten in terms of those with positive mass using the Kramers–Wannier duality transformation

$$m \to -m \qquad \sigma \to \mu \qquad \mu \to \sigma \qquad (36.35)$$

Taking these facts into account we get the following expressions for the '$-$' bosonic exponents:

$$\cos(\sqrt{\pi}\phi_-) = \mu_3 \sigma \qquad \sin(\sqrt{\pi}\phi_-) = \sigma_3 \mu$$
$$\cos(\sqrt{\pi}\theta_-) = \sigma_3 \sigma \qquad \sin(\sqrt{\pi}\theta_-) = \mu_3 \mu \qquad (36.36)$$

Combining (36.35) and (36.36), from (36.33) we get the following, manifestly SU(2)-invariant, expressions:

$$n_x^+ \sim \sigma_1 \mu_2 \sigma_3 \sigma \qquad n_y^+ \sim \mu_1 \sigma_2 \sigma_3 \sigma \qquad n_z^+ \sim \sigma_1 \sigma_2 \mu_3 \sigma \qquad (36.37)$$
$$n_x^- \sim \mu_1 \sigma_2 \mu_3 \mu \qquad n_y^- \sim \sigma_1 \mu_2 \mu_3 \mu \qquad n_z^- \sim \mu_1 \mu_2 \sigma_3 \mu \qquad (36.38)$$

It is instructive to compare them with two possible representations for the staggered magnetization operators for the $S = 1$ Heisenberg chain which can be derived from the SU(2)$_2$ WZNW model (see (31.63) and Tsvelik (1990)):

$$S^x \sim \sigma_1 \mu_2 \sigma_3 \qquad S^y \sim \mu_1 \sigma_2 \sigma_3 \qquad S^z \sim \sigma_1 \sigma_2 \mu_3 \qquad (36.39)$$

or

$$S^x \sim \mu_1 \sigma_2 \mu_3 \qquad S^y \sim \sigma_1 \mu_2 \mu_3 \qquad S^z \sim \mu_1 \mu_2 \sigma_3 \qquad (36.40)$$

The agreement is achieved if the singlet excitation band is formally shifted to infinity ($m_s m_t \to -\infty$). This implies substitutions $\sigma \simeq \langle \sigma \rangle \neq 0$, $\mu \simeq \langle \mu \rangle \simeq 0$ for $m_s < 0$ (which would correspond to antiferromagnetic interchain coupling at $u = 0$) or $\sigma \simeq \langle \sigma \rangle \simeq 0$, $\langle \mu \rangle \neq 0$ for $m_s > 0$ (a ferromagnetic interchain coupling). In the latter case, as expected, the staggered magnetization of the $S = 1$ chain is determined by the sum of staggered magnetizations of both chains.

Staggered susceptibility of the conventional (Haldane) spin liquid

As we have mentioned above, the correlation function of staggered magnetization depends on a mutual sign of the masses. As we shall see in a moment, the case $m_t m_s < 0$

corresponds to the conventional (we shall call it Haldane) spin liquid which exhibits a sharp single-magnon peak in $\chi''(\omega, q \sim \pi)$. The case $m_s m_t > 0$ corresponds to a *spontaneously dimerized* spin liquid where $\chi''(\omega, q \sim \pi)$ has a two-particle threshold. We emphasize that both these liquids have the same excitation spectra with spectral gaps. Moreover, the excitations carry the same quantum numbers (one triplet and one singlet modes). Nevertheless, as I have already mentioned and as I am going to demonstrate now, their correlation functions are quite distinct.

Let us consider the case $m_t m_s < 0$ and consider the asymptotic behaviour of the two-point correlation functions of staggered magnetizations in the two limits $r \to 0$ and $r \to \infty$. In the limit $r \to \infty$ they are given by (28.28):

$$\langle \sigma_a(r)\sigma_a(0)\rangle = G_\sigma(\tilde{r}) = \frac{A_1}{\pi} K_0(\tilde{r}) + \mathcal{O}(e^{-3\tilde{r}}) \tag{36.41}$$

$$\langle \mu_a(r)\mu_a(0)\rangle = G_\mu(\tilde{r})$$

$$= A_1 \left\{ 1 + \frac{1}{\pi^2} \left[\tilde{r}^2 \left(K_1^2(\tilde{r}) - K_0^2(\tilde{r}) \right) - \tilde{r} K_0(\tilde{r})K_1(\tilde{r}) + \frac{1}{2} K_0^2(\tilde{r}) \right] \right\} + \mathcal{O}(e^{-4\tilde{r}}) \tag{36.42}$$

where $\tilde{r} = rM$ ($M = m_t$ or $-m_s$), A_1 is a nonuniversal parameter, and it has been assumed that M is positive. If M is negative the correlation functions are obtained simply by interchanging σ and μ, and putting $M \to -M$ (the duality transformation (36.35)). Therefore, as might be expected, at large distances, a difference between the ladder and the $S = 1$ chains appears only in $\exp(-m_s r)$ terms due to the contribution of the excitation branch with $M = m_s$ absent in the $S = 1$ chain.

In the limit $\tilde{r} \to 0$ the correlation functions are of power law form:

$$G_\sigma(\tilde{r}) = G_\mu(\tilde{r}) = \frac{A_2}{\tilde{r}^{\frac{1}{4}}} \tag{36.43}$$

plus nonsingular terms. The ratio of the constants A_1 and A_2 is a universal quantity given by (28.30).

We conclude this section by writing down the exact expression for the staggered magnetization two-point correlation functions. The correlation function for spins on the same chain is given by

$$\langle n_1^a(\tau, x)n_1^a(0,0)\rangle = G_\sigma^2(m_t r)G_\mu(m_t r)G_\sigma(m_s r) + G_\mu^2(m_t r)G_\sigma(m_t r)G_\mu(m_s r) \tag{36.44}$$

The interesting asymptotics are

$$\langle n_1^a(\tau, x)n_1^a(0,0)\rangle = \frac{1}{2\pi r}\tilde{Z} \quad \text{at} \quad |m_t|r \ll 1 \tag{36.45}$$

$$= \frac{m}{\pi^2} Z K_0(m_t r)\left\{ 1 + \frac{2}{\pi^2}[(m_t r)^2 \left(K_1^2(m_t r) - K_0^2(m_t r)\right) - mr K_0(m_t r)K_1(m_t r)\right.$$

$$\left. + \frac{1}{2}K_0^2(m_t r)]\right\} + \mathcal{O}\left(\exp[-(2|m_t| + |m_s|)r]\right) \quad \text{at} \quad m_t r \gg 1 \tag{36.46}$$

where $r^2 = v_s^2 \tau^2 + x^2$ and

$$\frac{\tilde{Z}}{Z} = \frac{2^{4/3} e}{3^{1/4}} A^{-12} \approx 0.264 \tag{36.47}$$

The complete expressions for the functions $G_{\sigma,\mu}(\tilde{r})$ are given in Wu et al. (1976). For the interchain correlation function we get

$$\langle n_1^a(\tau, x) n_2^a(0, 0) \rangle = G_\sigma^2(m_t r) G_\mu(m_t r) G_\sigma(m_s r) - G_\mu^2(m_t r) G_\sigma(m_t r) G_\mu(m_s r) \tag{36.48}$$

At $m_t r \ll 1$ it decays as $(m_t r)^{-2}$; the leading asymptotic at $mr \gg 1$ is the same as (36.46) (up to the -1 factor). The difference appears only in terms of order $\exp[-(2|m_t| + |m_s|)r]$. The important point is that at $mr \gg 1$ the contribution from the singlet excitation appears only in a high order in $\exp(-m_t r)$. Therefore it is unobservable by neutron scattering at energies below $(2|m_t| + |m_s|)$.

Using the above expressions we can calculate the imaginary part of the dynamical spin susceptibility in two different regimes. For $|\pi - q| \ll 1$ we have (supposing that $J_\perp$ is antiferromagnetic):

$$\Im m \chi^{(R)}(\omega, \pi - q; q_\perp)$$

$$= Z \begin{cases} \sin^2(q_\perp/2) \left[\dfrac{m_t}{\pi |\omega|} \delta \left(\omega - \sqrt{v_s^2 q^2 + m_t^2} \right) + F(\omega, q) \right] & \omega < (2|m_t| + |m_s|) \\[3mm] \dfrac{0.264}{\sqrt{\omega^2 - v_s^2 q^2}} & \omega \gg (2|m_t| + |m_s|) \end{cases} \tag{36.49}$$

where the transverse 'momentum' $q_\perp$ takes values 0 and π. The factor Z is assumed to be m-independent so that at $m \to 0$ we reproduce the susceptibility of noninteracting chains. We have calculated the function $F(\omega, q)$ only near the $3m_t$ threshold where it is equal to

$$F(\omega, q) \approx \frac{1}{24\sqrt{3}\pi m_t} \frac{(\omega^2 - v_s^2 q^2 - 9m_t^2)}{m_t^2} \tag{36.50}$$

The easiest way to calculate $F(\omega, q)$ is the following. Let us expand the functions $K_0(r)$ and $K_1(r)$ in powers of r^{-1}:

$$\begin{aligned} K_0(r) &= \sqrt{\pi/2r}\, e^{-r}(1 - 1/8r + 9/128r^2 - 75/1024r^3 + \cdots) \\ K_1(r) &= \sqrt{\pi/2r}\, e^{-r}(1 + 3/8r - 15/128r^2 + 105/1024r^3 + \cdots) \end{aligned} \tag{36.51}$$

Substituting these expressions into (36.46) where we set $m = 1$ we get after many cancellations

$$\langle n_1^a(\tau, x) n_1^a(0, 0) \rangle = \frac{Z}{4\sqrt{2}\pi^{5/2}} r^{-5/2} e^{-r} [1 + \mathcal{O}(r^{-1})] \tag{36.52}$$

Since the correlation function is Lorentz invariant, we can calculate its ω-dependent part and then restore the q-dependence making the replacement $\omega^2 \to \omega^2 - v^2 q^2$ in the final expression. So for $q = 0$ we can write down the Fourier integral for the function (36.52) in

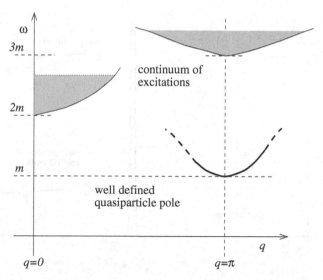

Figure 36.5. The area of the (ω, q) plane where the imaginary part of the dynamical magnetic susceptibility is finite.

polar coordinates

$$\chi = \frac{Z}{4\sqrt{2}\pi^{5/2}} \int d\phi \int_{\sim 1}^{\infty} dr\, r^{-3/2} \exp(-3r + i\omega_n r \cos\phi) \qquad (36.53)$$

The most singular ω-dependent part of the integral comes from large distances $r \sim (3 - i\omega)^{-1}$. With this in mind we can estimate the r-integral as follows:

$$\int d\phi \int_{\sim 1}^{\infty} dr\, r^{-3/2} \exp(-3r + i\omega_n r \cos\phi) = \text{reg} + \Gamma(-1/2)[3 - i\omega \cos\phi]^{1/2} \qquad (36.54)$$

where the abbreviation 'reg' stands for terms regular in $(\omega \cos\phi - 3)$. The regular terms do not contribute to the imaginary part of $\chi^{(R)}$ after the analytical continuation $i\omega_n \to \omega + i0$. Likewise the next terms in the expansion in $r-1$ will give regular terms together with higher powers of $[3 - i\omega \cos\phi]^{1/2}$. After the analytic continuation we arrive at the following expression for the leading singularity:

$$\Im m\chi^{(R)}(\omega) = -\Gamma(-1/2)\frac{Z}{4\sqrt{2}\pi^{5/2}} \Re e \int_0^{2\pi} d\phi \sqrt{\omega \cos\phi - 3} \qquad (36.55)$$

When $\omega \approx 3$ only small ϕ contribute and we can substitute $\cos\phi \approx 1 - \phi^2/2$ and obtain the expression (36.50).

For $|q| \ll 1$ we have

$$\Im m\chi^{(R)}(\omega, q; q_\perp) = [1 + \cos^2(q_\perp/2)]f(s, m) \qquad (36.56)$$

where $f(s, m)$ is given by (36.32).

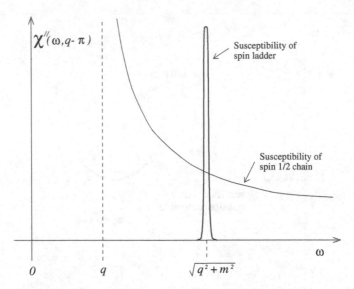

Figure 36.6. The optical magnon peak in the Haldane spin liquid.

Spontaneously dimerized spin liquid

The discussion of this section is based on results obtained by Nersesyan and Tsvelik (1997). If the masses m_s and m_t have opposite signs, and if the triplet branch of the spectrum remains the lowest, $|m_t| \ll |m_s|$, the two-chain spin ladder is in the Haldane liquid phase with short-range correlations of the staggered magnetization, but with coherent $S = 1$ and $S = 0$ single-magnon excitations. Since we have three interchain couplings, g_s, g_0 and u, the masses may vary independently, and we can ask how the properties of the spin ladder are changed in regions where $m_s m_t > 0$. The thermodynamic properties, being dependent on m_a^2, remain unchanged. The symmetry of the ground state and the behaviour of the correlation functions, however, experience a deep change: the ground state turns out to be spontaneously dimerized, and the spectral function of the staggered magnetization displays only incoherent background.

Transitions from the Haldane phase to spontaneously dimerized phases take place when either the triplet excitations become gapless, with the singlet mode still having a finite gap, or vice versa. The transition at $m_t = 0$ belongs to the universality class of the critical, exactly integrable, $S = 1$ spin chain (see (36.25)); the corresponding non-Haldane phase with $|m_t| < |m_s|$ represents the dimerized state of the $S = 1$ chain with spontaneously broken translational symmetry and doubly degenerate ground state. The critical point $m_s = 0$ is of the Ising type; it is associated with a transition to another dimerized phase ($|m_t| > |m_s|$), not related to the $S = 1$ chain.

To be specific, let us assume that $u < 0$ and $g_s < 0$. We start from the Haldane-liquid phase, $m_t < 0$, $m_s > 0$, increase $|u|$ and, passing through the critical point $m_t = 0$, penetrate into a new phase with $0 < m_t < m_s$. The change of the relative sign of the two masses amounts to the duality transformation in the singlet (ρ) Ising system, implying that in the definitions (36.37) and (36.38) the order and disorder parameters, σ and μ, must be interchanged. As a result, the spin correlation functions are now given by different expressions:

$$\langle \mathbf{n}^+(\mathbf{r}) \cdot \mathbf{n}^+(\mathbf{0}) \rangle \sim K_0^2(m_t r) \qquad \langle \mathbf{n}^-(\mathbf{r}) \cdot \mathbf{n}^-(\mathbf{0}) \rangle \sim K_0(m_t r) K_0(m_s r) \qquad (36.57)$$

The total and relative dimerization fields, $\epsilon_\pm = \epsilon_1 \pm \epsilon_2$, can be easily found to be

$$\epsilon_+ \sim \mu_1 \mu_2 \mu_3 \mu_0 \qquad \epsilon_- \sim \sigma_1 \sigma_2 \sigma_3 \sigma_0 \qquad (36.58)$$

and their correlation functions are

$$\langle \epsilon_+(\mathbf{r}) \epsilon_+(\mathbf{0}) \rangle \sim C \left[1 + \mathcal{O}\left(\frac{\exp(-2m_{t,s})}{r^2} \right) \right] \qquad \langle \epsilon_-(\mathbf{r}) \epsilon_-(\mathbf{0}) \rangle \sim K_0^3(m_t r) K_0(m_s r) \qquad (36.59)$$

From (36.59) it follows that the new phase is characterized by long-range dimerization ordering along each chain, with zero relative phase: $\langle \epsilon_1 \rangle = \langle \epsilon_2 \rangle = (1/2)\langle \epsilon_+ \rangle$ The onset of dimerization is associated with spontaneous breakdown of the Z_2-symmetry related to simultaneous translations by one lattice spacing on the both chains.

In the dimerized phase the spin correlations undergo dramatic changes. Using long-distance asymptotics of the MacDonald functions, from (36.57) we obtain:

$$\Im m \, \chi_+(\omega, \pi - q) \sim \frac{\theta \left(\omega^2 - q^2 - 4m_t^2 \right)}{m_t \sqrt{\omega^2 - q^2 - 4m_t^2}}$$

$$(36.60)$$

$$\Im m \, \chi_-(\omega, \pi - q) \sim \frac{\theta [\omega^2 - q^2 - (m_t + m_s)^2]}{\sqrt{m_t m_s} \sqrt{\omega^2 - q^2 - (m_t + m_s)^2}}$$

where $\pm$ signs refer to the case where the wave vector in the direction perpendicular to the chains is equal to 0 and π respectively. We observe the disappearance of coherent magnon poles in the dimerized spin fluid; instead we find two-magnon thresholds at $\omega = 2m_t$ and $\omega = m_t + m_s$, similar to the structure of $\chi''(\omega, q)$ at small wave vectors in the Haldane fluid phase. The fact that two massive magnons, each with momentum $q \sim \pi$, combine to form a two-particle threshold, still at $q \sim \pi$ rather than $2\pi \equiv 0$, is related to the fact that, in the dimerized phase with $2a_0$ periodicity, the new Umklapp is just π.

Exercise

Calculate the spectral function $\Im m \chi(q, \omega)$ at the Ising transition, $m_s = 0$, when $\langle \sigma(r)\sigma(0) \rangle = \langle \mu(r)\mu(0) \rangle \sim r^{-1/4}$. What happens with the pole at $\omega = \sqrt{q^2 + m_t^2}$?

Chirally stabilized spin liquid

On the phase diagram of a spin ladder there is a special line $g_s = 0$, $u = 0$, $g_0 > 0$ where the excitations carry spin $1/2$. At this point the only remaining relevant perturbation is the

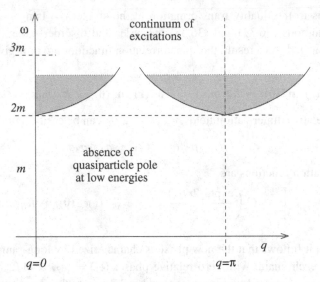

Figure 36.7. The area of the (ω, q) plane where χ'' does not vanish.

current-current interaction (36.61). The entire two-chain Hamiltonian density at this point acquires the following form:

$$\mathcal{H} = \frac{2\pi v}{3} \sum_{j=1,2} \left(: \mathbf{J}_j^2 : + : \bar{\mathbf{J}}_j^{\,2} : \right) + g_0 (\mathbf{J}_1 + \bar{\mathbf{J}}_1) \cdot (\mathbf{J}_2 + \bar{\mathbf{J}}_2) \qquad (36.61)$$

It can be rewritten as follows:

$$\mathcal{H} = \mathcal{H}_+ + \mathcal{H}_- + V \qquad (36.62)$$

$$\mathcal{H}_+ = \frac{2\pi v}{3} \left(: \mathbf{J}_1^2 : + : \bar{\mathbf{J}}_2^{\,2} : \right) + g_0 \mathbf{J}_1 \bar{\mathbf{J}}_2 \qquad (36.63)$$

$$\mathcal{H}_+ = \frac{2\pi v}{3} \left(: \mathbf{J}_2^2 : + : \bar{\mathbf{J}}_1^{\,2} : \right) + g_0 \mathbf{J}_2 \bar{\mathbf{J}}_1 \qquad (36.64)$$

$$V = g_0 (\mathbf{J}_1 \mathbf{J}_2 + \bar{\mathbf{J}}_1 \bar{\mathbf{J}}_2) \qquad (36.65)$$

Hamiltonians H_+ and H_- commute with each other and interaction V is irrelevant. Therefore the excitations possess an additional quantum number – parity. The spectrum in a sector with given parity is represented by massive spin-1/2 particles. The mass is exponentially small in $1/g_0$. The reader can find further details about this and related models in the publications by Allen *et al.* (2000) and Nersesyan and Tsvelik (2003).

Spin $S = 1$ antiferromagnets

The spin $S = 1$ Heisenberg chain can be understood as a limiting case of the spin $S = 1/2$ ladder with a ferromagnetic rung interaction. Realistic systems also include a single-ion anisotropy, which is a common occurrence in magnetic systems. This anisotropy leads to a splitting of the triplet mode observed experimentally. The corresponding calculations can be found in the original publication (Tsvelik, 1990) and also in the paper by Fujiwara *et al.* (1993).

References

Aijro, Y., Goto, T., Kikuchi, H., Sakakibara, T. and Inami, T. (1989). *Phys. Rev. Lett.*, **63**, 1424.

Allen, D., Essler, F. H. L. and Nersesyan, A. A. (2000). *Phys. Rev. B*, **61**, 8871.

Babujan, H. M. (1983). *Nucl. Phys. B*, **215**, 317.

Fujiwara, N., Goto, T., Maegawa, S. and Kohmoto, T. (1993). *Phys. Rev. B*, **47**, 11860.

Nersesyan, A. A. and Tsvelik, A. M. (1997). *Phys. Rev. Lett.*, **78**, 3939.

Nersesyan, A. A. and Tsvelik, A. M. (2003). *Phys. Rev. B*, **67**, 024422.

Nomura, K. (1989). *Phys. Rev. B*, **40**, 2421.

Renard, J. P., Verdaguer, M., Regnault, L. P., Erkelens, W. A. C., Rossat-Mignod, J. and Stirling, W. G. (1987). *Europhys. Lett.*, **3**, 945.

Schulz, H. J. (1986). *Phys. Rev. B*, **34**, 6372.

Takhtajan, L. A. (1982). *Phys. Lett. A*, **87**, 479.

Tsvelik, A. M. (1990). *Phys. Rev. B*, **42**, 10499.

Tsvelik, A. M. (2001). *Nucl. Phys. B*, **612**, 479.

Wu, T. T., McCoy, B., Tracy, C. A. and Barouch, E. (1976). *Phys. Rev. B*, **13**, 316.

37

Kondo chain

In this chapter I discuss the following model Hamiltonian describing an M-fold degenerate band of conduction electrons interacting with a periodic arrangement of local spins S:

$$H - \mu N = \sum_r \sum_{j=1}^{M} \left[-\frac{1}{2}(c^+_{r+1,\alpha,j} c_{r,\alpha,j} + c^+_{r,\alpha,j} c_{r+1,\alpha,j}) - \mu c^+_{r,\alpha,j} c_{r,\alpha,j} \right.$$

$$\left. + J(c^+_{r,\alpha,j} \hat{\sigma}_{\alpha\beta} c_{r,\beta,j}) \mathbf{S}_r \right] \tag{37.1}$$

The related problem is a long-standing problem of the Kondo lattice or, in more general words, the problem of the coexistence of conduction electrons and local magnetic moments. We have discussed this problem very briefly in Chapter 21, where it was mentioned that this remains one of the biggest unsolved problems in condensed matter physics. The only part of it which is well understood concerns a situation where localized electrons are represented by a single local magnetic moment (the Kondo problem). In this case we know that the local moment is screened at low temperatures by conduction electrons and the ground state is a singlet. The formation of this singlet state is a nonperturbative process which affects electrons very far from the impurity. The relevant energy scale (the Kondo temperature) is exponentially small in the exchange coupling constant. It still remains unclear how conduction and localized electrons reconcile with each other when the local moments are arranged regularly (Kondo lattice problem). Empirically, Kondo lattices resemble metals with very small Fermi energies of the order of several degrees. It is widely believed that conduction and localized electrons in Kondo lattices hybridize at low temperatures to create a single narrow band (see the discussion in Chapter 21). However, our understanding of the details of this process remains vague. The most interesting problem is how the localized electrons contribute to the volume of the Fermi sea (according to the large-N approximation, they do contribute). The most dramatic effect of this contribution is expected to occur in systems with one conduction electron and one spin per unit cell. Such systems must be insulators (the so-called Kondo insulator). The available experimental data apparently support this point of view: all compounds with an odd number of conduction electrons per spin are insulators (Aeppli and Fisk, 1992). At low temperatures they behave as semiconductors with very small gaps of the order of several degrees. The marked exception is FeSi where the size of the gap is estimated as $\sim$700 K (Schlesinger et $al.$, 1993). The conservative approach to Kondo insulators would be to calculate their band structure treating the on-site

Coulomb repulsion U as a perturbation. The advantage of this approximation is that one gets an insulating state already in the zeroth order in U. The disadvantage is that it contradicts the principles of perturbation theory which prescibe that the strongest interactions are taken into account first. It turns out also that the pragmatic sacrifice of principles does not lead to a satisfactory description of the experimental data: the band theory fails to explain many experimental observations (see the discussion in Schlesinger et al. (1993)).

In this chapter we study a one-dimensional model of the Kondo lattice (37.1) at half-filling ($\mu = 0$). It will be demonstrated that at least in this case the insulating state forms not due to a hybridization of conduction electrons with local moments, but as a result of strong antiferromagnetic fluctuations. The presence of these fluctuations is confirmed by the numerical calculations by Tsunetsugu et al. (1992) and Yu and White (1993) which demonstrate a sharp enhancement of the staggered susceptibility in one-dimensional Kondo insulators. The interaction of electrons with fluctuations of magnetization converts them into massive spin polarons with the spectral gap exponentially small in the coupling constant. The described scenario does not require a global antiferromagnetic order, just the contrary – the spin ground state remains disordered with a finite correlation length. At present it is not clear whether such a scenario can be generalized for higher dimensions. If this is the case, then Kondo insulators are either antiferromagnets (then they have a true gap), or spin liquids with a strongly enhanced staggered susceptibility. In the latter case instead of a real gap there is a pseudogap, a drop in the density of states on the Fermi level.

In what follows we shall use the path integral formalism. As we know from Chapter 16, in the path integral approach spins are treated as classical vectors $\mathbf{S} = S\mathbf{m}\,(\mathbf{m}^2 = 1)$ whose action includes the Berry phase. In the present case we have the following Euclidean action:

$$A = \int d\tau \left\{ \sum_r \left[iS \int_0^1 du(\mathbf{m}_r(u, \tau)[\partial_u \mathbf{m}_r(u, \tau) \times \partial_\tau \mathbf{m}_r(u, \tau)]) \right. \right.$$

$$\left. \left. + c^*_{r,\alpha,a} \partial_\tau c_{r,\alpha,a} \right] - H(c^*, c; S\mathbf{m}) \right\} \tag{37.2}$$

where the first term represents the spin Berry phase responsible for the correct quantization of local spins and the last term is the Hamiltonian (37.1) with $\mu = 0$.

Further, we shall follow the semiclassical approach assuming that all fields can be separated into fast and slow components. Then the fast components will be integrated out and we shall obtain an effective action for the slow ones. As will become clear later, this approach is self-consistent if the exchange integral is small, $JM \ll 1$. In this case there is a local antiferromagnetic order with the correlation length much larger than the lattice distance. The subsequent derivation essentially repeats the derivation of the O(3) nonlinear sigma model described in Chapter 16. The only marked difference is the region of validity: the semiclassical approach works well for spin systems only if the underlying spins are large; for the Kondo lattice we have another requirement: the smallness of the exchange coupling constant. The suggested decomposition of variables is

$$\mathbf{m}_r = a\mathbf{L}(x) + (-1)^r \mathbf{n}(x)\sqrt{1 - a^2 \mathbf{L}(x)^2} \quad (\mathbf{Ln}) = 0$$
$$c_r = i^r \psi_\mathrm{L}(x) + (-i)^r \psi_\mathrm{R}(x) \tag{37.3}$$

where $|L|/a \ll 1$ is a rapidly varying ferromagnetic component of the local magnetization. Substituting (37.3) into (37.2) and keeping only nonoscillatory terms, we get:

$$A = \int d\tau dx\, L$$

$$L = iS\left(L[\mathbf{n} \times \partial_\tau \mathbf{n}]\right) + 2\pi S \times (\text{top-term}) + \bar{\psi}_j \left[\gamma_\mu \partial_\mu + JS(\hat{\sigma}\mathbf{n}(x))\sqrt{1 - a^2 \mathbf{L}(x)^2}\right]\psi_j$$

$$(37.4)$$

where the topological term is given by (16.18). The first two terms come from the expansion of the spin Berry phase which is explained in detail in Chapter 16. In the fermionic action I have omitted the term describing the interaction of fermionic currents with the ferromagnetic component of the magnetization. This term is responsible for the Kondo screening and can be neglected because, as will be shown later, the Kondo temperature is smaller than the characteristic energy scale introduced by the antiferromagnetic fluctuations.

The fermionic determinant of (37.4) is a particular case of the more general determinant calculated in the last section of Chapter 32. It is reduced to the determinant (32.33) after the transformation $\psi_R \to i\psi_R$, $\psi_L \to \psi_L$ and the substitution $m = JS$, $g = i(\mathbf{n}\hat{\sigma})$. The final result is

$$L = iS\left(L[\mathbf{n} \times \partial_\tau \mathbf{n}]\right) + \frac{M}{2\pi}[(\partial_x \mathbf{n})^2 + (\partial_\tau \mathbf{n})^2] + \frac{M}{\pi}(JS)^2 \ln\frac{1}{JS}(\mathbf{L})^2$$

$$+ \pi(2S - M) \times (\text{top-term}) \qquad (37.5)$$

The Wess–Zumino term in (32.33) yields the topological term, but with the factor M in front. The term with $\mathbf{L}^2$ comes from the static part of the determinant. Integrating over fast ferromagnetic fluctuations described by $\mathbf{L}$, and rescaling time and space we get

$$A = \frac{M}{2\pi} \int d\tau dx [v^{-2}(\partial_\tau \mathbf{n})^2 + (\partial_x \mathbf{n})^2] + \pi(2S - M) \times (\text{top-term}) \qquad (37.6)$$

$$v^{-2} = 1 + \frac{\pi^2}{2J^2 M^2 \ln(1/JS)} \qquad (37.7)$$

This is the action of the O(3) nonlinear sigma model with the dimensionless coupling constant

$$g = \frac{\pi v}{M} = \frac{\pi}{\sqrt{M^2 + \pi^2/2J^2 \ln(1/JS)}} \qquad (37.8)$$

For $|M - 2S| = (\text{even})$ the topological term is always equal to $2\pi i \times$ integer and therefore gives no contribution to the partition function. In this case the model (37.6) is the ordinary O(3) nonlinear sigma model. As we know from Chapter 16, this model has a disordered ground state with a correlation length

$$\xi = a^{-1} g \exp(2\pi/g)$$

The spin excitations are massive triplets, particles with spin $S = 1$.

If $|M - 2S| = $ (odd) the topological term is essential. The model becomes critical and its low energy behaviour is the same as for the spin-$1/2$ antiferromagnetic Heisenberg chain. In this case the energy scale

$$\Delta_S(J) = \pi v/\xi \approx J g^{-1} \exp(-2\pi/g) \qquad (37.9)$$

marks the crossover to the critical regime where $\mathbf{n}$ has the same correlation functions as staggered magnetization of the Heisenberg chain (see the final section in Chapter 29). The specific heat is linear at low temperatures $T \ll \Delta_s(J)$ and comes from gapless spin excitations.

Despite the fact that the expression for the energy scale $\Delta_s(J)$ formally resembles the expression for the Kondo temperature

$$T_K = \sqrt{J} \exp\left(-2\pi/J\right)$$

$m(J)$ is always larger due to the presence of the large logarithm. Therefore at small J the antiferromagnetic exchange induced by the conduction electrons (the RKKY interaction) plays a stronger role than the Kondo screening. This justifies the neglect of the $\mathbf{L}(\psi_R^+ \sigma \psi_R + \psi_L^+ \sigma \psi_L)$ term in the evaluation of the fermionic determinant. Thus the leading contributions to the low energy dynamics come from antiferromagnetic fluctuations in agreement with the results of Tsunetsugu et al. (1992) and Yu and White (1993).

Our derivation includes one nontrivial element: the expansion of the fermionic determinant in (37.4) contains the topological term. This term cannot be obtained by the semiclassical expansion. The semiclassical approximation assumes adiabaticity, that is that eigenvalues of the fast system adjust to the slowly varying external field. Adiabaticity is violated when energy levels cross; the level crossing gives rise to topological terms. Apparently, level crossings occur in the fermionic action (37.4) when the $\mathbf{n}(\tau, x)$ field has a nonzero topological charge. However, I do not want to pursue this line of argument. Instead, I will justify the physical necessity of the topological term using the strong coupling limit $J/t \gg 1$. As usual in one dimension, we expect that the strong coupling limit correctly reproduces *qualitative* features of the solution. At $J/t \gg 1$ conduction electrons create bound states with spins. These bound states have spins $\mathbf{S}'_n$ with magnitude $|S - M/2|$. The next step in the t/J expansion gives the Heisenberg-type exchange between these spins with the exchange integral $j \sim t^2/J$. As we have learned from Chapter 16, the Heisenberg model in one dimension in the continuum limit is equivalent to the nonlinear sigma model with the topological term. That is what we got, see (37.6). Of course, the strong coupling limit does not give the correct J-dependence of the coupling constants at $J/t \ll 1$, but it gives the same coefficient $(S - 2M)$ in the topological term and now its meaning is clear.

It follows from our analysis that at half-filling the Kondo chain has very different properties in the charge and spin sectors. The characteristic energy scale for the spin excitations is given by (37.9); the scale for the charge excitations is obviously of the order of JS. This conclusion is also in agreement with the numerical calculations referred to above.

References

Aeppli, G. and Fisk, Z. (1992). *Comments Condens. Matter Phys.*, **16**, 155.

Schlesinger, Z., Fisk, Z., Zhang, H.-T., Maple, M. B., DiTusa, J. F. and Aeppli, G. (1993). *Phys. Rev. Lett.*, **71**, 1748.

Tsunetsugu, H., Hatsugai, Y., Ueda, K. and Sigrist, M. (1992). *Phys. Rev. B*, **46**, 3175.

Yu, C. C. and White, S. R. (1993). *Phys. Rev. Lett.*, **71**, 3866.

Gauge fixing in non-Abelian theories: $(1 + 1)$-dimensional quantum chromodynamics

The results of Chapter 33 can be applied to an interesting model problem of $(1 + 1)$-dimensional quantum chromodynamics. I restrict the discussion to massless quantum chromodynamics and will be using Minkovsky notation. The corresponding Lagrangian density in Minkovsky space-time is given by

$$L = \frac{1}{8\pi e^2} \mathrm{Tr}(F_{\mu\nu} F^{\mu\nu}) + \sum_{n=1}^{k} \bar{\psi}_{n\alpha} \gamma_\mu (i\partial_\mu \delta_{\alpha\beta} + A_\mu^{\alpha\beta}) \psi_{n\beta} \tag{38.1}$$

$$F_{\mu\nu} \equiv \tau^a F_{\mu\nu}^a = \partial_\mu A_\nu - \partial_\nu A_\mu + [A_\mu, A_\nu] \tag{38.2}$$

The more general problem with massive fermions is discussed in detail in the review paper of Frishman and Sonnenschein (1993). In the discussion which follows below I use the results on the gauged WZNW model derived by Bershadsky and Ooguri (1989) in a somewhat different context. In the absence of fermions $(1 + 1)$-dimensional quantum chromodynamics (or rather gluodynamics) is an exactly solvable problem. The gauge invariance eliminates all continuous degrees of freedom; only discrete ones remain.

The Lagrangian (38.1) has an important property of gauge invariance, namely it is invariant under the transformations

$$\begin{aligned} \bar{\psi} \rightarrow \bar{\psi} G \qquad \psi \rightarrow G^+ \psi \\ A_\mu \rightarrow G^+ A_\mu G + iG^+ \partial_\mu G \end{aligned} \tag{38.3}$$

where G is a matrix from the $SU(N)$ group. This gauge invariance leads to additional complications related to gauge fixing. This problem is so important and general that it deserves a separate discussion.

Let us consider the partition function of quantum chromodynamics:

$$Z = \int DA_\mu D\bar{\psi} D\psi \, \exp\{-iS[\bar{\psi}, \psi, A]\} \tag{38.4}$$

Obviously, if we do not place any restrictions on A_μ, the functional integral will diverge. This divergence occurs because we integrate over gauge transformations of the fields which do not change the action. The divergence comes exclusively from the measure and can be removed by its proper redefinition. To do this let us separate the Hilbert space of A_μ into 'slices'. Each slice is specified by some representative field $A_\mu^{(0)}$ and consists of all fields generated from it by the gauge transformations (38.3). The representative fields are specified by some condition which I shall find later. The idea is to separate the integral over A_μ into integrals over the representative field $A_\mu^{(0)}$ and the gauge group:

$$DA_\mu = DA_\mu^{(0)}[G^+DG]J\left(A_\mu^{(0)}\right) \tag{38.5}$$

where J is the Jacobian of this transformation and $[G^+DG]$ is the measure of integration on the gauge group. It is essential that due to the gauge invariance the Jacobian does not depend on G. Therefore omitting integration over G we get rid of all divergences. The remaining part of the measure includes integration over the representative fields with the corresponding Jacobian:

$$Z = \int DA_\mu^{(0)} J\left(A_\mu^{(0)}\right) D\bar{\psi} D\psi \exp\left\{-iS[\bar{\psi}, \psi, A_\mu^{(0)}]\right\} \tag{38.6}$$

The next problem is to calculate the Jacobian. Since it does not depend on G, we can calculate it at the point $G = 1$ where the calculation is easier. In the vicinity of this point we have

$$G = 1 + i\epsilon^a(x)\tau^a + \mathcal{O}(\epsilon^2) \tag{38.7}$$

where $\epsilon^a(x)$ are infinitely small functions. Substituting (38.7) into (38.3) we get the general expression for the gauge potential:

$$A_\mu^a = A_\mu^{(0),a} - D_\mu^{ab}\epsilon^b \qquad D_\mu^{ab} = \partial_\mu\delta^{ab} + f^{abc}A_\mu^c \tag{38.8}$$

In order to make the calculation of the Jacobian simpler, it is convenient to choose the representative fields in such a way that the gauge transformations are orthogonal to them:

$$\left(A_\mu^{(0),a}, D_\mu^{ab}\epsilon^b\right) = -\left(D_\mu^{ab}A_\mu^{(0),b}, \epsilon^a\right) = 0 \tag{38.9}$$

which means that

$$D_\mu^{ab} A_\mu^{(0),b} = 0 \tag{38.10}$$

This condition is called the Coulomb gauge. In the Coulomb gauge the integration separates:

$$DA = DA^{(0)}D\epsilon^a \det\left[D_\mu^{ab}\right] \tag{38.11}$$

The functional determinant can be written as a path integral over auxiliary Dirac fermion fields $\bar{\eta}, \eta$:

$$J\left[A^{(0)}\right] = \det\left[D_\mu^{ab}\right] \propto \det\left[i\gamma^\mu D_\mu^{ab}\right]$$
$$= \int D\bar{\eta}^a D\eta^a \exp\left[\int d^Dx\,\bar{\eta}^a\gamma^\mu\left(\partial_\mu\delta^{ab} + f^{acb}A_\mu^{(0)}\right)\eta^b\right] \tag{38.12}$$

These fields are called Faddeev–Popov ghosts, or simply ghosts.

The gauge fixing procedure, which I have just explained, has a general validity. Let us apply it to the $(1 + 1)$-dimensional quantum chromodynamics. I choose the same gauge as in the previous chapter:

$$\bar{A} = 0 \qquad A = -2ig^{-1}\partial g$$

(now $2i\partial = \partial_t - \partial_x$) with the additional Coulomb condition (38.10), which in this choice of gauge reads as

$$\bar{\partial}A = 0 \tag{38.13}$$

Thus quantum chromodynamics includes only chiral degrees of freedom! All right-moving fields are excluded by the condition (38.13). The integration over the representative field is the integration over the chiral component of g:

$$DA^{(0)} \propto [g^+(z)Dg(z)] \tag{38.14}$$

Integrating over the fermions we get the following expression for the partition function:

$$Z = \int [g^+(z)Dg(z)]J(g)\exp\left\{ikW[g] - \frac{i}{8\pi e^2}\int dt dx \operatorname{Tr}[\bar{\partial}(g^{-1}\partial g)]^2\right\} \tag{38.15}$$

The Jacobian itself is the fermionic determinant of the fermions transforming according to the adjoint representation of the group.[1] One can check directly that such Dirac fermions have $k = 2c_v$. Therefore we have

$$J[g] = \det(i\gamma_\mu \partial_\mu \delta_{ac} + if^{abc}A^b) = \exp\{2ic_v W[g]\} \tag{38.16}$$

Thus we have

$$Z_{QCD} = \int [g^+(z)Dg(z)] \exp\left\{i(k + 2c_v)W[g] - \frac{i}{8\pi e^2}\int dt dx \operatorname{Tr}[\bar{\partial}(g^{-1}\partial g)]^2\right\} \tag{38.17}$$

In fact the second term, having scaling dimensions $\Delta = \bar{\Delta} = 2$, is irrelevant. The quantum chromodynamics is governed by the *chiral* WZNW action with $\tilde{k} = -(k + 2c_v)$. The corresponding central charge is equal to

$$C_{QCD} = \frac{(k + 2c_v)G}{k + c_v} = 2G - \frac{kG}{k + c_v} \tag{38.18}$$

The discussed procedure includes an analytic continuation of the Wess–Zumino–Novikov–Witten model on negative k. It is for this reason we could not work in Euclidean space-time. As we have seen, the central charge remains positive; thus this procedure has a chance to be legitimate. The primary fields, however, acquire negative conformal dimensions. The only exceptions are the unit field and the chiral currents. These are the only well defined fields in $(1 + 1)$-dimensional massless quantum chromodynamics. Since these fields are chiral, i.e. depend only on one coordinate z, correlation functions of the field strength $\bar{\partial}J$ are short range, as in the Abelian quantum electrodynamics.

References

Bershadsky, M. and Ooguri, H. (1989). *Commun. Math. Phys.*, **126**, 49.
Frishman, Y. and Sonnenschein, J. (1993). *Phys. Rep.*, **223**, 309.

---

[1] Recall that the adjoint representation is defined as a representation composed of the structure constants.

Select bibliography

Abrikosov, A. A. (1988). *Fundamentals of the Theory of Metals*. North-Holland, Amsterdam.

Abrikosov, A. A., Gorkov, L. P. and Dzyaloshinsky, I. E. (1963). *Methods of Quantum Field Theory in Statistical Physics*, ed. R. A. Silverman, revised edn. Dover, New York.

Amit, D. J. (1984). *Field Theory, the Renormalization Group, and Critical Phenomena*. World Scientific, Singapore.

Anderson, P. W. (1984). *Basic Notions of Condensed Matter Physics*. Benjamin/Cummings.

Ashcroft, N. W. and Mermin, N. D. (1983). *Solid State Physics*. Holt-Saunders.

Auerbach, A. (1994). *Interacting Electrons and Quantum Magnetism*. Springer, New York.

Brezin, E. and Zinn-Justin, J. (eds.) (1990). *Fields, Strings and Critical Phenomena, Les Houches 1988, Session XLIX*. North-Holland, Amsterdam.

Cardy, J. (1996). *Scaling and Renormalization in Statistical Physics*. Cambridge University Press, Cambridge.

Di Francesco, Ph., Mathieu, P. and Sénéchal, D. (1999). *Conformal Field Theory*. Springer, Berlin.

Fradkin, E. (1991). *Field Theories of Condensed Matter Systems*. Addison-Wesley, New York.

Gogolin, A. O., Nersesyan, A. A. and Tsvelik, A. M. (1998). *Bosonization and Strongly Correlated Systems*. Cambridge University Press, Cambridge.

Itzykson, C. and Drouffe, J.-M. (1989). *Statistical Field Theory*. Cambridge University Press, Cambridge.

Itzykson, C., Saleur, H. and Zuber, J.-B. (eds.) (1988). *Conformal Invariance and Applications to Statistical Mechanics*. World Scientific, Singapore.

Polyakov, A. M. (1998). *Field Theories and Strings*. Harwood Academic, New York.

Popov, V. N. (1990). *Functional Integrals and Collective Excitations*. Cambridge University Press, Cambridge.

Sachdev, S. (1999). *Quantum Phase Transitions*. Cambridge University Press, Cambridge.

Schutz, B. (1980). *Geometrical Methods in Mathematical Physics*. Cambridge University Press, Cambridge.

Smirnov, F. A. (1992). *Form Factors in Completely Integrable Models of Quantum Field Theory*, Advanced Series in Mathematical Physics, Vol. 14. World Scientific, Singapore.

Wilczek, F. (1990). *Fractional Statistics and Anyon Superconductivity*. World Scientific, Singapore.

Zinn-Justin, J. (1993). *Quantum Field Theory and Critical Phenomena*, 2nd edn. Oxford University Press, Oxford.

Index